The Finite Element Method
for Engineers

The Finite Element Method for Engineers

Kenneth H. Huebner

Engineering Mechanics Department
General Motors Research Laboratories

A Wiley-Interscience Publication

JOHN WILEY & SONS

New York • Chichester • Brisbane • Toronto

Library of Congress Cataloging in Publication Data:

Huebner, Kenneth H   1942-
    The finite element method for engineers.

    "A Wiley-Interscience publication."
    Includes bibliographical references.
    1. Finite element method. I. Title.

TA347.F5H83 1975   624'.171   74-17452
ISBN 0-471-41950-8

Printed in the United States of America

10 9 8 7

To My Wife, Louise

PREFACE

One of the newest and most popular numerical techniques which engineers and scientists everywhere are using is the finite element method. With the help of high-speed digital computers, the finite element method has greatly enlarged the range of engineering problems amenable to numerical analysis.

The method of finite elements originated some 15 years ago in the aircraft industry as an effective means for analyzing complex airframe structures. It began as an extension of matrix methods for structural analysis, but now it is recognized as a powerful and versatile tool which permits a computer solution of almost all previously intractable problems in stress analysis. Many analysts agree that the finite element method represents a true breakthrough in solid mechanics.

An appealing feature of the finite element method is that it is not restricted to solid mechanics. Although this important fact was recognized about 8 years ago, the method has only recently been applied in other areas. Actually, it is applicable to almost all continuum or field problems. Hence it is not surprising that the method is receiving much attention today in engineering, physics, and mathematics.

This book is for persons who are involved in analysis in the physical sciences and want to learn about the nature and capabilities of finite element analysis. It is a "starting-point" text that presents the finite element method at an easy-to-understand, introductory level. Most previous treatments of the method center on structural mechanics problems. This book views the finite element method more generally as a numerical analysis tool for engineering mechanics problems. The approach has been to draw freely from the technical journals and to consolidate scattered information. The exposition, it is hoped, is a balanced mixture of theory and examples.

In this book, I have attempted to meet three objectives: (1) to describe as simply as possible the fundamentals of the finite element method; (2) to give the reader a coherent working familiarity with these fundaments so that he can, without difficulty, apply the method to his problems; and (3) to offer a useful overall view of the method to establish a convenient point of departure for further study of the special advanced topics continually appearing in the literature.

I have tried to begin the discussion of each topic at a relatively elementary level and to work up gradually to its more complex aspects. An elementary knowledge of engineering mechanics and the associated mathematical and computer skills is assumed. However, special aspects of matrix algebra and variational calculus that are employed in the method are treated in sufficient detail. Tensor notation (although it can be useful in this field) is omitted for simplicity.

The book is divided into two parts. The first part presents basic concepts and the fundamental theory of the method. It begins with a largely historical and discursive discussion which provides an overview and general orientation for the reader. Then, to provide a physical basis, a detailed development of the method (as it originated from structural mechanics) is given. Because matrices and variational calculus are important to an understanding of finite element analysis, these topics are also treated in the first two appendices. Once the physical basis has been established, the mathematical basis for the method is presented. Here the reader is shown how to apply the method to widely diverse problems in engineering mechanics. The fourth chapter in this section presents the most recent generalized finite element concepts. In summary, Part I enables the reader to comprehend the method, to discover how it extends from structures problems to general continuum problems, and to appreciate the limitations of the method.

The second part of the book offers five chapters on the applications of finite element methods in engineering mechanics. Chapter 6 presents finite element formulations for linear elasticity theory, and Appendix C provides a brief review of the relevant basic equations from solid mechanics. The formulations for the elasticity problems are based on the commonly used displacement method of analysis, though mention is also made of some references in which other possible formulations are treated. General field problems such as heat conduction, electromagnetics, and torsion are the subject of Chapter 7. Since there are many practical field problems in which time is an independent parameter, an extension of finite element concepts to the time domain is examined in some detail. An entire chapter is devoted to a particular type of field problem, namely, the fluid-film lubrication problem. Inclusion of this chapter reflects my special interest in this area.

The penultimate chapter of the book studies the application of finite element techniques in fluid dynamics. The basic equations and available variational principles are reviewed in Appendix D. A discussion of the diverse types of fluid mechanics problems amenable to finite element analysis is the central theme. Here the reader will find a treatment that takes him up to the latest applications of the method. Solving a problem by the finite element method ultimately reduces to writing a computer program to generate and solve a set of simultaneous equations. To guide the reader through this procedure, a typical program is presented and explained in detail in Chapter 10. Here other practical considerations associated with implementing the finite element method on a digital computer are also discussed.

I have attempted to make the book self-contained so that a person wishing to apply the finite element method to his particular problem need study only one text. Although this volume treats the most important aspects of finite element analysis, space limitations made it necessary to omit or considerably abridge the treatment of some specialized aspects. I believe, however, that the treatment is sufficiently comprehensive to meet the needs of students as well as those of most practicing engineers and scientists.

The references listed at the end of each chapter are those in which readers can find additional information or detailed developments of the more advanced topics. Because of the rapidly expanding literature on this subject, the reference lists are by no means exhaustive. Instead, they were selected to supplement material in the more fundamental research works.

KENNETH H. HUEBNER

Warren, Michigan
July 1974

ACKNOWLEDGMENT

The most important contributors to this book were all the researchers—too numerous to mention here—whose works are copiously cited throughout the book. These were the persons who provided the information which I have organized, summarized, and reported. To them, I am most grateful.

My special thanks go also to several of my colleagues. D. F. Hays granted some of the time and helped to create the atmosphere in which completion of the book was possible. A. O. "Butch" De Hart has a large measure of my gratitude for suggesting the writing of this book in the first place and for giving much active encouragement throughout the project. Professor J. F. Booker of Cornell University has been most helpful in reviewing parts of the manuscript, providing stimulating ideas, and making valuable suggestions. To Professor O. C. Zienkiewicz, whose friendship and research association were most treasured during this writing project, I express my sincere appreciation.

I am also indebted to a number of persons who helped to prepare the manuscript. In particular, I wish to thank Sallie Ellison and Jim Carter, staff members of the G. M. Research Library, for finding and providing many references, and Sue Moreau for her excellent typing assistance.

Finally, I am especially grateful to my wife Louise for her patience and understanding while enduring the "book widow syndrome" for over a year.

K. H. H.

CONTENTS

PART ONE

1

MEET THE FINITE ELEMENT METHOD

1.1 WHAT IS THE FINITE ELEMENT METHOD?

The finite element method is a numerical analysis technique for obtaining approximate solutions to a wide variety of engineering problems. Although originally developed to study the stresses in complex airframe structures, it has since been extended and applied to the broad field of continuum mechanics. Because of its diversity and flexibility as an analysis tool, it is receiving much attention in engineering schools and in industry.

Although this brief comment on the finite element method answers the question posed by the section heading, it does not give us the operational definition that we need to apply the method to a particular problem. Such an operational definition—along with a description of the fundamentals of the method—requires considerably more than one paragraph to develop. Hence Part 1 of this book is devoted to basic concepts and fundamental theory. Before discussing more aspects of the finite element method, we should first consider some of the circumstances leading to its inception, and we should briefly contrast it with other numerical schemes.

In more and more engineering situations today, we find that it is necessary to obtain approximate numerical solutions to problems rather than exact closed-form solutions. For example, we may want to find the load capacity of a plate which has several stiffeners and odd-shaped holes, the concentration of pollutants during nonuniform atmospheric conditions, or the rate of fluid flow through a passage of arbitrary shape. Without too much effort, we can write down the governing equations and boundary conditions for these problems, but we see immediately that no simple analytical solution can be found. The difficulty in these three examples lies in the fact that either the geometry or some other feature of the problem is irregular or "arbitrary." Analytical solutions to problems of this type seldom exist; yet these are the kinds of problems which engineers and scientists are called upon to solve.

The resourcefulness of the analyst usually comes to the rescue and provides several alternatives to overcome this dilemma. One possibility is to make simplifying assumptions—to ignore the difficulties and reduce the problem to one that can be handled. Sometimes this procedure works; but, more often than not, it leads to serious inaccuracies or wrong answers. Now that large-scale digital computers are widely available, a more viable alternative is to retain the complexities of the problem and try to find an approximate numerical solution.

Several approximate numerical analysis methods have evolved over the years—the most commonly used method is the general finite difference [1,2]† scheme. The familiar finite difference model of a problem gives a *pointwise* approximation to the governing equations. This model (formed by writing difference equations for an array of grid points) is improved as more points are used. With finite difference techniques, we can treat some fairly difficult problems; but, for example, when we encounter irregular geometries or unusual specification of boundary conditions, we find that finite difference techniques become hard to use.

In addition to the finite difference method, another, more recent numerical method (known as the "finite element method") has emerged. Unlike the finite difference method, which envisions the solution region as an array of grid points, the finite element method envisions the solution region as built up of many small, interconnected subregions or elements. A finite element model of a problem gives a *piecewise* approximation to the governing equations. The basic premise of the finite element method is that a solution region can be analytically modeled or approximated by replacing it with an assemblage of discrete elements. Since these elements can be put together in a variety of ways, they can be used to represent exceedingly complex shapes.

As an example of how a finite difference model and a finite element model might be used to represent a complex geometrical shape, consider the turbine blade cross section in Figure 1.1. For this device, we may want to find the distribution of displacements and stresses for a given force loading, or the distribution of temperature for a given thermal loading. The interior coolant passage of the blade, along with its exterior shape, gives it a non-simple geometry.

A uniform finite difference mesh would reasonably cover the blade (the solution region), but the boundaries must be approximated by a series of horizontal and vertical lines (or "stair steps"). On the other hand, the finite element model (using the simplest two-dimensional element—the triangle) gives a better approximation to the region and requires fewer nodes. Also, a

† Numbers in brackets denote references at the end of the chapter.

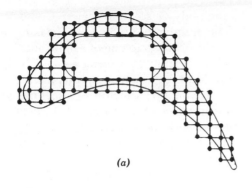

(a)

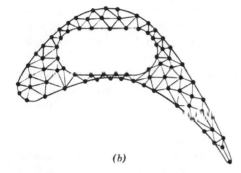

(b)

Figure 1.1. Finite difference and finite element discretizations of a turbine blade profile. (*a*) Typical finite difference model. (*b*) Typical finite element model.

better approximation to the boundary shape results because the curved boundary is represented by a series of straight lines. This example is not intended to suggest that finite element models are decidedly better than finite difference models for all problems. The only purpose of the example is to demonstrate that the finite element method is particularly well suited for problems with complex geometries.

1.2 HOW THE METHOD WORKS

We have been alluding to the essence of the finite element method, but now we shall discuss it in greater detail. In a continuum† problem of any dimension, the field variable (whether it is pressure, temperature, displacement, stress, or some other quantity) possesses infinitely many values because it

---

† We define a continuum to be a body of matter (solid, liquid, or gas) or simply a region of space in which a particular phenomenon is occurring.

is a function of each generic point in the body or solution region. Consequently, the problem is one with an infinite number of unknowns. The finite element discretization procedures reduce the problem to one of a finite number of unknowns by dividing the solution region into elements and by expressing the unknown field variable in terms of assumed approximating functions within each element. The approximating functions (sometimes called *interpolation functions*) are defined in terms of the values of the field variables at specified points called *nodes* or *nodal points*. Nodes usually lie on the element boundaries where adjacent elements are considered to be connected. In addition to boundary nodes, an element may also have a few interior nodes. The nodal values of the field variable and the interpolation functions for the elements completely define the behavior of the field variable within the elements. For the finite element representation of a problem, the nodal values of the field variable become the new unknowns. Once these unknowns are found, the interpolation functions define the field variable throughout the assemblage of elements.

Clearly, the nature of the solution and the degree of approximation depend not only on the size and number of the elements used, but also on the interpolation functions selected. As one would expect, we cannot choose functions arbitrarily because certain compatability conditions should be satisfied. Often functions are chosen so that the field variable or its derivatives are continuous across adjoining element boundaries. The essential guidelines for choosing interpolation functions are discussed in Chapters 3 and 5. These are applied to the formulation of different kinds of effective elements.

Thus far, we have briefly discussed the concept of modeling and an arbitrarily shaped solution region with an assemblage of discrete elements; and we have pointed out that interpolation functions must be defined for each element. We have not yet mentioned, however, an important feature of the finite element method which sets it apart from other approximate numerical methods. That feature is the ability to formulate solutions for individual elements before putting them together to represent the entire problem. This means, for example, that if we are treating a problem in stress analysis, we can find the force-displacement or stiffness characteristics of each individual element and then assemble the elements to find the stiffness of the whole structure. In essence, a complex problem reduces to considering a series of greatly simplified problems.

Another advantage of the finite element method is the variety of ways in which one can formulate the properties of individual elements. There are basically four different approaches. The first approach to obtaining element properties is called the *direct approach* because its origin is traceable to the direct stiffness method of structural analysis. Although the direct approach can be used only for relatively simple problems, it is presented in Chapter 2 because it is the easiest to understand when meeting the finite element method for the first time. The direct approach also suggests the need for matrix algebra (Appendix A) in dealing with the finite element equations.

Element properties obtained by the direct approach can also be determined by the more versatile and more advanced *variational approach*. The variational approach relies on the calculus of variations (Appendix B) and involves extremizing a *functional*. For problems in solid mechanics, the functional turns out to be the potential energy, the complementary potential energy, or some derivative of these, such as the Reissner variational principle. Knowledge of the variational approach (Chapter 3) is necessary to work beyond the introductory level and to extend the finite element method to a wide variety of engineering problems. Whereas the direct approach can be used to formulate element properties for only the simplest element shapes, the variational approach can be employed for both simple and sophisticated element shapes.

A third and even more versatile approach to deriving element properties has its basis entirely in mathematics and is known as the *weighted residuals approach* (Chapter 4). The weighted residuals approach begins with the governing equations of the problem and proceeds without relying on a functional or a variational statement. This approach is advantageous because it thereby becomes possible to extend the finite element method to problems where no functional is available. For some problems, we do not have a functional—either because one may not have been discovered or because one does not exist.

A fourth approach relies on the balance of thermal and/or mechanical energy of a system. The *energy balance approach* (like the weighted residuals approach) requires no variational statement and hence broadens considerably the range of possible applications of the finite element method. We will discuss each of these approaches in subsequent chapters.

Regardless of the approach used to find the element properties, the solution of a continuum problem by the finite element method always follows an orderly step-by-step process. To summarize in general terms how the finite element method works, we will succinctly list these steps now; they will be developed in detail later.

1. *Discretize the continuum.* The first step is to divide the continuum or solution region into elements. In the example of Figure 1.1, the turbine blade has been divided into triangular elements which might be used to find the temperature distribution or stress distribution in the blade. A variety of element shapes (such as those cataloged in Chapter 5) may be used, and, with care, different element shapes may be employed in the same solution region. Indeed, when analyzing an elastic structure that has different types of components such as plates and beams, it is not only desirable but also necessary

to use different types of elements in the same solution. Although the number and the type of elements to be used in a given problem are matters of engineering judgment, the analyst can rely on the experience of others for guidelines. The discussion of applications in Chapters 6–9 reveals many of these useful guidelines.

2. *Select interpolation functions.* The next step is to assign nodes to each element and then choose the type of interpolation function to represent the variation of the field variable over the element. The field variable may be a scalar, a vector, or a higher-order tensor. Often, although not always, polynomials are selected as interpolation functions for the field variable because they are easy to integrate and differentiate. The degree of the polynomial chosen depends on the number of nodes assigned to the element, the nature and number of unknowns at each node, and certain continuity requirements imposed at the nodes and along the element boundaries. The magnitude of the field variable as well as the magnitude of its derivatives may be the unknowns at the nodes.

3. *Find the element properties.* Once the finite element model has been established (that is, once the elements and their interpolation functions have been selected), we are ready to determine the matrix equations expressing the properties of the individual elements. For this task, we may use one of the four approaches just mentioned: the direct approach, the variational approach, the weighted residual approach, or the energy balance approach. The variational approach is often the most convenient, but for any application the approach used depends entirely on the nature of the problem.

4. *Assemble the element properties to obtain the system equations.* To find the properties of the overall system modeled by the network of elements, we must "assemble" all the element properties. In other words, we must combine the matrix equations expressing the behavior of the elements and form the matrix equations expressing the behavior of the entire solution region or system. The matrix equations for the system have the same form as the equations for an individual element except that they contain many more terms because they include all nodes.

The basis for the assembly procedure stems from the fact that, at a node where elements are interconnected, the value of the field variable is the same for each element sharing that node. Assembly of the element equations is a routine matter in finite element analysis and is usually done by electronic computer.

Before the system equations are ready for solution, they must be modified to account for the boundary conditions of the problem. In Chapter 2, we will see explicitly how the assembly process leads to the system equations and how the boundary conditions are introduced.

5. *Solve the system equations.* The assembly process of the preceding step gives a set of simultaneous equations which we can solve to obtain the unknown nodal values of the field variable. If the equations are linear, we can use a number of standard solution techniques such as those mentioned in Chapter 10; if the equations are nonlinear, their solution is more difficult to obtain. Several alternative approaches to nonlinear problems are presented in Chapter 10.

6. *Make additional computations if desired.* Sometimes we may want to use the solution of the system equations to calculate other important parameters. For example, in a fluid mechanics problem such as the lubrication problem, the solution of the system equations gives the pressure distribution within the system. From the nodal values of the pressure, we may then calculate velocity distributions and flows or perhaps shear stresses if these are desired.

1.3 A BRIEF HISTORY OF THE METHOD

Although the label "finite element method" first appeared in 1960, when it was used by Clough [3] in a paper on plane elasticity problems, the ideas of finite element analysis date back much further. In fact, the questions "Who originated the finite element method, and when did it begin?" have three different answers depending on whether one asks an applied mathematician, a physicist, or an engineer. Each of these specialists has some justification for claiming the finite element method as his own because each developed the essential ideas independently at different times and for different reasons. The applied mathematicians were concerned with boundary value problems of continuum mechanics; in particular, they wanted to find approximate upper and lower bounds for eigenvalues. The physicists were also interested in solving continuum problems, but they sought means to obtain piecewise approximate functions to represent their continuous functions. Faced with increasingly complex problems in the aeroelasticity field, the engineers were searching for a way in which to find the stiffness influence coefficients of shell-type structures reinforced by ribs and spars. The efforts of these three groups resulted in three sets of papers with distinctly different viewpoints.

The first efforts to use piecewise continuous functions defined over triangular domains appear in the applied mathematics literature with the work of Courant [4] in 1943. Motivated by Euler's [5] paper, Courant used an assemblage of triangular elements and the principle of minimum potential energy to study the St. Venant torsion problem. After Courant's work, nearly a decade passed before these discretization ideas were used again. The works of Polya [6,7], Hersch [8], and Weinberger [9,10], who focused their attention on bounding eigenvalues, mark a period of renewed interest.

In 1959, Greenstadt [11], motivated by a discussion in the book by Morse and Feshback [12], outlined a discretization approach involving "cells" instead of points; that is, he imagined the solution domain to be divided into a set of contiguous subdomains. In his theory, he describes a procedure for representing the unknown function by a series of functions, each associated with one cell. After assigning approximating functions and evaluating the appropriate variational principle in each cell, he uses continuity requirements to tie together the equations for all the cells. By this means, he reduces a continuous problem to a discrete one. Greenstadt's theory allows for irregularly shaped cell meshes and contains many of the essential and fundamental ideas that serve as the mathematical basis for the finite element method as we know it today.

In the early 1960's (about the same time that finite element concepts began to develop in the engineering community), the significant works of White [13] and Friedrichs [14] appeared. These authors, apparently unaware of the engineering activities at that time, used triangularly shaped elements to develop difference equations from variational principles. Although they used regular meshes, they recognized the need for making special provisions at irregular boundaries.

As the popularity of the finite element method began to grow in the engineering and physics communities, more applied mathematicians became interested in giving the method a firm mathematical foundation. As a result, a number of studies (notably refs. 15–36) were aimed at estimating discretization error, rates of convergence, and stability for different types of finite element approximations. These studies most often focused on the special case of linear elliptic boundary value problems. Although the finite element method has been and is being frequently applied to nonlinear problems [37], corresponding mathematical studies of convergence and accuracy for nonlinear problems have seldom appeared.

A fundamental consideration in the finite element method—the development of suitable function approximations to field variables—has been advanced by some of the mathematical literature on spline functions [38–46].

Since the late 1960's, the mathematical literature on the finite element method has grown more than in any previous period. Several books and monographs [47–50] are devoted to the mathematical foundations of the method. A survey paper by Oden [51] summarizes for the interested reader some of the recent and salient mathematical contributions. In this book, we shall not study the rigorous mathematical basis of the finite element method because such detailed knowledge is unnecessary for most practical applications. Instead, we shall call upon pertinent results when they are needed.

While the mathematicians were developing and using finite element concepts, the physicists were also busy with similar ideas. The work of Prager and Synge [52] leading to the development of the hypercircle method is a key example. As a concept in function space, the hypercircle method was originally developed in connection with classical elasticity theory to give its minimum principles a geometric interpretation. Outgrowths of the hypercircle method (such as the one suggested by Synge [53]) can be applied to the solution of continuum problems in much the same way as finite element techniques can be applied. McHahon [54], a student of Synge, demonstrated this in 1953 when he published an analysis incorporating tetrahedral elements and linear interpolation functions.

It was physical intuition which first brought finite element concepts to the engineering community. In the 1930's, when a structural engineer encountered a truss problem such as the one shown in Figure 1.2a, he immediately knew how to solve for component stresses and deflections as well as the overall strength of the unit. First, he would recognize that the truss was simply an assembly of beams or rods whose force-deflection characteristics he knew well. Then he would combine these individual characteristics according to the laws of equilibrium and solve the resulting system of equations for the unknown forces and deflections for the overall system.

This procedure worked well whenever the structure in question had a *finite* number of interconnection points, but then the following question arose: "What can we do when we encounter an elastic continuum structure such as a plate which has an *infinite* number of interconnection points?" For example, in Figure 1.2b, if a plate replaces the truss, the problem becomes considerably more difficult. Intuitively, Hrenikoff [55] reasoned that this difficulty could be overcome by assuming the continuum structure to be divided into elements or structural sections (beams) interconnected at only a finite number of node points. Under this assumption, the problem reduces to that of a conventional structures type, which could be handled by the old methods. Attempts to apply Hrenikoff's "framework method" were successful, and thus the seed of finite element techniques began to germinate in the engineering community.

Shortly after Hrenikoff, McHenry [56] and Newmark [57] offered further development of these discretization ideas, while Kron [58,59] studied topological properties of discrete systems. There followed a 10-year spell of inactivity, which was broken in 1954 when Argyris and his collaborators [60–66] began to publish a series of papers extensively covering linear structural analysis and efficient solution techniques well suited to automatic digital computation. The actual solution of plane stress problems by means of triangular elements whose properties were determined from the equations

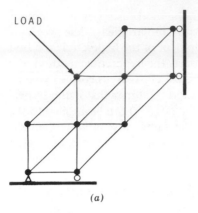

(a)

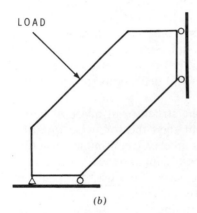

(b)

Figure 1.2. Example of a truss and a similarly shaped plate supporting the same load. (a) Truss. (b) Plate.

of elasticity theory was first given in the now classical paper of Turner, Clough, Martin, and Topp [67]. These investigators were the first to introduce what is now known as the direct stiffness method for determining finite element properties. Their studies, along with the advent of the digital computer at that time, opened the way to the solution of complex plane elasticity problems. After further treatment of the plane elasticity problem by Clough [68] in 1960, engineers began to recognize the efficacy of the finite element method.

Concepts of the method began to solidify after 1963 when Besseling [69], Melosh [70], Fraeijs de Veubeke [71], and Jones [72] recognized that the finite element method was a form of the Ritz method and confirmed it as a general technique to handle elastic continuum problems. In 1965, the finite element method received an even broader interpretation when Zienkiewicz and Cheung [73] reported that it is applicable to all field problems which can be cast into variational form. During the late 1960's and early 1970's (while mathematicians were working on establishing errors, bounds, and convergence criteria for finite element approximations), engineers and other appliers of the finite element method were also studying similar concepts for various problems in the area of solid mechanics. Several of these studies [74–84], although restricted to certain types of problems, have yielded useful results.

In the years since 1960, the finite element method has received widespread use in engineering. Hundreds of papers, the proceedings of several conferences and short courses [85–101], and seven books [102–108] have been published on the subject.† Although a major portion of this literature deals with static and dynamic structural analysis, there has been a continuing steady increase in the number of applications in other fields. Unquestionably, the finite element method is now a well-established and accepted engineering analysis tool.

1.4   RANGE OF APPLICATIONS

Applications of the finite element method can be divided into three categories, depending on the nature of the problem to be solved. In the first category are all the problems known as *equilibrium problems* or time independent problems. The majority of applications of the finite element method fall into this category. For the solution of equilibrium problems in the solid mechanics area, we need to find the displacement distribution or the stress distribution or perhaps the temperature distribution for a given mechanical or thermal loading. Similarly, for the solution of equilibrium problems in fluid mechanics, we need to find pressure, velocity, temperature, and sometimes concentration distributions under steady-state conditions.

In the second category are the so-called *eigenvalue problems* of solid and fluid mechanics. These are steady-state problems whose solution often requires the determination of natural frequencies and modes of vibration of solids and fluids. Examples of eigenvalue problems involving both solid and fluid mechanics appear in civil engineering when the interaction of lakes and dams is considered, and in aerospace engineering when the sloshing of liquid fuels in flexible tanks is involved. Another class of eigenvalue problems includes the stability of structures and the stability of laminar flows.

---

† A number of the recent survey articles [109–115] also serve as excellent sources for additional references.

In the third category is the multitude of time-dependent or *propagation problems* of continuum mechanics. This category is composed of the problems which result when the time dimension is added to the problems of the first two categories.

Just about every branch of engineering is a potential user of the finite element method. But the mere fact that this method can be used to solve a particular problem does not mean that it is the most practical solution technique. Often several attractive techniques are available to solve a given problem. Each technique has its relative merits, and no technique enjoys the lofty distinction of being "the best" for all problems. Consequently, when a designer or analyst has a continuum problem to solve, his first major step is to decide which method to use. This involves a study of the alternative methods of solution, the availability of computer facilities and computer packages, and, most important of all, the amount of time and money that can be spent to obtain a solution. These important aspects of the .finite element method are considered further throughout this book.

The range of possible applications of the finite element method extends to all engineering disciplines, but civil and aerospace engineers concerned with stress analysis are the most frequent users of the method. Major aircraft companies and other organizations involved in the design of structures have developed elaborate finite element computer programs. Many of these special-purpose programs are proprietary; however, some companies offer the use of their programs for a fee. Also, a number of large-scale stress analysis programs are available in the public domain; these are listed in Chapter 10.

1.5   THE FUTURE OF THE FINITE ELEMENT METHOD

Our brief look at the history of the finite element method shows us that its early development was sporadic. The applied mathematicians, the physicists, and the engineers all dabbled with finite element concepts; but they did not recognize at first the diversity and the multitude of potential applications. After 1960, this situation changed and the tempo of development increased markedly. By 1972, the finite element method had become the most active field of interest in the numerical solution of continuum problems.

As an analysis technique, the finite element method has reached the point where no additional dramatic developments or breakthroughs can be expected. Instead, future growth will involve broader applications to practical problems, increased understanding of special important aspects, and further refinement of the basic techniques.

Although in solid mechanics the finite element method can be used to solve a very large number of complex problems, there are still some areas where more work needs to be done. Some examples are the treatment of problems involving material failures, such as cracking, fracturing, and bond release in composites. Much needed attention must also be given to the modeling of nonlinear material behavior and the accurate characterization of material properties. All problems involving the determination of free boundaries should also be included in the list requiring further work.

Outside the field of solid mechanics, many extensions of the finite element method will continue to appear. In the general area of continuum mechanics, efforts will be made to refine the approach to propagation or time-dependent problems. Mathematicians will doubtless work to put the method on a broader theoretical foundation and to provide insight into problems of determining error bounds and rates of convergence for both linear and nonlinear problems. Several new types of elements will also be introduced, but many people believe that this aspect of the finite element method is already overworked.

Finally, from a practitioner's viewpoint, the finite element method, like any other numerical analysis technique, can always be made more efficient and easier to use. As the method is applied to larger and more complex problems, it becomes increasingly important that the solution process remains economical. This means that studies to find better ways to solve simultaneous linear and nonlinear equations will certainly continue. Also, since implementation of the finite element method usually requires a considerable amount of data handling, we can expect that ways to automate this process and make it more error-free will evolve.

REFERENCES

1. G. E. Forsythe and W. R. Wasow, *Finite Difference Methods for Partial Differential Equations*, John Wiley and Sons, New York, 1960.

2. R. D. Richtmyer and K. W. Morton, *Difference Methods for Initial-Value Problems*, 2nd ed., John Wiley-Interscience, New York, 1967.

3. R. W. Clough, "The Finite Element Method in Plane Stress Analysis," *Proceedings of 2nd ASCE Conference on Electronic Computation*, Pittsburgh, Pa., September 8 and 9, 1960.

4. R. Courant, "Variational Methods for the Solutions of Problems of Equilibrium and Vibrations," *Bull. Am. Math. Soc.*, Vol. 49, 1943.

5. L. Euler, *Methods Inveniendi Lineas Curvas Maximi Minimine Proprietate Gaudentes*, M. Bousquet, Lausanne and Geneva, 1774.

6. G. Polya, "Sur une Interpretation de la Methode des Differences Finies qui peut Fournir des Bornes Superieures ou Inferieures," *C. R. Acad. Sci.*, Vol. 235, 1952.

7. G. Polya, *Estimates for Eigenvalues: Studies Presented to Richard von Mises*, Academic Press, New York, 1954.

8. J. Hersch, "Equations Differentielles et Functions de Cellules," *C. R. Acad. Sci.*, Vol. 240, 1955.

9. H. F. Weinberger, "Upper and Lower Bounds for Eigenvalues by Finite Difference Methods," *Commun. Pure Appl. Math.*, Vol. 9, 1956.

10. H. F. Weinberger, "Lower Bounds for Higher Eigenvalues by Finite Difference Methods," *Pacific J. Math.*, Vol. 8, 1958.

11. J. Greenstadt, "On the Reduction of Continuous Problems to Discrete Form," *IBM J. Res. Dev.*, Vol. 3, 1959.

12. P. M. Morse and H. Feshback, *Methods of Theoretical Physics*, McGraw-Hill Book Company, New York, 1953, Section 9.4.

13. G. N. White, "Difference Equations for Plane Thermal Elasticity," LAMS-2745, Los Alamos Scientific Laboratory, Los Alamos, N. Mex., 1962.

14. K. O. Friedrichs, "A Finite Difference Scheme for the Neumann and the Dirichlet Problem," NYO-9760, Courant Institute of Mathematical Sciences, New York University, New York, 1962.

15. J. Cea, "Approximation Variationnelle des Problemes aux Limites," *Ann. Inst. Fourier*, Vol. 14, 1964.

16. R. B. Kellogg, "Difference Equations on a Mesh Arising from a General Triangulation," *Math. Comp.*, Vol. 18, 1964.

17. R. B. Kellogg, "Ritz Difference Equations on a Triangulation," *Proceedings of the Conference on the Application of Computing Methods to Reactor Problems*, Argonne National Laboratory, May 17–19, 1965.

18. P. G. Ciarlet, "Variational Methods for Non-Linear Boundary Value Problems," Thesis, Case Institute of Technology, Cleveland, Ohio, June 1966.

19. L. A. Oganesjan, "Convergence of Difference Schemes in Case of Improved Approximation of the Boundary" (in Russian), *Z. Vychisl. Mat. Mat. Fiz.*, Vol. 6, 1966.

20. R. S. Varga, "Hermite Interpolation-Type Ritz Methods for Two-Point Boundary Value Problems," in *Numerical Solutions of Partial Differential Equations*, J. H. Bramble (ed.), Academic Press, New York, 1967.

21. K. O. Friedrichs and H. B. Keller, "A Finite Difference Scheme for Generalized Neumann Problems," in *Numerical Solutions of Partial Differential Equations*, J. H. Bramble (ed.), Academic Press, New York, 1967.

22. J. P. Aubin, "Approximation des Espaces de Distributions et des Operateurs Differentiels," *Bull. Soc. Math. France*, Mem. 12, 1967.

23. P. G. Ciarlet, M. H. Schultz, and R. S. Varga, "Numerical Methods of High-Order Accuracy for Non-Linear Boundary Value Problems. I: One-Dimensional Problem," *Numer. Math.*, Vol. 9, 1967.

24. J. J. Göel, "Construction of Basic Functions for Numerical Utilization of Ritz's Method," *Numer. Math.*, Vol. 12, 1968.

25. M. Zlamal, "On the Finite Element Method," *Numer. Math.*, Vol. 12, 1968.

26. G. Birkhoff, M. H. Schultz, and R. S. Varga, "Piecewise Hermite Interpolation in One and Two Variables with Applications to Partial Differential Equations," *Numer. Math.*, Vol. 11, 1968.

27. M. H. Schultz, "L-Multivariate Approximation Theory," *SIAM J. Numer. Anal.*, Vol. 6, No. 2, June 1969.

28. A. Zenisek, "Interpolation Polynomials on the Triangle," *Numer. Math.*, Vol. 15, 1970.

29. J. H. Bramble and M. Zlamal, "Triangular Elements in the Finite Element Method," *Math. Comp.*, Vol. 24, No. 112, October 1970.

30. R. E. Carlson and C. A. Hall, "Ritz Approximations to Two-Dimensional Boundary Value Problems," *Numer. Math.*, Vol. 18, 1971.

31. I. Babuska, "The Rate of Convergence for the Finite Element Method," *SIAM J. Numer. Anal.*, Vol. 8, 1971.

32. I. Babuska, "Error Bounds for the Finite Element Method," *Numer. Math.*, Vol. 16, 1971.

33. G. Fix and N. Nassif, "On Finite Element Approximations to Time-Dependent Problems," *Numer. Math.*, Vol. 19, 1972.

34. Y. Yamamoto and N. Tukuda, "A Note on the Convergence of Finite Element Solutions," *Int. J. Numer. Methods Eng.*, Vol. 3, 1971.

35. I. Fried, "Discretization and Computational Errors in High-Order Finite Elements," *AIAA J.*, Vol. 9, No. 10, 1972.

36. I. Fried, "Accuracy of Complex Finite Elements," *AIAA J.*, Vol. 10, No. 3, 1972.

37. J. T. Oden, *Finite Elements of Non-linear Continua*, McGraw-Hill Book Company, New York, 1972.

38. G. Birkhoff and H. L. Garabedian, "Smooth Surface Interpolation," *J. Math. Phys.*, Vol. 39, 1960.

39. C. de Boor, "Bicubix Spline Interpolation," *J. Math. Phys.*, Vol. 41, 1962.

40. G. Birkhoff and C. de Boor, "Piece-wise Polynomial Interpolation and Approximation," in *Approximation of Functions*, H. L. Garabedian (ed.), Elsevier Publishing Company, Amsterdam, 1965.

41. J. H. Ahlberg, E. N. Nilson, and J. L. Walsh, *The Theory of Splines and Their Applications*, Academic Press, New York, 1967.

42. W. J. Gordon and D. H. Thomas, "Cardinal Functions for Spline Interpolation," *General Motors Res. Rept.* GMR-770, 1968.

43. W. J. Gordon, "Distributive Lattices and the Approximation of Multivariate Functions," *Proceedings of the Symposium on Approximation with Special Emphasis on Spline Functions*, held at the Mathematics Research Center, University of Wisconsin, May 5–7, 1969, Academic Press, New York, 1969.

44. I. J. Schoenberg (ed.), *Approximations with Special Emphasis on Spline Functions*, Academic Press, New York, 1969.

45. W. J. Gordon, "Spline-Blended Surface Interpolation Through Curve Networks," *J. Math. Mech.*, Vol. 18, No. 10, 1969.

46. T. N. E. Greville (ed.), *Theory and Applications of Spline Functions*, Academic Press, New York, 1969.

47. R. S. Varga, "Functional Analysis and Approximation Theory in Numerical Analysis," *SIAM (Regular Conference Series in Applied Mathematics)*, Philadelphia, Pa., 1971.

48. W. G. Strang and G. Fix, *An Analysis of the Finite Element Method*, Prentice-Hall, 1973.

49. I. Babuska and A. K. Aziz (eds.), *The Mathematical Foundations of the Finite Element Method—with Applications to Partial Differential Equations*, Academic Press, New York, 1973.

50. J. Whiteman (ed.), *The Mathematics of Finite Elements and Applications*, Academic Press, New York, 1973.

51.   J. T. Oden, "Some Aspects of Recent Contributions to the Mathematical Theory of Finite Elements," in *Advances in Computational Methods in Structural Mechanics and Design*, Proceedings of the 2nd U.S.-Japan Seminar, held in August 1972, University of Alabama Press, Huntsville, Ala., 1972.

52.   W. Prager and J. L. Synge, "Approximation in Elasticity Based on the Concept of Function Space," *Quart. Appl. Math.*, Vol. 5, 1947.

53.   J. L. Synge, "Triangulation in the Hypercircle Method for Plane Problems," *Proc. Roy. Irish Acad.*, Vol. 54A21, 1952.

54.   J. McMahon, "Lower Bounds for the Electrostatic Capacity of a Cube," *Proc. Roy. Irish Acad.*, Vol. 55A9, 1953.

55.   A. Hrenikoff, "Solution of Problems in Elasticity by the Framework Method," *J. Appl. Mech.*, Vol. 8, 1941.

56.   D. McHenry, "A Lattice Analogy for the Solution of Plane Stress Problems," *J. Inst. Civ. Eng.*, Vol. 21, 1943.

57.   N. M. Newmark, in *Numerical Methods of Analysis in Engineering*, L. E. Grinter (ed.), Macmillan Company, New York, 1949.

58.   G. Kron, "Tensorial Analysis and Equivalent Circuits of Elastic Structures," *J. Franklin Inst.*, Vol. 238, 1944.

59.   G. Kron, "Equivalent Circuits of the Elastic Field," *J. Appl. Mech.*, Vol. 66, 1944.

60.   J. H. Argyris, "Energy Theorems and Structural Analysis," *Aircraft Eng.*, Vol. 26, October–November 1954.

61.   J. H. Argyris, "Energy Theorems and Structural Analysis," *Aircraft Eng.*, Vol. 27, February–March–April–May 1955.

62.   J.H. Argyris, "The Matrix Analysis of Structures with Cut-Outs and Modifications," *Proceedings of the 9th International Congress on Applied Mechanics*, Section II: Mechanics of Solids, September 1956.

63.   J. H. Argyris and S. Kelsey, "Structural Analysis by the Matrix Force Method with Applications to Aircraft Wings," *Wiss. Ges. Luftfahrt Jahrb.*, 1956.

64.   J. H. Argyris, "The Matrix Theory of Statics," *Ing.-Arch.*, Vol. 25, 1957.

65.   J. H. Argyris and S. Kelsey, "The Analysis of Fuselages of Arbitrary Cross-Section and Taper," *Aircraft Eng.*, Vol. 31, 1959.

66.   J. H. Argyris and S. Kelsey, *Energy Theorems and Structural Analysis*, Butterworth and Company, London, 1960.

67.   M. J. Turner, R. W. Clough, H. C. Martin, and L. C. Topp, "Stiffness and Deflection Analysis of Complex Structures," *J. Aeronaut. Sci.*, Vol. 23, No. 9, 1956.

68.   R. W. Clough, "The Finite Element Method in Plane Stress Analysis," *Proceedings of 2nd ASCE Conference on Electronic Computation*, Pittsburgh, Pa., September 8 and 9, 1960.

69.   J. F. Besseling, "The Complete Analogy Between the Matrix Equations and the Continuous Field Equations of Structural Analysis," International Symposium on Analogue and Digital Techniques Applied to Aeronautics, Liege, Belgium, 1963.

70.   R. J. Melosh, "Basis for the Derivation of Matrices for the Direct Stiffness Method," *AIAA J.*, Vol. 1, 1963.

71.   B. Fraeijs de Veubeke, "Upper and Lower Bounds in Matrix Structural Analysis," in *AGARD-ograph 72*, B. F. de Veubeke (ed.), Pergamon Press, New York, 1964.

72.   R. E. Jones, "A Generalization of the Direct-Stiffness Method of Structural Analysis," *AIAA J.*, Vol. 2, 1964.

73. O. C. Zienkiewicz and Y. K. Cheung, "Finite Elements in the Solution of Field Problems," *Engineer*, Vol. 220, 1965.

74. S. W. Key, "A Convergence Study of the Direct Stiffness Method," Ph.D. Dissertation, University of Washington, Seattle, Wash., 1966.

75. R. W. McLay, "Completeness and Convergence Properties of Finite Element Displacement Functions—A General Treatment," AIAA 5th Aerospace Science Meeting (*AIAA Paper* 67-143), New York, 1967.

76. P. Tong and T. H. H. Pian, "The Convergence of the Finite Element Method in Solving Linear Elastic Problems," *Int. J. Solids Struct.*, Vol. 3, 1967.

77. E. R. de Arentes e Oliveira, "Completeness and Convergence in the Finite Element Method," *Proceedings of the 2nd Conference on Matrix Methods in Structural Mechanics* (AFFDL-TR-68-150), Wright-Patterson Air Force Base, Dayton, Ohio, October 1968.

78. M. W. Johnson and R. W. McLay, "Convergence of the Finite Element Method in the Theory of Elasticity," *J. Appl. Mech.*, Vol. 35, No. 2, June 1968.

79. J. E. Waltz, R. E. Fulton, and N. J. Cyrus, "Accuracy and Convergence of Finite Element Approximations," *Proceedings of the 2nd Conference on Matrix Methods in Structural Mechanics* (AFFDL-TR-68-150), Wright-Patterson Air Force Base, Dayton, Ohio, October 1968.

80. P. C. Dunne, "Complete Polynomial Displacement Fields for the Finite Element Method," *J. Roy. Aeronaut. Soc.*, Vol. 72, 1968.

81. H. Ramstad, "Convergence and Numerical Accuracy with Special Reference to Plate Bending," in *Finite Element Methods in Stress Analysis*, I. Holand and K. Bell (eds.), Tapir Press, Trondheim, Norway, 1969.

82. A. Zenisek and M. Zlamal, "Convergence of a Finite Element Procedure for Solving Boundary Value Problems of the Fourth Order," *Int. J. Numer. Methods Eng.*, Vol. 2, 1970.

83. M. Mikkola, "On the Convergence of the Finite Element Method," *4th Scandinavian Meeting on Strength of Materials*, Helsinki, 1971.

84. E. R. de Arantes e Oliveira, "Convergence Theorems in the Theory of Structures," in *The NATO Lectures on Finite Element Methods in Continuum Mechanics*, J. T. Oden and E. R. A. Oliveira (eds.), University of Alabama Press, Huntsville, Ala., 1973.

85. O. C. Zienkiewicz and G. Hollister (ed.), *Stress Analysis*, John Wiley and Sons, London, 1966.

86. J. Przmieniecki et al. (eds.), *Proceedings of the 1st Conference on Matrix Methods in Structural Mechanics* (AFFDL-TR-66-80), Wright-Patterson Air Force Base, Dayton, Ohio, October 1965.

87. L. Berke et al. (eds.), *Proceedings of the 2nd Conference on Matrix Methods in Structural Mechanics* (AFFDL-TR-68-150), Wright-Patterson Air Force Base, Dayton, Ohio, October 1968.

88. *Proceedings of the Third Conference on Matrix Methods in Structural Mechanics*, Wright-Patterson Air Force Base, Dayton, Ohio, October 1971 (in press).

89. *First Conference on Electronic Computation*, held in Kansas City, Mo., November 1958, American Society of Civil Engineers (special publication).

90. *Second Conference on Electronic Computation*, held in Pittsburgh, Pa., September 1960, American Society of Civil Engineers (special publication).

91. *Third Conference on Electronic Computation*, held in Boulder, Colo., June 1963, published as *ASCE Proc., J. Struct. Div.*, Vol. 89, No. ST4, August 1963.

92. *Fourth Conference on Electronic Computation*, held in Los Angeles, Calif., September 1966, published as *ASCE Proc., J. Struct. Div.*, Vol. 92, No. ST6, December 1966.

93. *Fifth Conference on Electronic Computation*, held at Purdue University, Lafayette, Ind., September 1970, published as *ASCE Proc., J. Struct. Div.*, Vol. 97, No. ST1, January 1971.

94. B. M. Fraeijs de Veubeke (ed.), "High-Speed Computing of Elastic Structures," *Proceedings of IVTAM Symposium on High-Speed Computing of Elastic Structures*, University of Liege, Liege, Belgium, August 1970.

95. B. M. Fraeijs de Veubeke (ed.), *Matrix Methods of Structural Analysis*, Pergamon Press, Oxford, 1964.

96. W. Rowan and R. Hackett (eds.), *Proceedings of the Symposium on Application of Finite Element Methods in Civil Engineering*, Vanderbilt University, Nashville, Tenn., November 1969.

97. R. H. Gallagher, Y. Yamada, and J. T. Oden (eds.), *Recent Advances in Matrix Methods of Structural Analysis and Design*, University of Alabama Press, Huntsville, Ala., 1971.

98. *Proceedings from the Symposium on Numerical and Computer Methods in Structural Mechanics*, held at the University of Illinois, Urbana, Ill., September 1971.

99. G. L. M. Gladwell (ed.), "Computer Aided Engineering," *Solid Mechanics Study* 5, University of Waterloo, Ontario, 1971.

100. I. Holand and K. Bell (eds.), *Finite Element Methods in Stress Analysis*, Tapir Press, Trondheim, 1969.

101. H. Tottenham and C. Brebbia (eds.), *Finite Element Techniques in Structural Mechanics*, Stress Analysis Publishers, Southampton, England, 1971.

102. O. C. Zienkiewicz, *The Finite Element Method in Engineering Science*, McGraw-Hill Book Company, London, 1971.

103. C. Desai and J. Abel, *Introduction to the Finite Element Method*, Van Nostrand-Reinhold, New York, 1971.

104. J. T. Oden, *Finite Elements of Non-linear Continua*, McGraw-Hill Book Company, New York, 1972.

105. R. H. Gallagher, *Finite Element Analysis: Fundamentals*, Prentice-Hall, 1975.

106. H. C. Martin and G. F. Carey, *Introduction to Finite Element Analysis*, McGraw-Hill Book Company, New York, 1973.

107. D. H. Norrie and G. de Vries, *The Finite Element Method*, Academic Press, New York, 1973.

108. J. Robinson, *Integrated Theory of Finite Element Methods*, John Wiley and Sons, London, 1973.

109. C. A. Felippa and R. W. Clough, "The Finite Element Method in Solid Mechanics," *Numerical Solution of Field Problems in Continuum Physics*, SIAM-AMS Proceedings, Vol. 2, American Mathematical Society, Providence, R.I., 1970.

110. A. C. Singhal, "775 Selected References on the Finite Element Method and Matrix Methods of Structural Analysis," *Rept.* S-12, Civil Eng. Dept., Laval University, Quebec, January 1969.

111. Z. Zudans, "Survey of Advanced Structural Design Analysis Techniques," *Nucl. Eng. Design*, Vol. 10, No. 4, 1969.

112. O. C. Zienkiewicz, "The Finite Element Method: From Intuition to Generality," *Appl. Mech. Rev.*, Vol. 23, No. 3, March 1970.

113. P. V. Marcal, "Finite Element Analysis with Material Non-Linearities—Theory and Practice," in *Recent Advances in Matrix Methods of Structural Analysis and Design*, R. H. Gallagher et al. (eds.), University of Alabama Press, Huntsville, Ala., 1971.

114. J. E. Akin, D. L. Fenton, and W. C. T. Stoddart, "The Finite Element Method—A Bibliography of Its Theory and Application," *Rept*. EM 72-1, Dept. of Eng. Mech., University of Tennessee, Knoxville, Tenn., 1971.

115. J. R. Whiteman, "A Bibliography for Finite Element Methods," *Brunel Univ. Rept.* TR/9, Dept. of Mathematics, Brunel University, Uxbridge, 1972.

2

THE DIRECT APPROACH:
A PHYSICAL INTERPRETATION

2.1 INTRODUCTION

The finite element method offers a way to solve a complex continuum problem by allowing us to subdivide it into a series of simpler interrelated problems. Essentially, it gives us a consistent technique for modeling the whole as an assemblage of discrete parts or finite elements. The "whole" may be a body of matter or a region of space in which some phenomenon of interest is occurring. The degree to which the assemblage of elements represents the whole usually depends on the number, size, and type of elements chosen for the representation. Sometimes it is possible to choose the elements in a way that leads to an exact representation, but this occurs only in special cases. Most often the choice of elements is a matter of engineering judgment based on accumulated experience. We will continue to elaborate on this important point throughout the book.

Since the fundamental ideas of discretization in the finite element method stem from the physical procedures used in network analysis and structural framework analysis, we shall begin our discussion of finite element concepts by considering simple examples from these areas. The advantage of this approach is that an understanding of the techniques and essential concepts is gained without much mathematical manipulation. Accordingly, we can develop an intuitive "feel" for the process before going on to more advanced topics. As we have seen, in any finite element analysis the first step is to replace a complex system by an equivalent idealized system consisting of individual elements connected to each other at specified points or nodes. Implicit in this procedure is the problem of defining or identifying the elements and then determining their properties. For some problems the part that should be chosen as one element immediately suggests itself, whereas

22

in other cases the choice is not so obvious. In this chapter we will consider several example problems for which we can easily identify discrete elements. Once the elements have been selected, we will use direct physical reasoning to establish the element equations in terms of pertinent variables. Then we will combine the element equations to form the governing equations for the complete system. By this means we will illustrate the so-called direct approach for determining element properties, and we will establish the general assembly procedures common to all finite element analyses.

Finally, in the concluding section of this chapter, we will point out that the early direct approaches to the finite element method lead to far broader interpretations of the method and open the way for applying it to many different kinds of problems in mathematical physics.

2.2 DEFINING ELEMENTS AND THEIR PROPERTIES

2.2.1 Linear spring systems

One of the most elementary systems that we can examine from a finite element viewpoint is the linear spring system shown in Figure 2.1. In this system we have two springs connected in series in, say, the x coordinate direction.

One end of the spring on the left-hand side is rigidly attached to a wall, while the spring on the right is free to move. We assume that both springs can experience either tension or compression. Forces, displacements, and spring stiffnesses are the only parameters in this system.

The way to subdivide this system into discrete elements is immediately obvious. If we define each spring to be an element, the system consists of two elements and three nodes (points of connection where forces can be transmitted and displacements can exist).

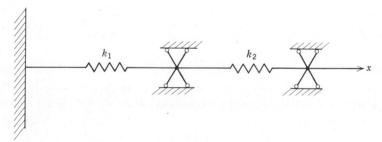

Figure 2.1. A simple linear spring system.

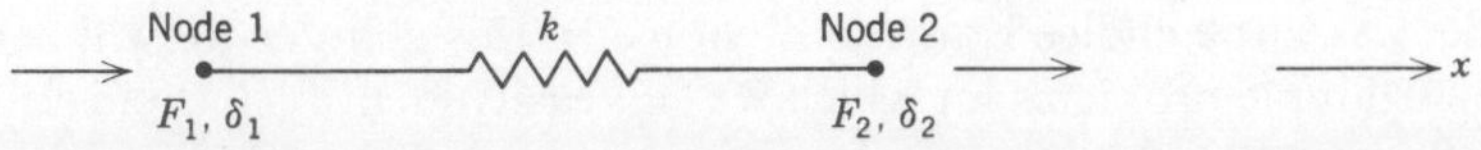

Figure 2.2. One element of a linear spring system.

To determine the properties of an element, in this case its force-displacement equations, we focus our attention on an isolated element shown in the free-body diagram of Figure 2.2. A force and a displacement are defined at each node, and for convenience we can take these forces and displacements in the positive x direction. The field variable for this trivial example is displacement. We do not need to choose an interpolation function to represent the variation of the field variable over the element because an exact representation is already available. Hooke's law in its simplest form provides the means to relate the nodal displacements and the applied nodal forces.

According to Hooke's law, when an elastic spring experiences an axial load F, it deflects an amount δ, given by

$$\delta = \frac{1}{k} F = cF \tag{2.1}$$

where k is the spring stiffness coefficient and c is the deflection coefficient. From equation 2.1 we may interpret k as the force required to produce unit deflection and c as the deflection caused by a unit force.

Using equation 2.1, we can now write two more equations, one for each node, expressing the equilibrium of forces in terms of displacements:

$$F_1 = k\,\delta_1 - k\,\delta_2 \tag{2.2}$$

$$F_2 = -k\,\delta_1 + k\,\delta_2 \tag{2.3}$$

If we use matrix notation,† equations 2.2 and 2.3 may be written as one equation expressing the force-displacement properties of the elements, that is,

$$\begin{bmatrix} k & -k \\ -k & k \end{bmatrix} \begin{Bmatrix} \delta_1 \\ \delta_2 \end{Bmatrix} = \begin{Bmatrix} F_1 \\ F_2 \end{Bmatrix} \tag{2.4a}$$

or

$$[K]\{\delta\} = \{F\} \tag{2.4b}$$

Elements of $[K]$ are usually subscripted as k_{ij} to denote their location in the ith row and jth column of $[K]$. The square matrix $[K]$ is known as the element stiffness matrix, the column vector $\{\delta\}$ is the nodal displacement

† Matrix notation and matrix manipulations used in this book are summarized in Appendix A.

vector, and the column vector $\{F\}$ is the resultant nodal force vector for the element.

Although equation 2.4 was derived for one of the simplest type of finite elements, namely, a linear spring, it possesses many of the characteristics of the equations expressing the properties of more complex elements. For instance, the *form*† of equation 2.4 remains the same regardless of the type of problem, the complexity of the element, or the way in which the element properties were derived. In this simple example, Hooke's law enabled us to determine exact values for the stiffness coefficients in the matrix $[K]$; but for the more complex situations that we will encounter later, the stiffness coefficients will be determined approximately by using assumed displacement functions and variational principles. Whether the stiffness coefficients of $[K]$ are determined exactly or approximately, their interpretation is the same—a typical stiffness coefficient of $[K]$, k_{ij}, is defined for this example as the force required at node i to produce a unit deflection at node j. This definition holds because only one force and one displacement exist at each node. We also note that our simple stiffness matrix obeys the Maxwell-Betti reciprocal theorem, which states that all stiffness matrices for linear structures referred to orthogonal coordinate systems must be symmetric.

The element properties given by equation 2.4 apply to either the right- or the left-hand element, depending on which value of spring stiffness we substitute into $[K]$. The fact that the left-hand element is constrained to have zero displacement at one node does not influence the derivation of the element properties. Constraint conditions are taken into account only after the element equations are assembled to form the system equations. We shall discuss the assembly procedure common to all finite element analysis later in this chapter after we have considered several other examples showing how element properties can be established by the physical direct method.

2.2.2 Flow systems

One-dimensional heat flow

Another simple system that we can conveniently study using finite element concepts is shown in Figure 2.3. Here we have a section of layered material through which heat is flowing in only the x direction. In this heat conduction

---

† We will call the form of equation 2.4 the "standard form" because element equations and system equations usually appear in this way. Even the matrix equations for nonlinear problems can be written in the standard form, but the influence coefficients in $[K]$ for this case can be variables instead of constants. As we will see later, the matrix equations for eigenvalue problems have a slightly different form.

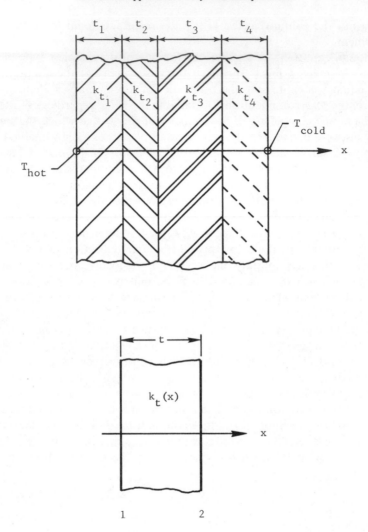

Figure 2.3. One-dimensional heat flow through a composite material.

problem we will assume that there is no internal heat generation, that the left-hand side of the wall is held at a uniform temperature higher than that of the right-hand side, and that each layer is an inhomogeneous solid whose thermal conductivity is a known function of position in the direction of heat flow. Heat flux, temperature, thermal conductivity, and layer thickness are the pertinent parameters.

This problem splits into a series of simpler ones if we consider each layer of the material as a finite element whose characteristics can be determined by the basic law of heat conduction. The field variable for this problem is the temperature.

The "nodes" for a typical element are now the bounding *planes* of the layer, and each node is characterized by a temperature which is uniform over the plane. Our system then consists of four elements and five nodes. A typical isolated element is also shown in Figure 2.3. We can find the exact heat flow behavior of an element by using the basic law between heat flow and temperature gradient established by the French mathematician J. B. J. Fourier. Again, we do not need an assumed interpolation function. According to Fourier's law, the quantity of heat crossing a unit area per unit time in the x-direction is given by

$$q = -k_t(x)\frac{dT}{dx} \tag{2.5}$$

where $k_t(x)$ is the thermal conductivity of the material. Integrating equation 2.5 across the thickness of the element gives

$$q\int_{x_1}^{x}\frac{dx}{k_t(x)} = -(T - T_1)$$

or, if we let

$$\frac{1}{\bar{k}_t} = \int_{x_1}^{x}\frac{dx}{k_t(x)}$$

we can write

$$q - q_1 = -\bar{k}_t(T - T_1)$$

Hence the heat flux entering nodes 1 and 2 is given by

$$Q_1 = -\bar{k}_t(T_2 - T_1) \tag{2.6}$$

$$Q_2 = -\bar{k}_t(T_1 - T_2) \tag{2.7}$$

respectively.

In matrix notation equation 2.6 and 2.7 may be written as

$$\begin{bmatrix} \bar{k}_t & -\bar{k}_t \\ -\bar{k}_t & \bar{k}_t \end{bmatrix}\begin{Bmatrix} T_1 \\ T_2 \end{Bmatrix} = \begin{Bmatrix} Q_1 \\ Q_2 \end{Bmatrix}$$

or, more concisely, as

$$[K_t]\{T\} = \{Q\} \tag{2.8}$$

where $[K_t]$ = matrix of thermal influence coefficients,
$\quad\quad\{T\}$ = column vector of nodal temperatures,
$\quad\quad\{Q\}$ = column vector of nodal heat fluxes.

We recognize that equation 2.8 has the standard form and that it completely defines the heat conduction properties of our simple thermal element. The thermal "stiffness" matrix is analogous to the structural stiffness matrix. Later, when we consider the general heat conduction problem, we will see that element properties will again be expressed in the form of equation 2.8. The only difference will be in the dimension of $[K_t]$ and the complexity of its terms.

Fluid and electrical networks

Figure 2.4 shows a simple fluid flow network that could represent the water distribution system in a small building. The system is composed of many individual flow paths, and the problem is to find how the pressure and flow emanating from a given source, such as a pump, are distributed among the various paths.

We can imagine this system to be a collection of finite elements if we define each flow path between two junctions to be an element. Using an arbitrary numbering scheme, we find that this system has 13 elements and 14 nodes. The element characteristics, that is, the pressure loss–flow rate relations, may then be derived for some cases from first principles of fluid mechanics and no assumed interpolation functions are needed. For example, if we may assume that the flow paths are circular pipes of constant cross-sectional area and that fully developed laminar flow exists in each pipe, then for any orientation of the pipe the pressure drop $(P_1 - P_2)$ between stations 1 and 2 due to a flow Q is given as [1]

$$P_1 - P_2 = \frac{128QL\mu}{\pi D^4} \tag{2.9}$$

where L is the length of the pipe, μ is the dynamic viscosity of the fluid, and D is the pipe diameter. For a typical element (Figure 2.4b) we can use equation 2.9 to write two equations for the flows at the nodes in terms of the pressures at the nodes. The flows *entering* the element at its ends are as follows:

Node 1: $Q_1 = \dfrac{\pi D^4}{128L\mu}(P_1 - P_2)$

Node 2: $Q_2 = \dfrac{\pi D^4}{128L\mu}(P_2 - P_1)$

In matrix notation these equations become

$$\frac{\pi D^4}{128L\mu}\begin{bmatrix} 1 & -1 \\ -1 & 1 \end{bmatrix}\begin{Bmatrix} P_1 \\ P_2 \end{Bmatrix} = \begin{Bmatrix} Q_1 \\ Q_2 \end{Bmatrix} \tag{2.10a}$$

or

$$[K_p]\{P\} = \{Q\} \tag{2.10b}$$

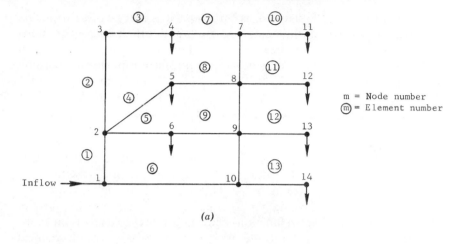

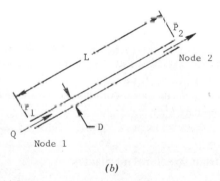

(b)

Figure 2.4. A fluid flow network. (a) A schematic representation of a flow system with numbered nodes and elements. (b) A typical element.

where $[K_p]$ = a "fluidity" matrix,

$\{P\}$ = column vector of nodal pressures,

$\{Q\}$ = column vector of nodal flows.

Again we see that equation 2.10 has the standard form. Our development of the element equations for the special case of laminar pipe flow does not include the pressure losses associated with the fittings and valves that are part of most fluid networks. These pressure losses are normally taken into account by introducing an equivalent length of pipe which can be lumped with the actual pipe length.

If the fluid flow in the network is turbulent, it is still possible to define an element as a length of fluid-carrying conduit, but the element equations are no longer linear. This can be seen by examining the Fanning equation, an empirical relation governing fully developed turbulent pipe flow. According to Fanning's equation, [1]

$$P_1 - P_2 = \frac{8 f_M L}{\pi D^5} Q^2$$

where L is the length of the pipe, D is its diameter, and f_M is the Moody friction factor, which is a function of the Reynolds number and the pipe roughness. Although it would be possible to use Fanning's equation to develop element equations similar to equation 2.10, the resulting fluidity matrix would contain known functions of the flow rate Q instead of constants. The same nonlinear character would prevail in the final equations assembled from the individual element equations, and hence special solution techniques would be required. Generally finite element techniques work best for linear networks.

The idea of defining a flow path as a finite element of a fluid flow network also applies to direct-current electrical networks. A current-carrying member of the electrical network can be taken as a finite element, and Ohm's law then provides the means for establishing the element characteristics. Procedures directly analogous to those used for the fluid flow networks may be used; voltages V_1 and V_2 play the same role as the nodal pressures, and a current I replaces the flow rate Q.

2.2.3   Simple elements from structural mechanics

In the engineering community the idea of modeling a structure as a series of finite elements began as an extension of the traditional methods used for analyzing framed structures such as trusses and bridges. These structures consist of bars in the form of spars or ribs which are distinct parts interconnected only at certain points where forces can be transmitted. Hence it seemed natural to an engineer to regard these structures as assemblages of individual components or elements.

To find the force-deflection characteristics or the stiffness of a complete framed structure, the engineer first determined the stiffness of its individual components, and the method he originally used is now known as the *direct stiffness method*. To illustrate the direct stiffness method we will derive the stiffness matrices for two simple elements, a pin-connected bar element and a triangular element for a thin plate.

A truss element

A pin-connected bar, the simplest type of truss element, provides a good starting point to illustrate how the direct method is used to calculate the stiffness matrix of a structural element. Consider the pin-connected truss member and coordinate system shown in Figure 2.5. We define each end of the element to be a node where two components of force and displacement may act. Since the nodes are pinned, no moments can exist there. This element is different from any we have discussed thus far because it has two degrees of freedom at each node instead of one. In total the element has four degrees of freedom, so we will need four equations to describe its force-deflection

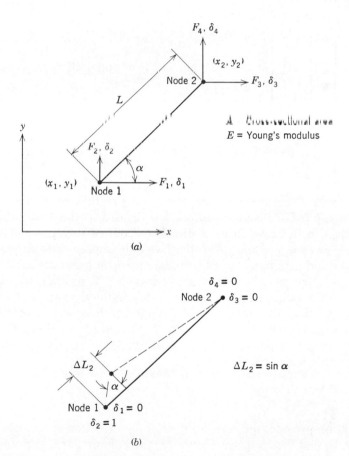

Figure 2.5. A pin-connected truss element with forces and displacements defined at its nodes. (*a*) A truss element and an assigned coordinate system. (*b*) Element configuration for $\delta_2 = 1$, $\delta_1 = \delta_3 = \delta_4 = 0$.

characteristics. In matrix notation these equations will have the standard form:

$$
\begin{bmatrix}
k_{11} & k_{12} & k_{13} & k_{14} \\
k_{21} & k_{22} & k_{23} & k_{24} \\
k_{31} & k_{32} & k_{33} & k_{34} \\
k_{41} & k_{42} & k_{43} & k_{44}
\end{bmatrix}
\begin{Bmatrix}
u_1 \\
v_1 \\
u_2 \\
v_2
\end{Bmatrix}
=
\begin{Bmatrix}
F_1 \\
F_2 \\
F_3 \\
F_4
\end{Bmatrix}
$$

or

$$[K]\{\delta\} = \{F\}$$

where, as before,

$[K]$ = element stiffness matrix,

$$
\{\delta\} =
\begin{Bmatrix}
\delta_1 \\
\delta_2 \\
\delta_3 \\
\delta_4
\end{Bmatrix}^{\dagger}
=
\begin{Bmatrix}
u_1 \\
v_1 \\
u_2 \\
v_2
\end{Bmatrix}
= \text{column vector of nodal displacements,}
$$

$$
\{F\} =
\begin{Bmatrix}
F_1 \\
F_2 \\
F_3 \\
F_4
\end{Bmatrix}
= \text{column vector of nodal forces.}
$$

We can use the operational definition of $[K]$ to find its various stiffness coefficients, k_{ij}. For instance, since k_{12} is defined as the force (associated with node 1 in this case) in the x direction required to produce a unit deflection $\delta_2 = v_1 = 1$ in the y direction, we may impose these displacement conditions and determine k_{12} either theoretically or experimentally.

If we assume that, when the beam is displaced at its ends, its change in length is small compared to its initial length, and that its angle of inclination remains essentially the same, we may apply linear theory to relate forces and displacements. Again, there is no need for assumed functions. From elementary strength of materials [2], the axial tensile force necessary to elongate the beam an amount ΔL is given by

$$F = \frac{AE}{L} \Delta L$$

This relation provides the basis for developing the stiffness coefficients. For example, to find the coefficients in the second column of $[K]$ we subject the

---

† The arrangement of nodal displacements in the column vector is arbitrary, but once it is specified, $[K]$ is defined with respect to that arrangement.

element to a unit displacement at node 1 in the y direction, while holding the other displacements equal to zero (Figure 2.5b). The compression force required to shorten the element by ΔL_2 is $-(AE/L)\,\Delta L_2 = -(AE/L)\sin\alpha$. The reaction components of this force in the x and y directions are the stiffness coefficients k_{21} and k_{22}, respectively. Hence

$$k_{12} = \frac{AE}{L}\sin\alpha\cos\alpha$$

$$k_{22} = \frac{AE}{L}\sin\alpha\sin\alpha$$

The other two coefficients in the second column are the reaction forces, which are equal in magnitude but opposite in sign. Thus

$$k_{32} = -\frac{AE}{L}\sin\alpha\cos\alpha$$

$$k_{42} = -\frac{AE}{L}\sin\alpha\sin\alpha$$

Continuing in a similar manner, we can find the remaining coefficients. The result is a symmetric matrix given by

$$[K] = \frac{AE}{L}\begin{bmatrix} \cos^2\alpha & \cos\alpha\sin\alpha & -\cos^2\alpha & -\cos\alpha\sin\alpha \\ \cos\alpha\sin\alpha & \sin^2\alpha & -\cos\alpha\sin\alpha & -\sin^2\alpha \\ -\cos^2\alpha & -\cos\alpha\sin\alpha & \cos^2\alpha & \cos\alpha\sin\alpha \\ -\cos\alpha\sin\alpha & -\sin^2\alpha & \cos\alpha\sin\alpha & \sin^2\alpha \end{bmatrix}$$

where

$$L = \sqrt{(x_1 - x_2)^2 + (y_1 - y_2)^2} \qquad (2.11)$$

$$\cos\alpha = \frac{x_2 - x_1}{L}, \qquad \sin\alpha = \frac{y_2 - y_1}{L}$$

A triangular element

At this point in our derivations of element properties by the direct method we have considered only one-dimensional elements. Since the number of practical problems which can be treated by one-dimensional formulations is limited, we need to turn our attention to multidimensional elements. In two-dimensional problems the simplest and probably the most widely used element is the three-node triangular one.

Use of the triangular element in elasticity problems stems from the earliest work and is now classical. In the following we will show how the direct method was first used by Turner et al. [3] to derive the stiffness matrix for a three-node triangular element for plane stress problems. There are six essential steps in this derivation:

1. Assume the functional form for the displacement field within an element.
2. Express the displacement field in terms of the displacements defined at the nodes.
3. Introduce the strain-displacement equations and thereby determine the state of element strain corresponding to the assumed displacement field.
4. Write the constitutive equations relating stress to strain and thereby introduce the influence of the material properties of the element.
5. Find the set of nodal forces which are the static equivalent of the stresses acting at the edges of the element.
6. Combine the results of steps 1 through 5 and identify the element stiffness matrix from the resulting expression.

Step 1

Figure 2.6 shows a typical triangular element of a thin disk experiencing plane stress. The element is referenced to an orthogonal coordinate system, and its three nodes have been arbitrarily numbered in the counterclockwise direction. If we define two components of displacement at each node, this element has six degrees of freedom, given by the column vector of displacements:

$$\{\delta\} = \begin{Bmatrix} \delta_1 \\ \delta_2 \\ \delta_3 \\ \delta_4 \\ \delta_5 \\ \delta_6 \end{Bmatrix} = \begin{Bmatrix} u_1 \\ v_1 \\ u_2 \\ v_2 \\ u_3 \\ v_3 \end{Bmatrix} \tag{2.12}$$

The displacement distribution within the element clearly must be a function of x and y, and it must be *uniquely* determined by the six nodal displacements. With this constraint, a linear displacement function is the only choice, that is,

$$\begin{aligned} u &= \alpha_1 + \alpha_2 x + \alpha_3 y \\ v &= \alpha_4 + \alpha_5 x + \alpha_6 y \end{aligned} \tag{2.13}$$

where α_1 through α_6 are constants yet to be determined. Equation 2.13 states that the displacements vary linearly within the element and along the element boundaries. Such behavior of the displacement functions ensures continuity of the displacement field throughout the structure modeled by the mesh of adjacent elements. This desirable feature derives from the

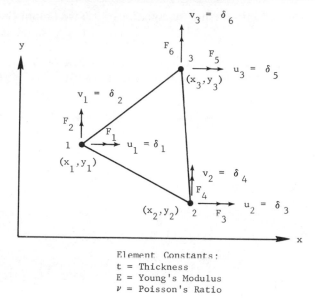

Element Constants:
t = Thickness
E = Young's Modulus
ν = Poisson's Ratio

Figure 2.6.   A triangular element in plane stress.

fact that the displacement along a boundary between two adjacent elements is uniquely specified by the displacements at the two nodes defining the ends of the boundary.

Step 2

Having defined the displacement field within an element, we are now ready to find the constants α_1 through α_6 in terms of the coordinates of the node points and the nodal displacements. Evaluating equation 2.13 at each of the nodes gives two sets of three equations each:

For node 1: $u_1 = \alpha_1 + \alpha_2 x_1 + \alpha_3 y_1$

For node 2: $u_2 = \alpha_1 + \alpha_2 x_2 + \alpha_3 y_2$ (2.14)

For node 3: $u_3 = \alpha_1 + \alpha_2 x_3 + \alpha_3 y_3$

Likewise, for the y component of displacement we have

$$v_1 = \alpha_4 + \alpha_5 x_1 + \alpha_6 y_1$$
$$v_2 = \alpha_4 + \alpha_5 x_2 + \alpha_6 y_2$$
$$v_3 = \alpha_4 + \alpha_5 x_3 + \alpha_6 y_3$$

With a little algebraic manipulation we can easily solve equations 2.14 for α_1, α_2, and α_3 in terms of u_1, u_2, u_3 and the nodal coordinates. When the result is substituted into the first of equations 2.13, we obtain the following expression for u:

$$u = \left(\frac{a_1 u_1 + a_2 u_2 + a_3 u_3}{2\Delta}\right) + \left(\frac{b_1 u_1 + b_2 u_2 + b_3 u_3}{2\Delta}\right) x$$

$$+ \left(\frac{c_1 u_1 + c_2 u_2 + c_3 u_3}{2\Delta}\right) y \tag{2.16}$$

where

$$\begin{aligned}
a_1 &= x_2 y_3 - x_3 y_2, & a_2 &= x_3 y_1 - x_1 y_3, & a_3 &= x_1 y_2 - x_2 y_1 \\
b_1 &= y_2 - y_3, & b_2 &= y_3 - y_1, & b_3 &= y_1 - y_2 \\
c_1 &= x_3 - x_2, & c_2 &= x_1 - x_3, & c_3 &= x_2 - x_1
\end{aligned} \tag{2.17}$$

and

$$\Delta = \text{area of the triangular element} = \frac{1}{2} \begin{vmatrix} 1 & x_1 & y_1 \\ 1 & x_2 & y_2 \\ 1 & x_3 & y_3 \end{vmatrix} \tag{2.18}$$

when the nodes of the element are numbered in the counterclockwise direction.

Following the same procedure for equations 2.15, we finally obtain

$$v = \left(\frac{a_1 v_1 + a_2 v_2 + a_3 v_3}{2\Delta}\right) + \left(\frac{b_1 v_1 + b_2 v_2 + b_3 v_3}{2\Delta}\right) x$$

$$+ \left(\frac{c_1 v_1 + c_2 v_2 + c_3 v_3}{2\Delta}\right) y \tag{2.19}$$

Step 3

Now that the displacement field is specified, we can proceed to find the element strains corresponding to the assumed displacements. For the plane stress case we are considering, the only strains that exist are those in the x-y plane $(\epsilon_x, \epsilon_y, \gamma_{xy})$, which for small displacements are given by[†]

$$\epsilon_x = \frac{\partial u}{\partial x}, \qquad \epsilon_y = \frac{\partial v}{\partial y}, \qquad \gamma_{xy} = \frac{\partial v}{\partial x} + \frac{\partial u}{\partial y} \tag{2.20}$$

[†] The reader who is unfamiliar with these relations may want to refer to the review of the basic relations of solid mechanics given in Appendix C.

An expression for the element strain can now be obtained by substituting equations 2.16 and 2.19 into equation 2.20 and differentiating. Using matrix notation, we may write

$$\epsilon_x = \frac{1}{2\Delta} \lfloor b_1 \quad 0 \quad b_2 \quad 0 \quad b_3 \quad 0 \rfloor \begin{Bmatrix} u_1 \\ v_1 \\ u_2 \\ v_2 \\ u_3 \\ v_3 \end{Bmatrix}$$

$$\epsilon_y = \frac{1}{2\Delta} \lfloor 0 \quad c_1 \quad 0 \quad c_2 \quad 0 \quad c_3 \rfloor \begin{Bmatrix} u_1 \\ v_1 \\ u_2 \\ v_2 \\ u_3 \\ v_3 \end{Bmatrix}$$

$$\gamma_{xy} = \lfloor c_1 \quad b_1 \quad c_2 \quad b_2 \quad c_3 \quad b_3 \rfloor \begin{Bmatrix} u_1 \\ v_1 \\ u_2 \\ v_2 \\ u_3 \\ v_3 \end{Bmatrix}$$

or

$$\{\epsilon\} = \begin{Bmatrix} \epsilon_x \\ \epsilon_y \\ \gamma_{xy} \end{Bmatrix} = \frac{1}{2\Delta} \begin{bmatrix} b_1 & 0 & b_2 & 0 & b_3 & 0 \\ 0 & c_1 & 0 & c_2 & 0 & c_3 \\ c_1 & b_1 & c_2 & b_2 & c_3 & b_3 \end{bmatrix} \begin{Bmatrix} u_1 \\ v_1 \\ u_2 \\ v_2 \\ u_3 \\ v_3 \end{Bmatrix} \qquad (2.21a)$$

If we let

$$[B] = \frac{1}{2\Delta} \begin{bmatrix} b_1 & 0 & b_2 & 0 & b_3 & 0 \\ 0 & c_1 & 0 & c_2 & 0 & c_3 \\ c_1 & b_1 & c_2 & b_2 & c_3 & b_3 \end{bmatrix} \qquad (2.21b)$$

then

$$\{\epsilon\}^{(e)} = [B]^{(e)}\{\delta\}^{(e)} \qquad (2.21c)$$

where the superscript (e) designates that the equation holds for one element. Equations 2.21 are the important relations that enable us to calculate the strain of an element once the displacements of its nodes are known. We see that the choice of a linear displacement field leads to a constant strain state in the element.

Step 4

Hooke's law provides the linear constitutive equations we need to relate stress to strain. In the simplest one-dimensional case, probably the most familiar one, Hooke's law states that stress is directly proportional to strain through a proportionality constant E called Young's modulus, that is,

$$\sigma = E\epsilon$$

In contrast to the one-dimensional case, we see from Appendix C that the three-dimensional version of Hooke's law has six components of stress related to six components of strain through a 6×6 proportionality matrix $[C]$, that is,

$$\begin{Bmatrix} \sigma_x \\ \sigma_y \\ \sigma_z \\ \tau_{xy} \\ \tau_{yz} \\ \tau_{xz} \end{Bmatrix} = \begin{bmatrix} C_{11} & C_{12} & \cdots & C_{16} \\ \vdots & & & \vdots \\ C_{61} & \cdots & \cdots & C_{66} \end{bmatrix} \begin{Bmatrix} \epsilon_x \\ \epsilon_y \\ \epsilon_z \\ \gamma_{xy} \\ \gamma_{yz} \\ \gamma_{xz} \end{Bmatrix} \tag{2.22a}$$

or

$$\{\sigma\} = [C]\{\epsilon\} \tag{2.22b}$$

The matrix $[C]$ in equation 2.22a contains 36 constants, but because of its symmetry there are actually only 21 distinct elastic constants. Equation 2.22b simplifies considerably for the special case of homogeneous and isotropic solids experiencing *plane stress*. For this case $[C]$ contains only 2 independent constants instead of 21. Two such constants are Young's modulus, E, and Poisson's ratio, v. In terms of these constants the stress-strain equations may be written as follows:

$$\begin{Bmatrix} \sigma_x \\ \sigma_y \\ \tau_{xy} \end{Bmatrix} = \frac{E}{1-v^2} \begin{bmatrix} 1 & v & 0 \\ v & 1 & 0 \\ 0 & 0 & \dfrac{1-v}{2} \end{bmatrix} \begin{Bmatrix} \epsilon_x \\ \epsilon_y \\ \gamma_{xy} \end{Bmatrix} \tag{2.23a}$$

or

$$\{\sigma\}^{(e)} = [C]^{(e)}\{\epsilon\}^{(e)} \tag{2.23b}$$

where, again, the superscript (e) signifies that the equation holds for a specific element.

Although equation 2.23 applies only when the material of the element is isotropic, we may use the element formulation we are deriving here to model inhomogeneous materials in plane stress. We can do this by assigning different E and v values to each element. Equation 2.23 reveals another important consequences of assuming a linear displacement field within the element—not only is the state of strain constant within the element, but the state of stress is constant also. A constant-stress state in the element is intuitively appealing because it ensures a state of equilibrium throughout the element

Step 5

The penultimate step in our derivation of the element stiffness matrix is to find a set of nodal forces which are statically equivalent to the constant stress field acting at the edges of the element. However, a difficulty common to the direct approach arises when we attempt to find nodal forces which are precise static equivalents. There are six unknown nodal forces, but only three static equilibrium equations with which to find them. To overcome this dilemma we find nodal forces which satisfy the "balance of forces" requirement and the "balance of moments" requirement, but the set we find is not unique. Figure 2.7 shows how the uniform stress distribution can be resolved into an "equivalent" set of forces acting at the midsides of the element. If the midside forces are then divided equally among the appropriate nodes and the x and y components summed, a *physically reasonable* set of "equivalent" nodal forces is obtained. This set of nodal forces is, in terms of edge stresses,

$$\{F\}^{(e)} = \begin{Bmatrix} F_1 \\ F_2 \\ F_3 \\ F_4 \\ F_5 \\ F_6 \end{Bmatrix} = -\frac{t}{2} \begin{bmatrix} y_3 - y_2 & 0 & x_2 - x_3 \\ 0 & x_2 - x_3 & y_3 - y_2 \\ y_1 - y_3 & 0 & x_3 - x_1 \\ 0 & x_3 - x_1 & y_1 - y_3 \\ y_2 - y_1 & 0 & x_1 - x_2 \\ 0 & x_1 - x_2 & y_2 - y_1 \end{bmatrix} \begin{Bmatrix} \sigma_x \\ \sigma_y \\ \tau_{xy} \end{Bmatrix} \tag{2.24a}$$

$$\{F\}^{(e)} = [A]^{(e)}\{\sigma\}^{(e)} \tag{2.24b}$$

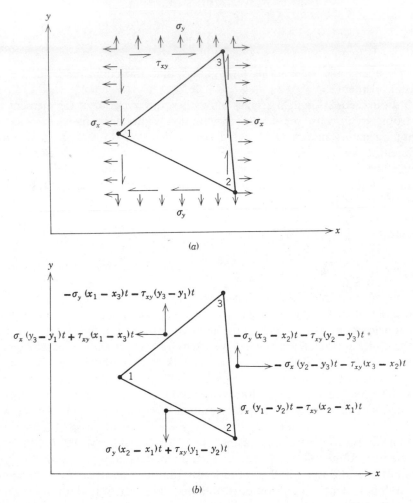

Figure 2.7. Resolving the uniform stress field of an element into equivalent nodal forces [3]. (*a*) Uniform stress distribution on the element. (*b*) Element midside forces which balance the stress distribution.

Step 6

Combining equations 2.22, 2.23, and 2.24 gives, after successive substitutions,

$$\{F\}^{(e)} = [A]^{(e)}[C]^{(e)}[B]^{(e)}\{\delta\}^{(e)} \tag{2.25}$$

The derivation of the element stiffness matrix is now complete. We recognize that equation 2.25 has the standard form (equation 2.4b) if we define the element stiffness as

$$[K]^{(e)} = [A]^{(e)}[C]^{(e)}[B]^{(e)} \tag{2.26}$$

Carrying out the matrix multiplications indicated by equation 2.26 leads to the following expression for the element stiffness matrix (from Turner et al. [3]):

$$[K] = \frac{Et}{2(1 - v^2)}$$

$$
\begin{bmatrix}
\dfrac{y_3}{x_2} + \dfrac{\lambda_1 x_{23}^{\,2}}{x_2 y_3} \\[2ex]
-\dfrac{\lambda_2 x_{32}}{x_2} & \dfrac{x_{23}^{\,2}}{x_2 y_3} + \dfrac{\lambda_1 y_3}{x_2} \\[2ex]
-\dfrac{y_3}{x_2} + \dfrac{\lambda_1 x_3 x_{23}}{x_2 y_3} & \dfrac{v x_{32}}{x_2} + \dfrac{\lambda_1 x_3}{x_2} & \dfrac{y_3}{x_2} + \dfrac{\lambda_1 x_3^{\,2}}{x_2 y_3} \\[2ex]
\dfrac{v x_3}{x_2} + \dfrac{\lambda_1 x_{32}}{x_2} & \dfrac{x_3 x_{23}}{x_2 y_3} - \dfrac{\lambda_1 y_3}{x_2} & -\dfrac{\lambda_2 x_3}{x_2} & \dfrac{x_3^{\,2}}{x_2 y_3} + \dfrac{\lambda_1 y_3}{x_2} \\[2ex]
-\dfrac{\lambda_1 x_{23}}{y_3} & -\lambda_1 & -\dfrac{\lambda_1 x_3}{y_3} & \lambda_1 & \dfrac{\lambda_1 x_2}{y_3} \\[2ex]
-v & -\dfrac{x_{23}}{y_3} & v & -\dfrac{x_3}{y_3} & 0 & \dfrac{x_2}{y_3}
\end{bmatrix}
$$

$$x_{ij} = x_i - x_j, \qquad y_{ij} = y_i - y_j$$

$$\lambda_1 = \frac{1-v}{2}, \qquad \lambda_2 = \frac{1+v}{2} \tag{2.27}$$

The stiffness matrix for a triangular element in plane strain may be derived using this same procedure with one exception—equation 2.23 must be changed to give the stress-strain relations for the plane strain case instead of the plane stress case.

Other examples of deriving stiffness matrices by means of the direct approach may be found in the book by Gallagher [4], who points out that the direct approach is useful for simple elements but unsatisfactory for more complex ones.

2.2.4 Coordinate transformations

In the preceding section we discussed the procedures for establishing the characteristics of various kinds of elements by the direct method. Before going on to discuss how element characteristics are assembled to obtain the characteristics of the entire system of elements, we need to consider the subject of coordinate transformations.

Regardless of the means used to find the characteristics of elements, it is sometimes more convenient or easier to derive the element characteristics in a local coordinate system—a coordinate system associated with the element. The local coordinate system may be different for each element in the assembly. If local coordinates are used, it is necessary, before the individual elements can be assembled, to transform the element equations so that all element characteristics are referred to a common global coordinate system.

In general, the element matrix equations to be transformed will have the standard form

$$[K]\{x^*\} = \{b^*\} \tag{2.28}$$

where the asterisk superscript designates a local reference system. If we assume that a transformation matrix $[\Phi]$ exists between the local and global systems, we may write

$$\{x^*\} = [\Phi]\{x\} \tag{2.29}$$

and

$$\{b^*\} = [\Phi]\{b\} \tag{2.30}$$

where the column vectors $\{x\}$ and $\{b\}$ are referenced to a global system.

When equations 2.29 and 2.30 are substituted into equation 2.28, there results

$$[K][\Phi]\{x\} = [\Phi]\{b\} \tag{2.31}$$

Premultiplying equation 2.31 by the inverse of the transformation matrix gives

$$[\Phi]^{-1}[K][\Phi]\{x\} = \{b\}$$

or

$$[\tilde{K}]\{x\} = \{b\} \tag{2.32}$$

where

$$[\tilde{K}] = [\Phi]^{-1}[K][\Phi] \tag{2.33}$$

Equation 2.33 gives the element matrix referenced to the global coordinate system provided that $[\Phi]^{-1}$ exists. If the column vectors $\{x^*\}$ and $\{b^*\}$ are directional quantities such as nodal displacements and forces, then the transformation matrix is simply the collection of the direction cosines relating the two coordinate systems. In this case, the transformation matrix $[\Phi]$ is an orthogonal matrix with the property that its inverse equals its transpose, that is,

$$[\Phi]^{-1} = [\Phi]^T \tag{2.34}$$

Hence we may write

$$[\tilde{K}] = [\Phi]^T[K][\Phi] \qquad (2.35)$$

In the solid mechanics problem these coordinate transformation rules can also be derived by using energy considerations and equating the work done by external forces in each coordinate system.

2.3 ASSEMBLING THE PARTS

Assuming that we have found by some means the necessary algebraic equations describing the characteristics of each element of our system, the next step in the finite element analysis of the system is to combine all these equations to form a complete set governing the composite of elements. The procedure for constructing the system equations from the element equations is the same regardless of the type of problem being considered or the complexity of the system of elements. Even if the system is modeled with a mixture of several different kinds of elements, the system equations are assembled from the element equations in the same way.

The system assembly procedure is based on our insistence of compatibility at the element nodes. By this we mean that at nodes where elements are connected the value (values) of the unknown nodal variable (or variables if more than one exists at the node) is (are) the same for all elements connecting at that node. The consequence of this rule is the basis for the assembly process. For example, in elasticity problems the nodal variables are usually generalized displacements which can be translations, rotations, or spatial derivatives of the translations. When these generalized displacements are matched at the nodes, the nodal stiffnesses and nodal loads for each of the elements sharing the node are added to obtain the net stiffness and the net load at the node. These assembly concepts, which originated in the matrix structural analysis and network analysis, are an essential part of every finite element solution.

2.3.1 Assembly rules derived from an example

Before we explicitly state the steps in the general assembly procedure and discuss an algorithm for its execution, we can introduce and illustrate its essential features with an elementary example. Consider finding the force-displacement behavior of the system of linear one-dimensional springs shown in Figure 2.8. We will assume that the spring elements in this system can experience only tension and compression. By definition this system has four elements and four nodes.

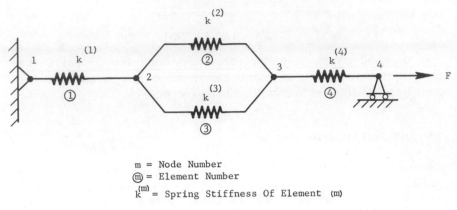

m = Node Number
ⓜ = Element Number
$k^{(m)}$ = Spring Stiffness Of Element (m)

Figure 2.8.   Linear spring system consisting of four elements.

An arbitrary global numbering scheme, as indicated in Figure 2.8, is established to identify these nodes and elements. In general, systematic numbering of all the nodes and elements in a finite element discretization is an essential part of the solution process. Once the numbering scheme† has been established for a finite element mesh, we must create the system's topology—the record of which nodes belong to each of the elements. This topology, given as input to the computer program or generated internally by the program, serves to define the connectivity of the element mesh. In other words, it tells how the elements are joined together. On the element level the topology is simply the ordered numbering of the nodes. Table 2.1 illustrates the system topology that we have established for our four-element spring system. For example, Table 2.1 tells us that element (4) has nodes 3 and 4, and that node 1 of the *element* actually is node 3 of the *system* while node 2 of the *element* is node 4 of the *system*.

Having specified the topology, we proceed to the element equations. From our first derivation in Section 2.2 we recall that for a typical spring element (see Figure 2.2) the relations expressing its stiffness are

$$\begin{bmatrix} k_{11} & -k_{12} \\ -k_{21} & k_{22} \end{bmatrix} \begin{Bmatrix} \delta_1 \\ \delta_2 \end{Bmatrix} = \begin{Bmatrix} F_1 \\ F_2 \end{Bmatrix} \tag{2.4}$$

where

$$k_{11} = k_{12} = k_{21} = k_{22} = k$$

† Node and element numbering may be done by hand or by a digital computer operating from a programed algorithm.

Since these element equations were derived in a local coordinate reference, which is the same as the global coordinates for this example, we do not need to transform the element stiffness and loads.

Table 2.1.   System topology—the correspondence between local and global numbering schemes

	Scheme	
Element	Local	Global
1	1	1
	2	2
2	1	2
	2	3
3	1	2
	2	3
4	1	3
	2	4

We recognize that under a given loading condition each element as well as the system of elements must be in equilibrium. If we impose this equilibrium condition at a particular node i, we find that

$$\sum_e F_i^{(e)} = F_i^{(1)} + F_i^{(2)} + F_i^{(3)} + \cdots = R_i \qquad (2.36)$$

which states that the sum of all the nodal forces in one direction at node i equals the resultant external load applied at node i.

Before evaluating the various $F_i^{(e)}$ and applying equation 2.36 to find the system equations, we review the standard subscript notation for stiffness coefficients. Each coefficient in a stiffness matrix is assigned a double subscript, say ij; the number i is the subscript designating the force F_i produced by a unit value of the displacement whose subscript is j. The force F_i is that which exists when $\delta_j = 1$ and all the other displacements are fixed. A displacement and a resultant force in the direction of the displacement carry the same subscript.

Consider evaluating equation 2.36 at each node in our linear spring system. We shall use the subscript notation based on the global numbering scheme. Since we insist upon the condition of displacement compatibility, that is, the displacement at each node is the same for all elements sharing that node, we can write the following force balance:

At node 1:

$$k_{11}{}^{(1)}\delta_1 + k_{12}{}^{(1)}\delta_2 = R_1$$

At node 2:

$$k_{21}{}^{(1)}\delta_1 + (k_{22}{}^{(1)} + k_{22}{}^{(2)} + k_{22}{}^{(3)})\delta_2 + (k_{23}{}^{(2)} + k_{23}{}^{(3)})\delta_3 = 0$$

At node 3:

$$(k_{32}{}^{(2)} + k_{32}{}^{(3)})\delta_2 + (k_{33}{}^{(2)} + k_{33}{}^{(3)} + k_{33}{}^{(4)})\delta_3 + k_{34}{}^{(4)}\delta_4 = 0$$

At node 4:

$$k_{43}{}^{(4)}\delta_3 + k_{44}{}^{(4)}\delta_4 = F$$

(2.37)

Using matrix notation, we can write these system equilibrium equations as

$$
\begin{bmatrix}
k_{11}{}^{(1)} & k_{12}{}^{(1)} & 0 & 0 \\
k_{21}{}^{(1)} & (k_{22}{}^{(1)} + k_{22}{}^{(2)} + k_{22}{}^{(3)}) & (k_{23}{}^{(2)} + k_{23}{}^{(3)}) & 0 \\
0 & (k_{32}{}^{(2)} + k_{32}{}^{(3)}) & (k_{33}{}^{(2)} + k_{33}{}^{(3)} + k_{33}{}^{(4)}) & k_{34}{}^{(4)} \\
0 & 0 & k_{43}{}^{(4)} & k_{44}{}^{(4)}
\end{bmatrix}
$$

$$
\times
\begin{Bmatrix}
\delta_1 \\
\delta_2 \\
\delta_3 \\
\delta_4
\end{Bmatrix}
=
\begin{Bmatrix}
R_1 \\
0 \\
0 \\
F
\end{Bmatrix}
$$

or

$$[K]\{\delta\} = \{R\} \qquad (2.38)$$

Equations 2.38 are the assembled force-displacement characteristics for the complete system, and $[K]$ is the assembled stiffness matrix. As these equations stand, they cannot be solved for the nodal displacements until they have been modified to account for the boundary conditions. Element stiffness matrices and the assembled stiffness matrix are always singular, that is, their inverse cannot be found because their determinants are zero. In our linear spring system, as well as in other general elasticity problems, the displacement field or the set of nodal displacements cannot be found unless enough nodal displacements are fixed to prevent the structure from moving as a rigid body when external loads are applied. As we see in Figure 2.8, this requirement is satisfied by the condition $\delta_1 = 0$, node 1 being rigidly fixed to some foundation. Since we are concerned here only with assembly procedures, we shall delay discussion of the procedures for treating boundary conditions until the next section.

An alternative approach to deriving these system equilibrium equations from first principles is to employ the principle of minimum potential energy.† According to this principle, the set of nodal displacements that minimizes the system's potential energy is the set existing when the system is at equilibrium. Hence, if we wrote the expression for the potential energy of our spring system in terms of spring stiffnesses, nodal displacements, and the exteral load F, and then minimized this expression with respect to $\{\delta\}$, equations 2.38 would result. We will not carry out this exercise since the result is the same as that already obtained; however, we mention this approach because it parallels the variational approaches to be discussed later.

Inspection of equation 2.38 reveals that the assembled stiffness matrix contains stiffness coefficients obtained by directly adding the individual element stiffness coefficients in the appropriate locations in the global stiffness matrix. The resultant load vector for the system is also obtained by adding individual element loads at the appropriate locations in the column matrix of resultant nodal loads. This result suggests that the element matrices may be thought of as submatrices for the entire system, and that the system matrices may be obtained by simple addition of the element submatrices. This observation is indeed valid, and it is the essence of the general assembly procedure for all finite element analyses. The element matrices are, of course, not added in a random fashion. Matrix addition is defined only for matrices of the same size. Thus, before we add the various element matrices, we first expand them to the dimension of the system matrix. The system matrix allows for the possibility that each nodal unknown (degree of freedom) can be related to each nodal "action"; thus, if the system has n nodal unknowns (degrees of freedom) the system matrix $[K]$ will be a square matrix of dimension $n \times n$. Our spring system has four nodes and only one nodal unknown (displacement) per node; hence, as we have seen, the system matrix $[K]$ has the dimensions 4×4. Expanded element matrices are constructed from the original element matrices by inserting the known stiffness coefficients into their proper locations in the enlarged matrices. It is helpful to think of the expanded element matrices initially as sets of $n \times n$ null matrices (all zero entries), which are then partially filled in by the insertion of the element submatrices.‡

Consider developing the expanded element submatrices for our linear spring system. For element (1) the local and global numbering schemes are,

---

† The principle of minimum potential energy may be applied to the linear spring system of Figure 2.8 because it is a conservative system.

‡ This is only a way to visualize the process. As we will see, we do not actually do this in practice.

by coincidence, the same (see Table 2.1), so that the subscripts of the element stiffness coefficients remain unchanged and we have

$$[\bar{K}]^{(1)} = \begin{bmatrix} k_{11}^{(1)} & k_{12}^{(1)} & 0 & 0 \\ k_{21}^{(1)} & k_{22}^{(1)} & 0 & 0 \\ 0 & 0 & 0 & 0 \\ 0 & 0 & 0 & 0 \end{bmatrix} \tag{2.39}$$

where $[\bar{K}]^{(1)}$ designates the expanded stiffness matrix for element (1).

For element (2) the correspondence between local and global numbering schemes indicates that the following holds:

Local	Global
$k_{11}^{(2)}$ $\longrightarrow$	$k_{22}^{(2)}$
$k_{12}^{(2)}$ $\longrightarrow$	$k_{23}^{(2)}$
$k_{21}^{(2)}$ $\longrightarrow$	$k_{32}^{(2)}$
$k_{22}^{(2)}$ $\longrightarrow$	$k_{33}^{(2)}$

Hence, when these coefficients are inserted into the expanded matrix, we have

$$[\bar{K}]^{(2)} = \begin{bmatrix} 0 & 0 & 0 & 0 \\ 0 & k_{22}^{(2)} & k_{23}^{(2)} & 0 \\ 0 & k_{32}^{(2)} & k_{33}^{(2)} & 0 \\ 0 & 0 & 0 & 0 \end{bmatrix} \tag{2.40}$$

Similarly, for the remaining two elements the expanded element matrices are

$$[\bar{K}]^{(3)} = \begin{bmatrix} 0 & 0 & 0 & 0 \\ 0 & k_{22}^{(3)} & k_{23}^{(3)} & 0 \\ 0 & k_{32}^{(3)} & k_{33}^{(3)} & 0 \\ 0 & 0 & 0 & 0 \end{bmatrix} \tag{2.41}$$

and

$$[\bar{K}]^{(4)} = \begin{bmatrix} 0 & 0 & 0 & 0 \\ 0 & 0 & 0 & 0 \\ 0 & 0 & k_{33}^{(4)} & k_{34}^{(4)} \\ 0 & 0 & k_{43}^{(4)} & k_{44}^{(4)} \end{bmatrix} \tag{2.42}$$

Now we observe that the master stiffness matrix of equation 2.38 can be obtained by simply adding equations 2.39–2.42, representing the contribution from each element. The mathematical statement of this assembly procedure is:

$$[\bar{K}] = \sum_{e=1}^{M} [\bar{K}]^{(e)} = [\bar{K}]^{(1)} + [\bar{K}]^{(2)} + [\bar{K}]^{(3)} + \cdots \tag{2.43}$$

where M is the total number of elements in the assemblage. In our example $M = 4$, but in an actual problem there might be several hundred elements. Even if the assemblage contains many different kinds of elements, equation 2.43 still holds—each individual element matrix is expanded (according to the global numbering scheme) to the dimension of the system matrix, and then these matrices are added.

The same expansion and summation principle applies for finding the column vectors of resultant external nodal actions† (forces in our example) from the element subvectors:

$$\{R\} = \sum_{e=1}^{M} \{\overline{R}\}^{(e)} \tag{2.44}$$

where $\{\overline{R}\}^{(e)}$ is the expanded column vector for element (e), and M is the total number of elements.

2.3.2 General assembly procedure

The preceding assembly procedure although developed from a simple example, is actually in principle the general procedure that applies to all finite element systems. We have assembled systems matrices by hand for a simple problem with four elements, but for real problems involving hundreds of elements the process is done by computer. The general procedure is summarized in the following steps.‡ For clarity we omit the special considerations that sometimes improve computing efficiency.

1. Set up $n \times n$ and $n \times 1$ null matrices (all zero entries), where $n =$ number of system nodal variables.

2. Starting with one element, transform the element equations from local to global coordinates if these two coordinates systems are not coincident.

3. Perform any necessary matrix operations on the element matrices.§

4. Using the established correspondence between local and global numbering schemes, change to the global indices (*a*) the subscript indices of the coefficients in the square matrix, and (*b*) the single subscript index of the terms in the column matrix.

---

† The nodal actions are the forcing mechanisms of the problem. These may be generalized forces (applied loads, torques, stresses, pressures, etc.) or generalized fluxes (velocities, heat flow, concentration, etc.).

‡ We assume that the element matrices have already been determined or can be determined as they are needed in the assembly process.

§ Sometimes elements have one or more nodes which have no connectivity. When this occurs, it is necessary to eliminate the nodal unknowns or degrees of freedom associated with these nodes. This concept is discussed in Chapter 5, where we treat different types of elements.

5. Insert these terms into the corresponding $n \times n$ and $n \times 1$ master matrices in the locations designated by their indices. Each time that a term is placed in a location where another term has already been placed, it is added to whatever value is there.

6. Return to step 2 and repeat this procedure for one element after another until all elements have been treated. The result will be an $n \times n$ master matrix $[K]$ of influence coefficients and a $n \times 1$ column matrix $\{R\}$ of resultant nodal actions. The complete system equations are then

$$\overset{n \times n}{[K]} \; \overset{n \times 1}{\{x\}} = \overset{n \times 1}{\{R\}} \tag{2.45}$$

where $\{x\}$ is the column matrix of nodal unknowns for the assemblage.

The generality of this assembly process for the finite element method offers a definite advantage: Once a computer program for the assembly process has been developed for the solution of one particular class of problems by the finite element method, it may be used again for the finite element solution of other classes of problems.

2.3.3 Features of the assembled matrix

Before discussing how equations 2.45 may be modified to account for boundary conditions, we will point out several important and useful features of the overall system matrix. The simple matrix of equations 2.38, for example, displays these features. First we note that the system matrix has its nonzero terms clustered about its main diagonal, while locations distant from the diagonal zero terms. Essentially the nonzero terms are contained in a band centered on the diagonal, and outside the band all terms are zero. Consequently, the matrix is said to be banded as well as *sparse* because it is not fully populated. Bandedness and sparseness reflect the connectivity of a finite element mesh. In a general finite element mesh, the system matrices are sparse because each element has relatively few nodes compared to all the system nodes and only a few elements share each node. Numbering of the nodes causes the system matrices to be banded. If an efficient nodal numbering scheme is used, the bandwidth can be minimized. Figure 2.9 shows schematically a typical coefficient matrix for a finite element system. The region of the matrix enclosed by dotted lines is the band which contains mostly nonzero coefficients. The width of this band is directly related to the maximum difference between any two global node numbers for an element. Another important feature is that system coefficient matrices are usually symmetric— a characteristic that can often be used to advantage in storing the matrices.

N x N

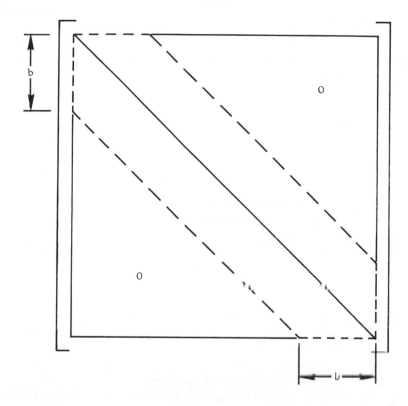

b = Half-Band Width
2b-1 = Band Width

Figure 2.9.   The structure of the coefficient matrix for a typical finite element system.

Many of the practical aspects of handling system matrices and numbering node points to obtain well-conditioned matrices are discussed in Chapter 10 on computational considerations.

2.3.4 Introducing boundary conditions

Regardless of the type of problem for which a set of system equations has been assembled, the final equations will have the form

$$\overset{n \times n}{[K]} \overset{n \times 1}{\{x\}} = \overset{n \times 1}{\{R\}} \tag{2.45}$$

These equations already take into account the boundary conditions involving specified external or applied nodal "actions" because these are included in the resultant column vector $\{R\}$ during assembly. However, for a unique solution of equations 2.45, at least one and sometimes more than on nodal variable must be specified and $[K]$ must be modified to render it nonsingular. The required number of specified nodal variables is dictated by the physics of the problem. Nodal variables may be specified for either interior or boundary nodes, but for nodes, i, whose coordinates are fixed, it is physically impossible to specify both x_i and R_i. However, at every node i we know either x_i or R_i. There are a number of ways to apply the boundary conditions to equations 2.45, and when they are applied, the number of nodal unknowns and the number of equations to be solved are effectively reduced. However, it is most convenient to introduce the known nodal variables in a way that leaves the original number of equations unchanged and avoids major restructuring of computer storage. Two straightforward means to accomplish this are described in the following discussion.

One way to include prescribed nodal variables in equation 2.45 while retaining an $n \times n$ system of equations is to modify the matrices $[K]$ and $\{R\}$ as follows [6]. If i is the subscript of a prescribed nodal variable, the ith row and the ith column of $[K]$ are set equal to zero and k_{ii} is set equal to unity. The term R_i of the column vector $\{R\}$ is replaced by the known value of x_i. Each of the $n - 1$ remaining terms of $\{R\}$ is modified by subtracting from it the value of the prescribed nodal variable multiplied by the appropriate column term from the original $[K]$ matrix. This procedure is repeated for each prescribed x_i until all of them have been included.

To illustrate this procedure for entering boundary conditions, we consider a simple example with only four system equations. Equation 2.45 expands to the form

$$\begin{bmatrix} k_{11} & k_{12} & k_{13} & k_{14} \\ k_{21} & k_{22} & k_{23} & k_{24} \\ k_{31} & k_{32} & k_{33} & k_{34} \\ k_{41} & k_{42} & k_{43} & k_{44} \end{bmatrix} \begin{Bmatrix} x_1 \\ x_2 \\ x_3 \\ x_4 \end{Bmatrix} = \begin{Bmatrix} R_1 \\ R_2 \\ R_3 \\ R_4 \end{Bmatrix}$$

Suppose that for this hypothetical system nodal variables x_1 and x_3 are specified as

$$x_1 = \beta_1, \qquad x_3 = \beta_3$$

When these boundary conditions are inserted, the equations become

$$\begin{bmatrix} 1 & 0 & 0 & 0 \\ 0 & k_{22} & 0 & k_{24} \\ 0 & 0 & 1 & 0 \\ 0 & k_{42} & 0 & k_{44} \end{bmatrix} \begin{Bmatrix} x_1 \\ x_2 \\ x_3 \\ x_4 \end{Bmatrix} = \begin{Bmatrix} \beta_1 \\ R_2 - k_{21}\beta_1 - k_{23}\beta_3 \\ \beta_3 \\ R_4 - k_{41}\beta_1 - k_{43}\beta_3 \end{Bmatrix}$$

This set of equations, unaltered in dimension, is now ready to be solved for all the nodal variables. The solution, of course, yields $x_1 = \beta_1$ and $x_3 = \beta_3$ along with the actual unknowns, x_2 and x_4.

Another way to insert prescribed nodal variables into the system equations is to modify certain diagonal terms of $[K]$ according to the easy method suggested by Payne and Irons [5]. The diagonal term of $[K]$ associated with a specified nodal variable is multiplied by a large number, say 1×10^{15}, while the corresponding term in $\{R\}$ is replaced by the specified nodal variable multiplied by the same large factor times the corresponding diagonal term. This procedure is repeated until all prescribed nodal variables have been treated. Effectively, this procedure makes the unmodified terms of $[K]$ very small compared to the modified terms (those associated with the specified nodal variables). After these modifications have been made, we proceed with the simultaneous solution of the complete set of n equations.

If we use this procedure to modify the original equations of our previous example, we obtain

$$\begin{bmatrix} k_{11} \times 10^{15} & k_{12} & k_{13} & k_{14} \\ k_{21} & k_{22} & k_{23} & k_{24} \\ k_{31} & k_{32} & k_{33} \times 10^{15} & k_{34} \\ k_{41} & k_{42} & k_{43} & k_{44} \end{bmatrix} \begin{Bmatrix} x_1 \\ x_2 \\ x_3 \\ x_4 \end{Bmatrix} = \begin{Bmatrix} \beta_1 k_{11} \times 10^{15} \\ R_2 \\ \beta_3 k_{33} \times 10^{15} \\ R_4 \end{Bmatrix}$$

To see that this procedure gives the desired result consider the first equation of the set:

$$k_{11} \times 10^{15} x_1 + k_{12} x_2 + k_{13} x_3 + k_{14} x_4 = \beta_1 \times k_{11} \times 10^{15}$$

For all practical purposes, this equation expresses the fact that

$$x_1 = \beta_1$$

since

$$k_{11} \times 10^{15} \gg k_{1j}, \quad j = 2, 3, 4$$

Computer coding is a littler easier for the second method than for the first one. Both methods preserve the sparse, banded, and usually symmetric properties of the original master matrix.

After the modified equations have been solved for the unknown nodal variables, we may return to the original set of equations and find the unknown nodal reactions if desired. An orderly, though impractical, way to do this is as follows [6]. Starting with the original equations, we may rearrange and partition them to obtain

$$\underset{n \times n}{\begin{bmatrix} [K_{11}] & [K_{12}] \\ [K_{12}]^T & [K_{22}] \end{bmatrix}} \underset{n \times 1}{\begin{Bmatrix} \{\tilde{x}_1\} \\ \{\tilde{x}_2\} \end{Bmatrix}} = \underset{n \times 1}{\begin{Bmatrix} \{\tilde{R}_1\} \\ \{\tilde{R}_2\} \end{Bmatrix}} \tag{2.46}$$

where $\{\tilde{R}_1\}$ is the column vector of *k*nown nodal reactions and $\{\tilde{x}_2\}$ is the column vector of *known* nodal variables. The matrix $[K_{11}]$ now possesses an inverse, so we may find the unknown $\{\tilde{x}_1\}$ by solving

$$[K_{11}]\{\tilde{x}_1\} = \{\tilde{R}_1\} - [K_{12}]\{\tilde{x}_2\} \tag{2.47}$$

With $\{\tilde{x}_1\}$ known from equation 2.47, the vector $\{\tilde{R}_2\}$ can be found from

$$\{\tilde{R}_2\} = [K_{12}]^T\{\tilde{x}_1\} + [K_{22}]\{\tilde{x}_2\} \tag{2.48}$$

The procedure represented symbolically by equations 2.46, 2.47, and 2.48 is usually impractical because the reordering process requires much additional programming effort and tends to destroy the bandedness property of the original matrix. Once the full set of equations has been assembled and the prescribed boundary conditions have been introduced, the next step is to solve the equations for the unknowns. Many schemes are available for this purpose (see, for example, Forsythe and Moler [7]). The most efficient schemes take advantage of the possible symmetry, sparseness, and bandedness of $[K]$. Although a detailed treatment of equation-solving routines is beyond the scope of this book, the more popular solution techniques are mentioned and references are cited in Chapter 10.

Numerical Example†

To help solidify the fundamental ideas presented in this chapter, we consider a trivial example problem in one dimension. Suppose that we have the column structure (Figure 2.10*a*) composed of two bars of different cross-sectional area. The column is loaded at one end by a compressive force F, and its displacement at the other end is constrained to be zero by an immovable wall. What are the forces and displacements at each end of the column and at the interface between the two bars? We can use the finite element method to answer this question in a systematic fashion.

The steps of the solution procedure are as follows.

1. *Discretize the structure.* The objective of the finite element procedure is to divide the problem into elements whose properties (in this case, the force-deflection characteristics) either are known or can be easily found. If we choose each of the uniform bar sections of the column to be an element, the system can be conveniently discretized in terms of axial force members, as shown in Figures 2.10*b* and 2.10*c*.

2. *Find element properties.* Now that we have identified the individual elements of the structure, the next step is to determine the element properties

† This example was suggested by Professor J. F. Booker of Cornell University.

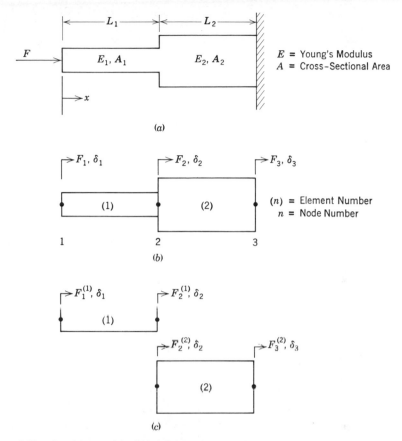

Figure 2.10.   A column and its finite element model. (*a*) Nonuniform column loaded in compression. (*b*) Finite element model. (*c*) Individual elements (axial force members).

or force-deflection characteristics. In this case, the task is particularly simple. Consider the typical isolated element shown in Figure 2.11. For an axial force member, we know that the stress σ_x at any station along the length of the member is constant; hence

$$\sigma_x = c_1$$

But from Hooke's law (Appendix C), the strain ϵ_x is given by

$$\epsilon_x = \frac{d\delta}{dx} = \frac{\sigma_x}{E} = \frac{c_1}{E}$$

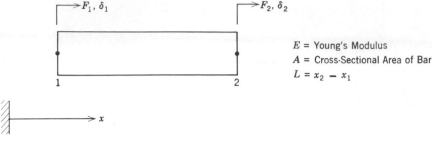

Figure 2.11. An axial force member as a finite element.

and by integrating we find

$$\delta(x) = \frac{1}{E}(c_1 x + c_2)$$

where E is Young's modulus. This linear variation of $\delta(x)$ is exact under the assumption that $\sigma_x = $ constant. We could have obtained the same result by simply assuming that displacement varies linearly along the bar; that is, we could have linearly interpolated the nodal displacements and written

At node 1: $\quad \delta_1 = \delta(x_1) = \frac{1}{E}(c_1 x_1 + c_2)$

At node 2: $\quad \delta_2 = \delta(x_2) = \frac{1}{E}(c_1 x_2 + c_2)$

or, in matrix form,

$$\begin{Bmatrix} \delta_1 \\ \delta_2 \end{Bmatrix} = \frac{1}{E} \begin{bmatrix} x_1 & 1 \\ x_2 & 1 \end{bmatrix} \begin{Bmatrix} c_1 \\ c_2 \end{Bmatrix}$$

Thus

$$\begin{Bmatrix} c_1 \\ c_2 \end{Bmatrix} = E \begin{bmatrix} x_1 & 1 \\ x_2 & 1 \end{bmatrix}^{-1} \begin{Bmatrix} \delta_1 \\ \delta_2 \end{Bmatrix}$$

Now we note that the nodal forces are given by

$$F_1 = -\sigma_x A = -c_1 A$$
$$F_2 = \quad \sigma_x A = \quad c_1 A$$

or

$$\begin{Bmatrix} F_1 \\ F_2 \end{Bmatrix} = \begin{bmatrix} -A & 0 \\ A & 0 \end{bmatrix} \begin{Bmatrix} c_1 \\ c_2 \end{Bmatrix}$$

Combining the above results gives the desired relationship between nodal forces and nodal displacements, that is,

$$\begin{Bmatrix} F_1 \\ F_2 \end{Bmatrix} = \frac{-E}{x_2 - x_1} \begin{bmatrix} -A & 0 \\ A & 0 \end{bmatrix} \begin{bmatrix} 1 & -1 \\ -x_2 & x_1 \end{bmatrix} \begin{Bmatrix} \delta_1 \\ \delta_2 \end{Bmatrix}$$

$$\begin{Bmatrix} F_1 \\ F_2 \end{Bmatrix} = \frac{AE}{x_2 - x_1} \begin{bmatrix} 1 & -1 \\ -1 & 1 \end{bmatrix} \begin{Bmatrix} \delta_1 \\ \delta_2 \end{Bmatrix}$$

or

$$\{F\}^{(i)} = [K]^{(i)} \{\delta\}^{(i)}, \quad i = 1, 2$$

Having found the general element equations, we can now evaluate them for a particular case. In some consistent set of units, suppose that we have

$$F = 10, \qquad E_1 = 1, \qquad A_1 = 1, \qquad L_1 = 1$$

and

$$E_2 = 1, \qquad A_2 = 2, \qquad L_2 = 1$$

Then in terms of the global numbering scheme (Figure 2.10b) the element equations are for element (1)

$$\begin{bmatrix} 1 & -1 & 0 \\ -1 & 1 & 0 \\ 0 & 0 & 0 \end{bmatrix} \begin{Bmatrix} \delta_1 \\ \delta_2 \\ \delta_3 \end{Bmatrix} = \begin{Bmatrix} F_1 \\ F_2 \\ F_3 \end{Bmatrix}^{(1)}$$

and for element (2)

$$\begin{bmatrix} 0 & 0 & 0 \\ 0 & 2 & -2 \\ 0 & -2 & 2 \end{bmatrix} \begin{Bmatrix} \delta_1 \\ \delta_2 \\ \delta_3 \end{Bmatrix} = \begin{Bmatrix} F_1 \\ F_2 \\ F_3 \end{Bmatrix}^{(2)}$$

3. *Assemble the element equations.* The system equations have the form

$$\begin{bmatrix} k_{11} & k_{12} & k_{13} \\ k_{21} & k_{22} & k_{23} \\ k_{31} & k_{32} & k_{33} \end{bmatrix} \begin{Bmatrix} \delta_1 \\ \delta_2 \\ \delta_3 \end{Bmatrix} = \begin{Bmatrix} F_1 \\ F_2 \\ F_3 \end{Bmatrix}$$

or

$$[K]\{\delta\} = \{F\}$$

Applying the assembly procedure gives

$$\begin{Bmatrix} F_1 \\ F_2 \\ F_3 \end{Bmatrix}^{(1)} + \begin{Bmatrix} F_1 \\ F_2 \\ F_3 \end{Bmatrix}^{(2)} = \begin{Bmatrix} F_1 \\ F_2 \\ F_3 \end{Bmatrix} = \{F\}$$

and

$$\begin{bmatrix} 1 & -1 & 0 \\ -1 & 1 & 0 \\ 0 & 0 & 0 \end{bmatrix} + \begin{bmatrix} 0 & 0 & 0 \\ 0 & 2 & -2 \\ 0 & -2 & 2 \end{bmatrix} = \begin{bmatrix} -1 & -1 & 0 \\ -1 & 3 & -2 \\ 0 & -2 & 2 \end{bmatrix}$$

4. *Solve the system equations.* We have three equations

$$\begin{bmatrix} 1 & -1 & 0 \\ -1 & 3 & -2 \\ 0 & -2 & 2 \end{bmatrix} \begin{Bmatrix} \delta_1 \\ \delta_2 \\ \delta_3 \end{Bmatrix} = \begin{Bmatrix} F_1 \\ F_2 \\ F_3 \end{Bmatrix}$$

in three unknowns: F_3, δ_1, and δ_2

$$\begin{Bmatrix} F_1 \\ F_2 \\ F_3 \end{Bmatrix} = \begin{Bmatrix} 10 \\ 0 \\ ? \end{Bmatrix}, \qquad \begin{Bmatrix} \delta_1 \\ \delta_2 \\ \delta_3 \end{Bmatrix} = \begin{Bmatrix} ? \\ ? \\ 0 \end{Bmatrix}$$

Segregating these equations according to equation 2.47 gives

$$\begin{bmatrix} 1 & -1 \\ -1 & 3 \end{bmatrix} \begin{Bmatrix} \delta_1 \\ \delta_2 \end{Bmatrix} = \begin{Bmatrix} F_1 \\ F_2 \end{Bmatrix} - \begin{bmatrix} 0 \\ -2 \end{bmatrix} \times 0 = \begin{Bmatrix} 10 \\ 0 \end{Bmatrix}$$

and solving simultaneously gives

$$\begin{Bmatrix} \delta_1 \\ \delta_2 \end{Bmatrix} = \begin{Bmatrix} 15 \\ 5 \end{Bmatrix}$$

From equation 2.48

$$F_3 = \lfloor 0 \quad -2 \rfloor \begin{Bmatrix} 15 \\ 5 \end{Bmatrix} + 2 \times 0$$

$$F_3 = -10$$

5. *Return to element equations (if desired).* The individual element loads, stress, and strain can be found by returning to the element equations. Hence

$$\begin{Bmatrix} F_1 \\ F_2 \end{Bmatrix}^{(1)} = \begin{bmatrix} 1 & -1 \\ -1 & 1 \end{bmatrix} \begin{Bmatrix} 15 \\ 5 \end{Bmatrix} = \begin{Bmatrix} 10 \\ -10 \end{Bmatrix}$$

$$\sigma_{x_1} = \frac{F_1}{A_2}, \qquad \sigma_{x_2} = \frac{F_2}{A_2}$$

$$\delta^{(1)}(x) = \frac{1}{E}(c_1 x + c_2)$$

$$\begin{Bmatrix} c_1 \\ c_2 \end{Bmatrix}^{(1)} = \frac{-E_1}{L_1} \begin{bmatrix} 1 & -1 \\ -L_2 & 0 \end{bmatrix} \begin{Bmatrix} 15 \\ 5 \end{Bmatrix}$$

and similarly for the other element.

2.4 CLOSURE

In this chapter we introduced the concepts of the finite element method for the elementary cases. We used the so-called direct approach or physical approach to derive equations expressing the behavior of a few simple elements. By this means several important features of the finite element method were exposed, and the physical interpretations of modeling a real system by a network of finite elements were introduced. As we progressed, the ideas of an element and a node took on more generalized meanings.

The utility of the direct approach, however, is severely limited because it cannot be generalized to study the behavior of more complex useful elements. But, with the element equations we did derive, we proceeded to develop the mathematical procedure by which element equations are assembled to form system equations. The assembly concepts we established are general and apply to all finite element analyses regardless of the means used to derive element equations.

We concluded the chapter with a discussion of how the assembled system equations should be modified before solution to account for boundary conditions. Having treated the assembly procedures and the methods for handling boundary conditions, we may now focus our attention in succeeding chapters on the formulation of element equations for many types of problems.

The direct approach is very helpful in explaining some of the finite element concepts, but to work beyond the elementary concepts we have to consider the mathematical foundations of the method. As we noted in Section 1.3, engineers began to recognize the mathematical foundations of the finite element method in 1963. For finite elements of an elastic continuum, civil engineers realized that the matrix equations they had previously derived by direct physical reasoning or by virtual work concepts could also be derived by directly minimizing an energy functional with respect to the nodal values of the field variable. If the field variable was the displacement field, the appro-private functional for the derivation was the integral expression for the total potential energy of the system. The discovery that the physically formulated finite method was actually a mathematical minimization process, and that element properties could be derived from variational principles governing the particular problem of interest, opened the way to far broader applications of the method. In addition, this mathematical interpretation of the method helped to establish continuity requirements for the assumed interpolation functions and facilitated the study of convergence properties and associated error estimates (bounds).

In the next chapter we will examine the variational basis of the finite element method and show how the method can be applied to almost all problems governed by an appropriate variational principle.

REFERENCES

1.   I. H. Shames, *Mechanics of Fluids*, McGraw-Hill Book Company, New York, 1962.
2.   S. Timoshenko, *Strength of Materials*, Part I: *Elementary Theory and Problems*, D. Van Nostrand, Princeton, N.J., 1955.
3.   M. J. Turner, R. W. Clough, H. C. Martin, and L. C. Topp, "Stiffness and Deflection Analysis of Complex Structures," *J. Aeronaut. Sci*, Vol. 23, No. 9, September 1956.
4.   R. H. Gallagher, *Correlation Study of Methods of Matrix Structural Analysis*, Pergamon Press, New York, 1964.
5.   N. A. Payne and B. Irons, Private communication with O. C. Zienkiewicz, 1963.
6.   C. A. Felippa and R. W. Clough, "The Finite Element Method in Solid Mechanics," *Numerical Solution of Field Problems in Continuum Physics*, SIAM-AMS Proceedings, Vol. 2, American Mathematical Society, Providence, R.I., 1970.
7.   G. Forsythe and C. B. Moler, *Computer Solution of Linear Algebraic Systems*, Prentice-Hall, Englewood Cliffs, N.J., 1967.

3

THE MATHEMATICAL APPROACH:
A VARIATIONAL INTERPRETATION

3.1 INTRODUCTION

In Chapter 2 we focused our attention on the direct approach to finite element analysis. This approach is helpful in gaining an understanding of the finite element method, but insurmountable difficulties arise when we try to apply it to complex problems. In this chapter we take a broader view and interpret the finite element method as an approximate means for solving variational problems.† At this point, however, we cannot fruitfully discuss the many specific techniques that are useful for particular types of problems. These specialized aspects of the finite element method will be introduced in the chapters of Part 2, where applications in solid and fluid mechanics are discussed.

To set the stage for the introduction of the mathematical concepts and to give them a place in the overall collection of solution techniques, we begin with a general discussion of the continuum problems of mathematical physics.

After briefly mentioning some of the more popular solution techniques for different classes of problems, we establish the necessary terminology and definitions to show how variational problems and the finite element method are related. The variational basis of the finite element method, the basis most often emphasized in the literature, dictates the criteria to be satisfied by the element interpolation functions, and it enables us to make definitive statements about convergence of the results as we use an ever-increasing number of smaller and smaller elements.

† We shall assume in this chapter that the reader is familiar with the calculus of variations and the nature of variational problems. To review the aspects of this subject relevant to the finite element method, the reader may refer to Appendix B.

A treatment of the rigorous mathematical considerations pertaining to error bounds and convergence is beyond the scope of this book, but some of the recent work in this area was mentioned in Chapter 1.

After a discussion of the variational approach to the formulation of element equations, we consider a detailed example. The last section of this chapter treats the problem of how to find variational principles for use in the finite element method.

3.2   CONTINUUM PROBLEMS

3.2.1   Introduction

Problems in engineering and science fall into two fundamentally different categories, depending on which point of view we adopt. One point of view is that all matter consists of single particles that retain their identity and nature as they move through space. Their position in space at any instant is given by the coordinates in some reference system, and these coordinates are functions of time—the only independent variable for any process. This viewpoint, known as the *Lagrangian* viewpoint, is the basis of Newtonian particle mechanics.

The other viewpoint, the one that we shall use, stems from the *continuum* rather than the molecular or particle approach to nature. In the continuum (sometimes called the *Eulerian*) viewpoint we say that all processes are characterized by field quantities which are defined at every point in space. The independent variables in *continuum problems* are the coordinates of space and time. The Eulerian viewpoint allows us to focus our attention on one point in space and then observe the phenomena occurring there.

Continuum problems are concerned with fields of temperature, stress, mass concentration, displacement, and electromagnetic and acoustic potentials, to name just a few examples. These problems arise from phenomena in nature that are *approximately* characterized by partial differential equations and their boundary conditions.

We shall briefly discuss the nature of continuum problems typically encountered and some of the possible means of solution. Then we will turn to the topic of solving these problems by the finite element method. Continuum problems of mathematical physics are often referred to as *boundary value problems* because their solution is sought in some domain defined by a given boundary, on which certain conditions called *boundary conditions* are specified. Except for free-boundary problems,† the shape of the boundary

---

† In free-boundary problems part of the problem is to determine the shape and location of the boundary. Finite element methods have been used to solve some special free-boundary problems, but, in general, a great deal of work remains to be done in this problem area.

and its location are always known. The boundary may be defined by a curve
or a surface in *n*-dimensional space, and the domain it defines may be
finite or infinite, depending on the extremities of the boundary. The boundary
is said to be *closed* if conditions *affecting the solution of the problem* are
specified everywhere on the boundary (even though part of the boundary
may extend to infinity), and *open* if part of the boundary extends to infinity
and no boundary conditions are specified on the part at infinity [1]. It is
important to note that our definition of a boundary value problem departs
from the usual one. The usual definition distinguishes between boundary
value problems and *initial value problems*, where time is an independent
variable. Because of our definition of the boundary of the domain, we may
describe all partial differential equations and their boundary conditions as
boundary value problems. Hence our definition groups the usual boundary
value problems and initial value problems into one category.

3.2.2 Problem statement

The kinds of continuum problems that we wish to solve are usually formulated in general terms as follows. Consider some domain D bounded
by the surface Σ. Let ϕ be a scalar function† defined in the interior of D such
that the behavior of ϕ in D is given by

$$\mathscr{L}(\phi) - f = 0 \qquad (3.1)$$

where f is a known scalar function of the independent variables, and $\mathscr{L}$ is a
linear or nonlinear differential operator. We will assume that the physical
parameters in the differential operator are known constants or known
functions. The simplest example of a linear differential operator is

$$\mathscr{L}(\ ) = \frac{\partial(\ )}{\partial x}$$

Two other examples of linear operators are as follows.
    Laplace operator in two dimensions:

$$\mathscr{L}(\ ) = \frac{\partial^2(\ )}{\partial x^2} + \frac{\partial^2(\ )}{\partial y^2}$$

Biharmonic operator in two dimensions:

$$\mathscr{L}(\ ) = \frac{\partial^4(\ )}{\partial x^4} + \frac{\partial^4(\ )}{\partial^2 x\, \partial^2 y} + \frac{\partial^4(\ )}{\partial y^4}$$

† Without loss of generality we restrict our attention to scalar functions. If we considered vector
functions, we would simply have a scalar equation like equation 3.1 for each component of the
vector.

In n dimensions, second-order differential operators can usually be reduced, by a suitable transformation, to the form

$$\mathscr{L}(\ ) = \sum_{i=1}^{n} A_i \frac{\partial^2(\ )}{\partial x_i^2} + \sum_{i=1}^{n} B_i \frac{\partial(\ )}{\partial x_i} + (\ )C + D \qquad (3.2)$$

where coefficients A_i, B_i, and C and the term D may be functions. The operator as given in equation 3.2 is *linear* if A_i, B_i, C_i, and D are functions only of the independent variables $(x_1, x_2, x_3, \ldots, x_n)$ and *quasi-linear* if A_i, B_i, C_i, and D are functions of x_i and the dependent parameter, as well as first derivatives of the dependent parameter. In general, an operator is linear only if

$$\mathscr{L}(f + g) = \mathscr{L}(f) + \mathscr{L}(g)$$

The general definition we have given the operator $\mathscr{L}(\ )$ in equation 3.1 precludes a discussion of appropriate boundary conditions. However, without boundary conditions, equation 3.1 does not describe a specific problem. For some cases it may be possible to integrate equation 3.1; but if this is done, the result will always contain arbitrary constants if $\mathscr{L}(\ )$ is an ordinary differential operator, or arbitrary functions if $\mathscr{L}(\ )$ is a *partial* differential operator. These constants or functions can be found only if the "proper" boundary conditions are specified.

3.2.3   Classification of differential equations

The important question of what kinds of conditions constitute "proper" boundary conditions for a given differential operator, or, equally important, what kinds of conditions are improper and cannot be satisfied, can be answered by investigating the classification of the given operator. All partial differential equations of the form of equation 3.1 can be classified as either *elliptic*, *hyperbolic*, *parabolic*, or some combination of these three categories, such as ultrahyperbolic, elliptically parabolic, or hyperbolically parabolic. For example, we consider the following very general partial differential equation in two independent variables:

$$A \frac{\partial^2 \phi}{\partial x^2} + 2B \frac{\partial^2 \phi}{\partial x\, \partial y} + C \frac{\partial^2 \phi}{\partial y^2} = D\left(x, y, \phi, \frac{\partial \phi}{\partial x}, \frac{\partial \phi}{\partial y}\right) \qquad (3.3)$$

where A, B, and C are functions of only x and y.

Equation 3.3 can be linear or nonlinear, depending on the form of the function D.

If $B^2 - AC < 0$, the equation is elliptic.

If $B^2 - AC = 0$, the equation is parabolic.

If $B^2 - AC > 0$, the equation is hyperbolic.

For parabolic and hyperbolic equations the solution domains are usually *open*, whereas for elliptic equations the solution domains are defined by a *closed* boundary. Because A, B, and C may be functions of each generic point in the solution region, the classification of equation 3.3 may change from one point in the solution region to another.

To determine the classification of more complex partial differential equations, it is necessary to study their *characteristics*. At the risk of oversimplifying the concepts, we can say that the characteristics of an equation are the lines (perhaps in n-dimensional space) along which the highest-order derivatives in the equation may be discontinuous. Once the characteristics of an equation are known, we have the following rules:

Elliptic equations have no real characteristics.
Parabolic equations have mixed real and imaginary characteristics.
Hyperbolic equations have all real and distinct characteristics.

The classification of partial differential equations and the role played by characteristics in determining proper boundary conditions for a differential equation are certainly beyond the scope of our central theme. The reader is referred to the books by Petrovsky [?] and Courant and Friedrichs [?] for further details on these subjects.

Our purpose in mentioning the classification of partial differential equations is to point out that the application of finite element techniques to the solution of differential equations and their boundary conditions depends on the classification of a given equation only insofar as the boundary conditions are concerned. If the proper boundary conditions are specified, the classification of the equation does not enter explicitly into consideration. When proper boundary conditions are given,† in principle the finite element method is applicable to linear and nonlinear partial differential equations valid over domains of *any* geometrical shape. By far the most frequent application of the finite element method has been to the solution of linear elliptic boundary value problems with irregular solution domains.

3.3 SOME METHODS FOR SOLVING CONTINUUM PROBLEMS

3.3.1 An overview

Returning to equation 3.1, we see that our general problem is to find the unknown function ϕ that satisfies equation 3.1 and the associated boundary conditions specified on Σ.

---

† In this book we shall assume that we are dealing with well-posed problems whose statements contain the proper boundary conditions.

There are many alternative approaches to the solution of linear and non-linear boundary value problems, and they range from completely analytical to completely numerical. Of these, the following deserve attention:

1. Direct integration (exact solutions)
    a. Separation of variables
    b. Similarity solutions
    c. Fourier and Laplace transformations
2. Approximate solutions
    a. Perturbation
    b. Power series
    c. Probability schemes
    d. Method of weighted residuals (MWR)
    e. Finite difference techniques
    f. Ritz method
    g. Finite element method

For a few problems it is possible to obtain an exact solution by direct integration of the differential equation. This is accomplished sometimes by an obvious separation of variables or by applying a transformation which makes the variables separable and leads to a similarity solution. Occasionally, a Fourier or Laplace transformation of the differential equation leads to an exact solution. However, the number of problems with exact solutions is severely limited, and most of these have already been solved. They are the classical problems.

Because regular and singular perturbation methods are applicable primarily when the nonlinear terms in the equation are small in relation to the linear terms, their usefulness is limited. The power series method is powerful and has been employed with some success, but since the method requires generation of a coefficient for each term in the series it is relatively tedious. Also, it is difficult, if not impossible, to demonstrate that the series converges.

The probability schemes, usually classified under the heading of Monte Carlo methods [4], are used for obtaining a statistical estimate of a desired quantity by random sampling. These methods work best when the desired quantity is a statistical parameter and sampling is done from a selective population.

With the advent of high-speed digital computers it appears that the three currently outstanding methods for obtaining approximate solutions of high accuracy are the method of weighted residuals, the finite difference method, and the finite element method. These methods are related, as we shall see, and in some cases the finite difference and the finite element methods can be shown to be special cases of the method of weighted residuals.

More than one volume would be needed to do justice to all of the important aspects of these methods. Since this book deals exclusively with the finite element method, readers interested in the other two approximate methods are referred to three notable treatments in the literature. Forsythe and Wasow [5] present a basic theoretical background for the finite difference method. Their treatment of the necessary algorithms for obtaining and solving the difference equations on computers is especially useful. They also discuss the derivation of difference equations from variational principles. Although in Chapter 4 we shall discuss aspects of the method of weighted residuals that are useful for the finite element method, more complete treatments of the weighted residuals technique can be found in the paper by Finlayson and Scriven [6] and the book by Finlayson [7].

3.3.2   The variational approach

Often continuum problems have different but equivalent formulations—a differential formulation and a variational formulation. In the differential formulation as we have seen, the problem is to integrate a differential equation or a system of differential equations subject to given boundary conditions. In the *classical variational formulation,*† the problem is to find the unknown function or functions which *extremize* (maximize, minimize) or make *stationary* a functional or system of functionals subject to the same given boundary conditions. The two problem formulations are equivalent because the functions that satisfy the differential equations and their boundary conditions also extremize or make stationary the functionals. This equivalence is apparent from the calculus of variations, which shows that the functionals are extremized or made stationary only when one or more Euler equations and their boundary conditions are satisfied. And these equations are precisely the governing differential equations of the problem. The classical variational formulation of a continuum problem often has advantages over the differential formulation from the standpoint of obtaining an approximate solution. First, the functional, which may actually represent some physical quantity in the problem, contains lower-order derivatives than the differential operator, and consequently an approximate solution can be sought in a larger class of functions. Second, the problem may possess reciprocal variational formulations, that is, one functional to be minimized and another functional of a different form to be maximized. In such cases we have a means for finding upper and lower bounds on the functional, and this capability may have significant engineering value. Third, the variational formulation allows

† Generally, we shall be concerned only with classical variational formulations in this book.

us to treat very complicated boundary conditions as *natural boundary conditions*.† And, fourth, from a purely mathematical standpoint the variational formulation is helpful because with the calculus of variations it can sometimes be used to prove the existence of a solution.

In the past, when engineers used the finite element method to solve their particular continuum problems, they most often relied on variational methods to derive the finite element equations. This approach is especially convenient when it is applicable; but before it can be used, a variational statement for the continuum problem must be obtained, that is, we must pose the problem in variational form. The topic of obtaining variational formulations for continuum problems is treated in Section 3.5. For convenience in our discussion here we will assume that the variational formulation for a given problem is known.

Historically, variational methods are among the oldest means of obtaining solutions to problems in physics and engineering. One general method for obtaining approximate solutions to problems expressed in variational form is known as the *Ritz method*. Since this method is basically a forerunner of the finite element procedure, we introduce some of its features before discussing the finite element approach. Actually, the finite element method is a special case of the Ritz method when the interpolation functions obey certain continuity requirements.

3.3.3   The Ritz method

The Ritz method consists of *assuming the form* of the unknown solution in terms of known functions (trial functions) with unknown adjustable parameters. (The trial functions are sometimes called coordinate functions.) From the family of trial functions we select the function which renders the functional stationary. The procedure is to substitute the trial functions into the functional and thereby express the functional in terms of the adjustable parameters. The functional is then differentiated with respect to each parameter, and the resulting equation is set equal to zero. If there are n unknown parameters, there will be n simultaneous equations to be solved for these parameters. By this means the approximate solution is chosen from the family of assumed solutions.

The procedure does nothing more than give us the "best" solution from the family of assumed solutions. Clearly, then, the accuracy of the approxi-

† As noted in Appendix B, boundary conditions may be either *geometric* or *natural*. In the finite element procedure we explicitly impose geometric boundary conditions via the procedure in Chapter 2, whereas the variational statement implicitly imposes the natural boundary conditions.

mate solution depends on the choice of trial functions. We require that the trial functions be defined over the whole solution domain and that they satisfy at least some and usually all of the boundary conditions. Sometimes, if we know the general nature of the desired solution, we can improve the approximation by choosing the trial functions to reflect this nature. If, by chance, the exact solution is contained in the family of trial solutions, the Ritz procedure gives the exact solution. Generally, the approximation improves as the size of the family of trial functions and the number of adjustable parameters increase. If the trial functions are part of an infinite set of functions that are capable of representing the unknown function to any degree of accuracy, the process of including more and more trial functions leads to a series of approximate solutions which converges to the true solution. Often a family of trial functions is constructed from polynomials of successively increasing degree, but in certain cases other kinds of functions may offer advantages [8].

Example: The Ritz Method

To illustrate the Ritz method we shall consider a simple example. Suppose that we want to find the function $\phi(x)$ satisfying

$$\frac{d^2\phi}{dx^2} = -f(x)$$

with the following boundary conditions: $\phi(a) = A$, $\phi(b) = B$.
We assume that $f(x)$ is a continuous function in the closed interval $[a, b]$. This problem is equivalent to finding the function $\phi(x)$ that minimizes the functional

$$J(\phi) = \int_a^b \left[\frac{1}{2}\left(\frac{d\phi}{dx}\right)^2 - f(x)\phi(x) \right] dx$$

We shall ignore the fact that this problem has an exact solution, and proceed to find an approximate solution. According to the Ritz method we assume that the desired solution can be approximately represented in $[a, b]$ by a combination of selected trial functions of the form

$$\phi(x) \approx C_1\psi_1(x) + C_2\psi_2(x) + \cdots + C_n\psi_n(x), \quad a \le x \le b$$

where the n constants C_i are adjustable parameters to be determined. The trial functions should be selected so that the expression for $\phi(x)$ satisfies the boundary conditions regardless of the choice of the constants C_i. Using polynomials is a simple and convenient way to construct the trial functions. thus if A = B = O for example, we can write

$$\phi(x) \approx (x - a)(x - b)(C_1 + C_2 x + C_3 x^2 + \cdots + C_n x^{n-1})$$

as a possible series of trial functions. When we substitute this approximate expression for $\phi(x)$ into the functional to be minimized, we obtain, after carrying out the integration,

$$J = J(C_1, C_2, \ldots, C_n)$$

Now we require that the C_i be chosen to minimize J. Hence, from differential calculus, we have

$$\frac{\partial J}{\partial C_1} = 0, \qquad \frac{\partial J}{\partial C_2} = 0, \qquad \ldots, \qquad \frac{\partial J}{\partial C_n} = 0$$

When these n equations are solved for the n parameters C_i, the approximate solution to $\phi'' = -f$ is obtained as shown in Figure 3.1. The accuracy of the approximate solution depends on the number of C's used in the trial function.

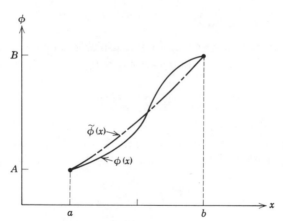

Figure 3.1. Rayleigh-Ritz approximation to the solution, $\phi(x)$. $\phi(x)$ = exact solution. $\tilde{\phi}(x)$ = an approximate Rayleigh-Ritz solution.

Generally, as n increases the accuracy improves. To assess the improvement in accuracy as more C's are used, we can solve the problem repeatedly by taking successively more terms in the approximation, that is, we can use

$$\phi_1(x) \approx (x - a)(x - b)C_1$$
$$\phi_2(x) \approx (x - a)(x - b)(C_1 + C_2 x)$$
$$\phi_3(x) \approx (x - a)(x - b)(C_1 + C_2 x + C_3 x^2)$$

and so on. By comparing the results at the end of each calculation we can estimate the effect of adding more terms on accuracy.

3.4   THE FINITE ELEMENT METHOD

3.4.1   Relation to the Ritz method

The finite element method and the Ritz method are essentially equivalent. Each method uses a set of trial functions as the starting point for obtaining an approximate solution; both methods take linear combinations of these trial functions; and both methods seek the combination of trial functions that makes a given functional stationary. The major difference between the methods is that the assumed trial functions in the finite element method are not defined over the whole solution domain, and they have to satisfy no boundary conditions, but only certain continuity conditions and then only sometimes. Because the Ritz method uses functions defined over the whole domain, it can be used only for domains of relatively simple geometric shape. In the finite element method the same geometric limitations exist, but only for the elements. Since elements with simple shapes can be assembled to represent exceedingly complex geometries, the finite element method is a far more versatile tool than the Ritz method. From a strict mathematical standpoint the finite element method is a special case of the Ritz method only when the piecewise trial functions obey certain continuity and completeness conditions to be discussed later.

3.4.2   Generalizing the definition of an element

The basic idea of the finite element method is to divide the solution domain into a finite number of subdomains (elements). These elements are connected only at node points in the domain and on the element boundaries. In this way, the solution domain is discretized and represented as a patchwork of elements. Frequently the boundaries of the finite elements are straight lines or planes, so if the solution domain has curved boundaries, these are approximated by a series of straight or flat segments.† In Chapter 2, when we considered a physical interpretation of the finite element method, we imagined the elements to be individual segments or parts of the actual system, for example, a spring, a fluid- or current-carrying conduit, a bar, or a triangular plate.

The nodes for these elements were part of the elements, and in cases where displacement was the unknown field variable the nodes could move as the

---

† An exception to this statement occurs when isoparametric elements are used. We shall introduce the concept of isoparametric elements in Chapter 5.

element deformed under load. In our example of a fluid-carrying conduit the nodes were simply places or locations in the system where pressure and flow were defined—the finite element and its nodes represented, not a part of the flowing fluid, but only a region through which the flow passed.

The mathematical interpretation of the finite element method requires us to generalize our definition of an element, and to think of elements in less physical terms. Instead of viewing an element as a physical part of the system we view the element as a part of the solution domain where the phenomena of interest are occurring. We imagine the solution domain to be sectioned by lines (or general planes in n dimensions) that define the boundaries of the elements. The elements are interconnected only at imaginary node points on the boundaries or surfaces of the elements. For solid mechanics problems we no longer need to imagine that the elements deform or change shape; rather we define them as regions of space where a displacement field exists. The nodes of an elements are then simply locations in space where the displacement and possibly its derivatives are known or sought. Likewise, for fluid mechanics problems the elements are regions over which a pressure field exists and through which fluid is flowing. The mathematical interpretation of a finite element mesh is that it is a *spatial subdivision* rather than a material subdivision [9]. This broader interpretation of an element allows us to carry over many of the basic ideas from one problem area to another.

In the finite element procedure, once the element mesh for the solution domain has been decided, the behavior of the unknown field variable over each element is approximated by continuous functions expressed in terms of the nodal values of the field variable and sometimes the nodal values of its derivatives up to a certain order. The functions defined over each finite element are called *interpolation functions*, *shape functions*, or field variable models. The collection of interpolation functions for the whole solution domain provides a piecewise approximation to the field variable.

3.4.3 Example of piecewise approximation

To illustrate the nature of this piecewise approximation we consider the representation of a two-dimensional field variable, $\phi(x, y)$. We will show how the nodal values of ϕ can uniquely and continuously define $\phi(x, y)$ throughout the domain of interest in the x-y plane, and we will introduce the notation for an interpolation function.

Suppose that we have the domain shown in Figure 3.2, and we section it into triangular elements with nodes at the vertices of the triangles. With this type of domain discretization we can allow ϕ to vary linearly over each

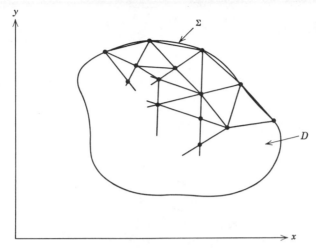

Figure 3.2. Two-dimensional domain divided into triangular elements.

element (see Figure 3.3). The plane passing through the three nodal values of ϕ associated with element (e) is described by the equation

$$\phi^{(e)}(x, y) = \beta_1^{(e)} + \beta_2^{(e)}x + \beta_3^{(e)}y \qquad (3.4)$$

We can express the constants $\beta_1^{(e)}$, $\beta_2^{(e)}$, and $\beta_3^{(e)}$ in terms of the coordinates of the element's nodes and the nodal values of ϕ by evaluating equation 3.4 at each node. Hence

$$
\begin{aligned}
\phi_i^{(e)} &= \beta_1^{(e)} + \beta_2^{(e)}x_i + \beta_3^{(e)}y_i \\
\phi_j^{(e)} &= \beta_1^{(e)} + \beta_2^{(e)}x_j + \beta_3^{(e)}y_j \\
\phi_k^{(e)} &= \beta_1^{(e)} + \beta_2^{(e)}x_k + \beta_3^{(e)}y_k
\end{aligned}
\qquad (3.5)
$$

Solving yields

$$\beta_1^{(e)} = \frac{\phi_i(x_jy_k - y_jx_k) + \phi_j(x_ky_i - x_iy_k) + \phi_k(x_iy_j - x_jy_i)}{2\Delta}$$

$$\beta_2^{(e)} = \frac{\phi_i(y_j - y_k) + \phi_j(y_k - y_i) + \phi_k(y_i - y_j)}{2\Delta} \qquad (3.6)$$

$$\beta_3^{(e)} = \frac{\phi_i(x_k - x_j) + (x_i - x_k) + \phi_k(x_j - x_i)}{2\Delta}$$

where (as in equation 2.18)

$$2\Delta = \begin{vmatrix} 1 & x_i & y_i \\ 1 & x_j & y_j \\ 1 & x_k & y_k \end{vmatrix} = 2\begin{bmatrix} \text{area of triangle whose} \\ \text{vertices are } i, j, k \end{bmatrix}$$

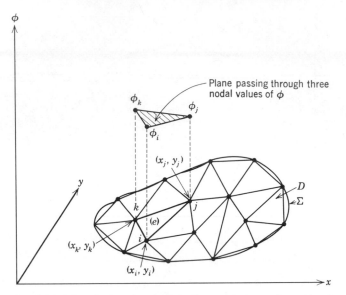

Figure 3.3. Subdivided domain D and a piecewise linear solution surface, $\phi(x, y)$.

Substituting equation 3.6 into equation 3.4 and rearranging terms, we have

$$\phi^{(e)}(x, y) = \frac{a_i + b_i x + c_i y}{2\Delta} \phi_i + \frac{a_j + b_j x + c_j y}{2\Delta} \phi_j + \frac{a_k + b_k x + c_k y}{2\Delta} \phi_k$$

(3.7)

where

$$a_i = x_j y_k - x_k y_j, \qquad b_i = y_j - y_k, \qquad c_i = x_k - x_j$$

and the other coefficients are obtained through a cyclic permutation of the subscripts i, j, k. If $i = 1, j = 2$, and $k = 3$, the coefficients are given explicitly by equation 2.17.

Now we define

$$N_n^{(e)} = \frac{a_n + b_n x + c_n y}{2\Delta}, \quad n = i, j, k$$

and let

$$\{\phi^{(e)}\} = \begin{Bmatrix} \phi_i \\ \phi_j \\ \phi_k \end{Bmatrix}, \qquad \lfloor N^{(e)} \rfloor = \lfloor N_i^{(e)}, N_j^{(e)}, N_k^{(e)} \rfloor$$

In general, the functions $N^{(e)}$ are called shape functions or interpolation functions, and they play a most important role in all finite element analysis. (Here the $N^{(e)}$ are linear interpolation functions for the three-node triangular element.) In matrix notation, we can write equation 3.7 as

$$\phi^{(e)}(x, y) = \lfloor N^{(e)} \rfloor \{\phi^{(e)}\} = N_i \phi_i + N_j \phi_j + N_k \phi_k \tag{3.8}$$

If the domain contains M elements, the complete representation of the field variable over the whole domain is given by

$$\phi(x, y) = \sum_{e=1}^{M} \phi^{(e)}(x, y) = \sum_{e=1}^{M} \lfloor N^{(e)} \rfloor \{\phi^{(e)}\} \tag{3.9}$$

From equation 3.9 we see that, if the nodal values of ϕ are known, we can represent the complete solution surface $\phi(x, y)$ as a series of interconnected, triangular-shaped planes. This many-faceted surface has no discontinuities or "gaps" at interelement boundaries because the values of ϕ at any two nodes defining an element boundary uniquely determine the linear variation of ϕ along that boundary. Figure 3.4 shows a wire model of the piecewise representation of the field variable expressed by equation 3.9.

Although we obtained equation 3.8 and 3.9 for a particular interpolation function (linear) and a particular element type (three node triangle), these equations have general validity. For more complex interpolation functions and other element types such as those discussed in Chapter 5, the form of equations 3.8 and 3.9 remains the same. Only the numbers of terms in the row and column matrices are different. Hence, if a solution domain is subdivided into elements, we can represent the unknown field variable in each element as

$$\phi^{(e)} = \overset{1 \times r}{\lfloor N^{(e)} \rfloor} \overset{r \times 1}{\{\phi\}} \tag{3.10}$$

where $\lfloor N^{(e)} \rfloor$ is the row vector of interpolation functions which are functions of the coordinates of the nodes, and $\{\phi\}^{(e)}$ is the column vector which is the collection of r discrete values consisting of nodal values of ϕ associated with the element *and* perhaps some other parameters that characterize the element and are not identified with any node. Such "nodeless" [10] variables may appear when constraints are imposed on the field variable, or when parameters are assigned to an element via modes of the interpolation functions that vanish on the element boundaries. These nodeless variables appear as *Lagrange multipliers* in the augmented functional. The nature of Lagrange multipliers and their relation to variational principles are discussed in Appendix B, but their application in finite element analysis is beyond the scope of this book.

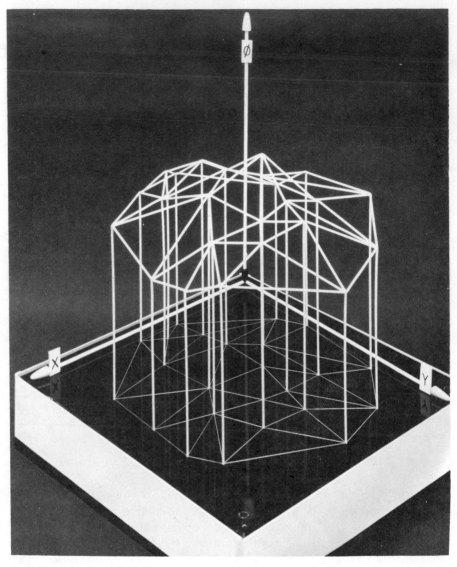

Figure 3.4. Wire model illustrating a piecewise linear representation of $\phi(x, y)$.

We shall now assume that we have the field variable, ϕ, completely represented in the solution domain in terms of a collection of nodal values of ϕ. Under this assumption, if these discrete values were known, the problem of finding an approximation to ϕ would be solved.

3.4.4 Element equations from a variational principle

The finite element solution to the problem involves picking the nodal values of ϕ so as to make stationary the functional, $I(\phi)$. To stationarize $I(\phi)$ with respect to the nodal values of ϕ, we require that

$$\delta I(\phi) = \sum_{i=1}^{n} \frac{\partial I}{\partial \phi_i} \delta \phi_i = 0 \tag{3.11}$$

where n is the total number of discrete values of ϕ assigned to the solution domain. Since the $\delta \phi_i$'s are independent, equation 3.11 can hold only if

$$\frac{\partial I}{\partial \phi_i} = 0, \quad i = 1, 2, \ldots, n \tag{3.12}$$

If the interpolation functions giving our piecewise representation of ϕ obey certain continuity and compatibility conditions, which we will discuss shortly, the functional $I(\phi)$ may be written as a sum of individual functionals defined for all elements of the assemblage, that is,

$$I(\phi) = \sum_{e=1}^{M} I^{(e)}(\phi^{(e)}) \tag{3.13}$$

where M is the total number of elements, and the superscript (e) denotes an element. Hence, instead of working with the functional defined over the whole solution region, we may focus our attention on the functionals defined for the individual elements. From equation 3.13 we have

$$\delta I = \sum_{e=1}^{M} \delta I^{(e)} = 0 \tag{3.14}$$

where the variation of $I^{(e)}$ is taken only with respect to the nodal values associated with element (e). Equation 3.14 implies that

$$\left\{ \frac{\partial I^{(e)}}{\partial \phi} \right\} = \frac{\partial I^{(e)}}{\partial \phi_j} = 0, \quad j = 1, 2, \ldots, r \tag{3.15}$$

where r is the number of nodes assigned to element (e). Equations 3.15 comprise a set of r equations that characterizes the behavior of element (e). The fact that we can represent the functional for the assemblage of elements as the sum of the functionals for all individual elements provides the key to formulating individual element equations from a variational principle. If, for example, the governing differential equations and boundary conditions for a problem are linear and self-adjoint, the corresponding variational statement of the problem involves a *quadratic functional*. When $I(\phi)$ is quadratic, $I^{(e)}(\phi)$ is also quadratic, and equations 3.15 for element (e) can always be written as [10]

$$\underset{r \times 1}{\left\{ \frac{\partial I^{(e)}}{\partial \phi} \right\}} = \underset{r \times r}{[K]^{(e)}} \underset{r \times 1}{\{\phi\}^{(e)}} - \underset{r \times 1}{\{F\}^{(e)}} = \underset{r \times 1}{\{0\}} \tag{3.16}$$

where $[K]^{(e)}$ is a square matrix of constant influence coefficients, $\{\phi\}^{(e)}$ is the column vector of nodal values, and $\{F\}$ is the vector of resultant nodal actions. The complete set of finite element equations for the problem is assembled by adding all the differentials of I as given by equations 3.15 for all the elements. The assembly procedure is identical to that discussed in Chapter 2. Symbolically, we can write the complete set of equations as

$$\frac{\partial I}{\partial \phi_i} = \sum_{e=1}^{M} \frac{\partial I^{(e)}}{\partial \phi_i} = 0, \quad i = 1, 2, \ldots, n \tag{3.17}$$

or

$$\left\{ \frac{\partial I}{\partial \phi_i} \right\} = \{0\}$$

Our problem is solved when the set of n equations 3.17 is solved simultaneously for the n nodal values of ϕ. If there are q nodes in the solution domain where ϕ is specified by boundary conditions, there will be $n - q$ equations to be solved for the $n - q$ unknowns. Note that the summation indicated in equation 3.17 contains many zero terms because only elements sharing node i will contribute to $\partial I / \partial \phi_i$. If node i does not belong to element e, $\partial I / \partial \phi_i = 0$. This fact is manifested in the "narrow band" and "sparseness" properties of the resulting matrix of influence coefficients.

We have now established the means for formulating individual finite element equations from a variational principle. The procedure can be summarized as follows.

If the functional for a given problem can be expressed as the sum of functionals evaluated for all elements we may focus our attention on an isolated element without regard for its eventual location in the assemblage. To derive the equations governing the element's behavior, we first use interpolation functions to define the unknown field variable ϕ in terms of its nodal values associated with the element; then we evaluate the functional $I^{(e)}$ by substituting the assumed form for $\phi^{(e)}$ and its derivatives and carry out the integration over the domain defined by the element boundaries. Finally, we perform the differentiations indicated by equation 3.15, and the result is the set of equations defining the element behavior. Since the differentiations are with respect to the discrete nodal values, we employ only the calculus, not the calculus of variations.

3.4.5 Requirements for interpolation functions

Our procedure for formulating the individual element equations from a variational principle and our privilege to assemble these equations to obtain the system equations rely on the assumption that the interpolation functions satisfy certain requirements. The requirements we place on the choice of the interpolation functions stem from the need to ensure that equation 3.13 holds and that our approximate solution converges to the correct solution when we use an increasing number of smaller elements, that is, when we refine the element mesh. Mathematical proofs of convergence assume that the process of mesh refinement occurs in a regular fashion defined by three conditions [11]: (1) the elements must be made smaller in such a way that every point of the solution domain can always be within an element, regardless of how small the element may be; (2) all previous minima must be contained in the refined meshes; and (3) the form of the interpolation functions must remain unchanged during the process of mesh refinement.

These three conditions are illustrated in Figure 3.5, where a simple two-dimensional solution domain in the form of an equilateral triangle is discretized with an increasing number of three-node triangles. We note that when elements with straight boundaries are used to model solution domains with curved boundaries the first two conditions are not satisfied and rigorous mathematical proofs of convergence may not be obtainable. In spite of this limitation many applications of the finite element method to problems with nonpolygonal solution domains yield acceptable engineering solutions.

To guarantee monotonic convergence in the sense just described and to make the assembly of the individual element equations meaningful, we require that the interpolation functions $N^{(e)}$ in the expressions

$$\phi^{(e)} = \lfloor N^{(e)} \rfloor \{\phi\}^{(e)}, \quad e = 1, 2, \ldots, M$$

be chosen so that the following general requirements are met:

1. At element interfaces (boundaries) the field variable ϕ and any of its partial derivatives up to one order less than the highest-order derivative appearing in $I(\phi)$ must be continuous.
2. All uniform states of ϕ and its partial derivatives up to the highest order appearing in $I(\phi)$ should have representation in $\phi^{(e)}$ when, in the limit, the element size shrinks to zero.

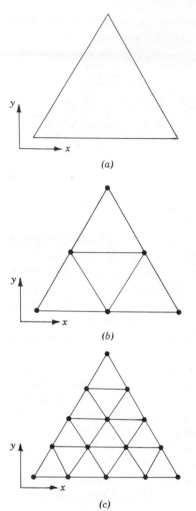

Figure 3.5.   Example of successive mesh refinement. (*a*) Original solution domain. (*b*) Discretization with 4 triangular elements. (*c*) Discretization with 16 triangular elements.

These requirements were given by Felippa and Clough [11] and justified by Oliveira [12]. The first one is known as the *compatibility* requirement, and the second as the *completeness* requirement. Elements whose interpolation functions satisfy the first requirement are called *compatible* or *conforming elements*; those satisfying the second requirement, *complete elements*. The definitions of an *incompatible element* is obvious.

When, for example, our field variable ϕ is defined as the displacement field in a body, these requirements on the interpolation functions have particular meanings. For instance, the compatibility requirement ensures the continuity of displacement between adjacent elements. Clearly, we could not expect a good approximation to reality if our displacement field representation were discontinuous across element boundaries. The second requirement, the completeness requirement, ensures that our displacement field representation allows for the possibility of rigid body displacements and constant strain states within the element. These conditions are discussed more fully in Part 2 of this book.

The compatibility requirement helps to ensure that the integral in $I(\phi)$ is well defined. Without interelement continuity we cannot be sure that the integral in $I(\phi)$ is unique. Uncertain contributions may arise from the "gaps" between elements. If the compatibility condition is violated, it is sometimes possible to add special boundary integrals for compensation [13]. It is always desirable when carrying out a finite element analysis to be sure that mesh refinement will lead to answers that are converging to the correct solution. For some problems, however, choosing interpolation functions that meet all the requirements may be difficult and may involve excessive numerical computation. For this reason some investigators have ventured to formulate interpolation functions for elements that do not meet all the compatibility and completeness requirements. In some instances acceptable convergence has been obtained, whereas in others no convergence or convergence to an incorrect solution has occurred. In solid mechanics applications we shall encounter a number of cases where complete but not compatible elements have been successfully used. Some applications [14] have shown that the compatibility condition does not always lead to the most rapid convergence. However, when extending the finite element method for use in other areas where far less experience is available, the safest approach is to pick functions and formulate elements that satisfy all the requirements. In Chapter 5 we shall discuss the formulation of many types of elements using polynomial interpolation functions.

Example

To show how the element equations can be derived from a variational principle we consider the same simple example we used to illustrate the Rayleigh-Ritz method in Section 3.3. The problem was to find the function $\phi(x)$ satisfying

$$\frac{d^2\phi}{dx^2} = -f(x) \quad \text{with boundary conditions } \phi(a) = A, \ \phi(b) = B$$

Equivalently, we seek the function $\phi(x)$ that minimizes

$$J(\phi) = \int_a^b \left[\frac{1}{2}\left(\frac{d\phi}{dx}\right)^2 - f(x)\phi(x) \right] dx$$

Since this is a one-dimensional problem (one independent variable), our elements will be line segments along the x axis (line elements). The number of nodes we assign to each element (and, hence, the type of interpolation function we can use for the element) is open to choice. Since the highest-order derivative appearing in $J(\phi)$ is a first-order derivative, the simplest interpolation functions satisfying our compatibility and completeness requirements are linear functions. Two nodal values of ϕ are required to define uniquely a linear variation of ϕ over an element; hence a typical element must have two nodes (Figure 3.6a). A linear variation of ϕ over element (e)

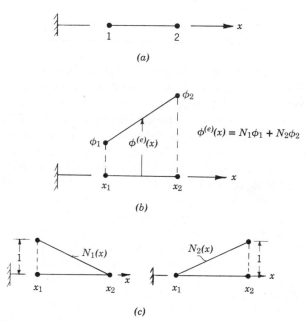

Figure 3.6. Linear representation of a field variable over a one-dimensional element. (a) One-dimensional line element. (b) Linear variation of $\phi(x)$ over element (e). (c) Linear interpolation functions for $\phi^{(e)}(x)$.

can be written in terms of its nodal values and the coordinates of the nodes (Figure 3.6b) as

$$\phi^{(e)}(x) = \left(\frac{x_2 - x}{x_2 - x_1}\right)\phi_1 + \left(\frac{x - x_1}{x_2 - x_1}\right)\phi_2$$

$$\phi^{(e)} = \lfloor N_1, N_2 \rfloor \begin{Bmatrix} \phi_1 \\ \phi_2 \end{Bmatrix} = \lfloor N \rfloor \{\phi\}^{(e)}$$

where

$$N_1(x) = \frac{x_2 - x}{x_2 - x_1}, \qquad N_2(x) = \frac{x - x_1}{x_2 - x_1}$$

The interpolation functions N_1 and N_2 are shown in Figure 3.6c. Since our interpolation functions for modeling the behavior of ϕ satisfy the compatibility and completeness requirements, we may write

$$J(\phi) = \sum_{e=1}^{M} J^{(e)}(\phi^{(e)})$$

where M is the number of elements we choose to model the domain $a \leq x \leq b$. We also have the mathematical assurance that, as M increases, our approximate solutions will converge to the correct solution. To find $J^{(e)}(\phi^{(e)})$ we substitute the equation for $\phi^{(e)}$ into the expression for $J(\phi)$ and obtain (integrating only over the domain of the element)

$$J^{(e)}(\phi^{(e)}) = \int_{x_1}^{x_2} \left[\frac{1}{2}\left(\lfloor N_1' \, N_2' \rfloor \begin{Bmatrix} \phi_1 \\ \phi_2 \end{Bmatrix} \right)^2 - f(x)\lfloor N_1 \, N_2 \rfloor \begin{Bmatrix} \phi_1 \\ \phi_2 \end{Bmatrix} \right] dx$$

where the prime denotes differentiation with respect to x. Carrying out the minimization of $J^{(e)}(\phi^{(e)})$ with respect to the nodal values of ϕ gives the equations for an element:

$$\frac{\partial J^{(e)}}{\partial \phi_1} = \int_{x_1}^{x_2} \left[N_1' \lfloor N_1' \, N_1' \rfloor \begin{Bmatrix} \phi_1 \\ \phi_2 \end{Bmatrix} - f N_1 \right] dx = 0$$

and

$$\frac{\partial J^{(e)}}{\partial \phi_2} = \int_{x_1}^{x_2} \left[N_2' \lfloor N_1' \, N_2' \rfloor \begin{Bmatrix} \phi_1 \\ \phi_2 \end{Bmatrix} - f N_2 \right] dx = 0$$

These equations may be written concisely as

$$[K]^{(e)}\{\phi\}^{(e)} = \{F\}^{(e)}$$

where

$$[K]^{(e)} = \int_{x_1}^{x_2} \begin{bmatrix} N_1' N_1' & N_1' N_2' \\ N_2' N_1' & N_2' N_2' \end{bmatrix}^{(e)} dx\dagger$$

$$\{\phi\}^{(e)} = \begin{Bmatrix} \phi_1 \\ \phi_2 \end{Bmatrix}, \qquad \{F\}^{(e)} = \int_{x_1}^{x_2} \begin{Bmatrix} f N_1 \\ f N_2 \end{Bmatrix}^{(e)} dx$$

---

† Integration of each term of the matrix is implied in this expression. We also note that $[K]^{(e)}$ is symmetric.

The components of $\{F\}^{(e)}$ may be viewed as the nodal forcing functions. With these definitions we see that the functional for one element can be written in quadratic form as

$$J^{(e)}(\phi^{(e)}) = \tfrac{1}{2}\lfloor\phi\rfloor^{(e)}[K]^{(e)}\{\phi\}^{(e)} - \lfloor F\rfloor^{(e)}\{\phi\}^{(e)}$$

The equations $[K]^{(e)}\{\phi\}^{(e)} = \{F\}^{(e)}$ give the characteristics of a particular finite element of our one-dimensional continuum. The same procedure we used to derive these equations can be employed to obtain the characteristics of more complicated elements. For instance, in this example the first *higher-order element* would be formulated as a line element with three nodes (two exterior nodes and one exterior node) and with quadratic interpolation functions such as those shown in Figure 3.7. It is easy to see how we can

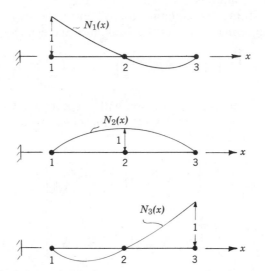

Figure 3.7.   Quadratic interpolation functions for three-node line elements.

formulate still higher-order elements with more nodes and subsequently higher-order interpolation functions. Regardless of the type of element we choose to formulate for this problem, the element equations will have the form $[K]^{(e)}\{\phi\}^{(e)} = \{F\}^{(e)}$.

To complete the finite element solution of this problem it is necessary to derive the equations for all the elements in the assemblage and then to assemble these algebraic equations according to the prescriptions in Chapter 2. After introducing the boundary conditions, we solve the resulting set of equations for the nodal values of ϕ. The result of the solution will be a piecewise representation of ϕ over $[a, b]$, such as that shown in Figure 3.7. By comparing Figures 3.1 and 3.8 we can see the primary difference between the Ritz method and the finite element method.

In view of the preceding example several general comments are in order. The task of deriving the equations for the individual elements for a problem may appear to be tedious, but actually the process is routine. If the same type of element is used throughout the solution domain, the element equations are given by one general expression. The expression contains nodal coordinates and other known parameters; hence it can serve as the computational algorithm for generating all the equations by computer. If several different types of elements are used, we simply have several different algorithms. Usually, however, for a given problem we use only one type of element.

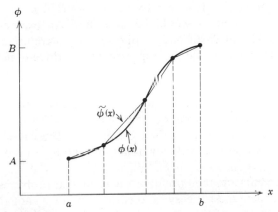

Figure 3.8.   Finite element approximation to the solution $\phi(x)$. $\phi(x)$ = exact solution. $\tilde{\phi}(x)$ = an approximate finite element solution.

As we saw in Chapter 2, the assembly procedure is automatically done within the computer also. Usually, the element matrices are inserted in the master matrix as they are formulated.

Before leaving our simple example we note a curious fact [15]. A rigorous mathematical proof exists to show that, if the problem $\phi'' = -f$, $\phi(a) = A$, $\phi(b) = B$ is solved by the finite element method using *linear* interpolation functions over two-node elements, and if at least two elements are used in the solution, then, regardless of the number of elements, the discrete nodal values of ϕ found from the solution lie on the exact solution, $\phi(x)$. This is one *isolated* case in which the finite element method gives exact nodal values.

3.4.6　Domain discretization

The first task in a finite element solution consists of discretizing the continuum by dividing it into a series of elements. Underlying the discretization process is the goal of achieving a good representation of the phenomena under study. There are no set rules for reaching this goal because of the vastly different circumstances encountered from one problem to another, but some helpful guidelines emerge from the large amount of available experience in finite element analysis.

Frequently, the first questions an analyst asks are, "What type of element should I use, and should I mix several different types of elements?" The answers to these questions depend on the problem being considered. Often only one type of element is used to represent the continuum unless circumstances dictate otherwise. It is easy to imagine a problem for which several different types of elements would be necessary. An example from solid mechanics would be an elastic body supported by pin-connected bars. In this case the elastic body would be represented by three-dimensional solid elements such as "bricks," while the bars would be approximated by one-dimensional elements like those we considered in Chapter 2. The most popular and versatile elements, because of the ease with which they can be assembled to fit complex geometries, are triangular elements in two dimensions and tetrahedral elements in three dimensions. These elements can have any number of exterior and interior nodes, depending on the type of interpolation functions defined for them. In Chapter 5 we will discuss many different types of elements and their usefulness in particular problems.

A uniform element mesh is easy to construct, but it may not always provide a good representation of the continuum. In regions of the solution domain where the gradient of the field variable is expected to vary relatively fast, a finer element mesh should be used. Also, it is most convenient to place nodes and element boundaries at locations where point external actions (such as forces in structural problems) are applied and where there are abrupt changes in the continuum. If the boundary of the region has any corners, nodes should also be placed at these corners. More elements should be used in regions where the boundary is irregular than in regions where it is smooth.

Another guideline is that the continuum should be discretized so as to give the element a well-proportioned shape. In other words, the ratio of an element's smallest dimension to its largest dimension should be near unity. Long, narrow elements should be avoided because they lead to a solution with directional bias that may not be correct. For a given element in a given location in the solution domain the best or optimum ratio of extreme dimensions depends on the local gradient of the field variable, and this is unknown a priori. Hence the conservative procedure is to use elements whose shape proportions are well balanced, although this is not always possible.

Provided that the elements obey the requirements for a convergent solution, we may expect that the more elements we use to model the solution domain, the better the accuracy of our results. Since increasing the number of elements leads to higher computational expenses, the analyst must base his decision on rational compromise as usual. When solving a particular type of problem for the first time, it is good practice to obtain several solutions with different numbers of elements. By comparing the results it is then possible to see whether enough elements are being used in the solution.

A similar trial-and-error procedure is used for determining satisfactory mesh representation of domains of *infinite* extent. The procedure is to construct a *finite* mesh encompassing the regions of the solution domain where the phenomena are occurring. By comparing solutions obtained for meshes of increasing extent, we can determine the point beyond which the location of the boundary no longer has a significant effect on the solution. Usually, the elements used to represent the extremities of the domain so determined are considerably larger than those in the interior.

3.4.7 Example of a complete finite element solution

In Chapter 1 we discussed in general terms the six-step finite element solution procedure. Now that we have seen how the finite element equations for a continuum problem can be derived from variational principles, we can return to these basic steps and illustrate them in some detail with an example. From Chapter 1 we recall the six basic steps of the method:

1. Discretize the continuum.
2. Select interpolation functions.
3. Find the element properties.
4. Assemble the element properties.
5. Solve the system equations.
6. Make additional computations if desired.

Using each of these steps, we shall now carry out a complete solution of a specific problem.

Suppose that we have two parallel flat plates of equal size and shape in the form of equilateral triangles. Imagine that the plates are separated by a thin fluid film. If these plates are aligned directly above one another and they are moving together with a velocity V while the incompressible fluid is being squeezed out from between them, the pressure field generated within the space between the plates satisfies the Reynolds-Poisson equation:

$$\frac{\partial^2 P}{\partial x^2} + \frac{\partial^2 P}{\partial y^2} = \frac{12\mu V}{h^3} = G = \text{constant} \qquad (a)$$

while the pressure vanishes along the boundary described by the lines (see Figure 3.9)

$$y = 0, \qquad y = 2 - \sqrt{3}x, \qquad y = \sqrt{3}x \qquad (b)$$

We choose this particular problem because it has an exact solution that we can compare with our finite element solution. It is easy to show by substitution that the pressure distribution

$$p(x, y) = -\frac{G}{4}(y - 2 + \sqrt{3}x)(y - \sqrt{3}x)y \qquad (c)$$

satisfies equation (a), is zero along the boundaries defined by equations (b), and hence is the unique solution to the problem. We will see in Section 3.5

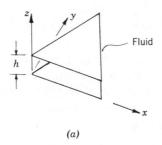

(a)

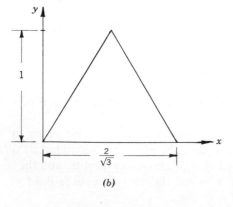

(b)

Figure 3.9. A simple two-dimensional squeeze film problem. (a) Triangular plates generating a squeeze film. (b) Geometry of solution domain.

that this problem is equivalent to finding the pressure distribution that satisfies the boundary conditions and minimizes the functional

$$J(P) = \int_D \int \left[\frac{1}{2} \left(\frac{\partial P}{\partial x} \right)^2 + \frac{1}{2} \left(\frac{\partial P}{\partial x} \right)^2 + GP \right] dx \, dy \qquad (d)$$

Consider now the application of the finite element method to obtain a solution. The first step is to subdivide the solution domain into elements. Since our solution domain has the shape of a triangle, an assemblage of triangular elements will give an exact representation. For simplicity we choose three triangular elements with nodes placed at the vertices of the triangles (see Figure 3.10). An actual finite element analysis may involve hundreds of nodes and elements, but we shall consider only a three-element

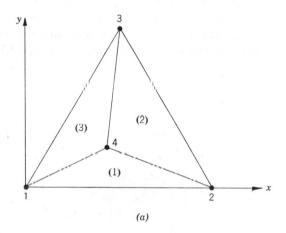

(a)

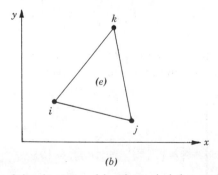

(b)

Figure 3.10. The finite element model and a typical three-node triangular element. (a) Discretization of the solution domain and the system numbering scheme (topology). (b) Notation for a typical element.

mesh, so that we can carry out the computations by hand. To avoid having to make coordinate transformations, we will derive the element equations in the global coordinate system. For this example we have no reason to choose a local coordinate system different from the global system. The next step in our solution is to choose interpolation functions. As we saw in Section 3.4.3, if we choose linear interpolation functions for each element, we can uniquely and continuously represent the field variable $p(x, y)$ in the solution domain. From equation 3.8 we can write for one element

$$P^{(e)}(x, y) = \lfloor N^{(e)} \rfloor \{P\}^{(e)} = N_i P_i + N_j P_j + N_k P_k \qquad (e)$$

where $\{P\}^{(e)}$ is the column vector of nodal pressures for element (e), and the interpolation functions $N^{(e)}$ are those defined in Section 3.4.3. Since these interpolation functions satisfy the interelement compatibility requirement, equation 3.13 holds and we may focus our attention on an isolated element.

To derive the element equations we substitute equation (e) into equation (d) and perform the differentiations indicated in equation 3.15:

$$\frac{\partial J^{(e)}}{\partial P_i} = \frac{\partial J(P^{(e)})}{\partial P_i}$$

$$= \iint_{A^{(e)}} \left[\frac{\partial P^{(e)}}{\partial x} \frac{\partial}{\partial P_i} \left(\frac{\partial P^{(e)}}{\partial x} \right) + \frac{\partial P^{(e)}}{\partial y} \frac{\partial}{\partial P_i} \left(\frac{\partial P^{(e)}}{\partial y} \right) + G \frac{\partial P^{(e)}}{\partial P_i} \right] dx \, dy = 0$$

$$= \iint_{A^{(e)}} \left[\left(\frac{\partial N_i}{\partial x} P_i + \frac{\partial N_j}{\partial x} P_j + \frac{\partial N_k}{\partial x} P_k \right) \frac{\partial N_i}{\partial x} \right.$$

$$\left. + \left(\frac{\partial N_i}{\partial y} P_i + \frac{\partial N_j}{\partial y} P_j + \frac{\partial N_k}{\partial y} P_k \right) \frac{\partial N_i}{\partial y} + G N_i \right] dx \, dy = 0$$

We may write similar expressions for $\partial J^{(e)}/\partial P_j$ and $\partial J^{(e)}/\partial P_k$. The three resulting equations can be written in the standard form:

$$\begin{Bmatrix} \dfrac{\partial J^{(e)}}{\partial P_i} \\[2mm] \dfrac{\partial J^{(e)}}{\partial P_j} \\[2mm] \dfrac{\partial J^{(e)}}{\partial P_k} \end{Bmatrix} = [K]^{(e)}\{P\}^{(e)} + \{R\}^{(e)} = \{0\} \qquad (f)$$

where the terms of the matrices $[K]^{(e)}$ and $[R]^{(e)}$ are given by

$$k_{ij}^{(e)} = \iint_{A^{(e)}} \left(\frac{\partial N_i}{\partial x} \frac{\partial N_j}{\partial x} + \frac{\partial N_i}{\partial y} \frac{\partial N_j}{\partial y} \right) dx\, dy \qquad (g)$$

$$R_i^{(e)} = \iint_{A^{(e)}} G N_i \, dx\, dy$$

Equations (f) and (g) may now be explicitly evaluated using our definition of the interpolation functions. Referring to Section 3.4.3, we have

$$\frac{\partial N_i}{\partial x} = \frac{b_i}{2\,\Delta_{(e)}}, \qquad \frac{\partial N_j}{\partial x} = \frac{b_j}{2\,\Delta_{(e)}}$$

$$\frac{\partial N_i}{\partial y} = \frac{c_i}{2\,\Delta_{(e)}}, \qquad \frac{\partial N_j}{\partial y} = \frac{c_j}{2\,\Delta_{(e)}}$$

where

$$b_i = y_j \quad y_{k} \quad 0_i \quad u_{k} \quad u_{j} \quad \Delta_{(e)} = \text{area of element } (e)$$

Hence

$$k_{ij}^{(e)} = \iint_{A^{(e)}} \left(\frac{b_i b_j}{4\,\Delta_{(e)}^2} + \frac{c_i c_j}{4\,\Delta_{(e)}^2} \right) dx\, dy$$

$$\qquad\qquad (h)$$

$$k_{ij}^{(e)} = \frac{1}{4\,\Delta_{(e)}} (b_i b_j + c_i c_j)$$

We note that, since the b's and c's are constant, the integrand is also constant and the surface integral over the triangular element is simply the area of the element, $\Delta_{(e)}$. Evaluating the nodal actions R_i for an element is not as simple because it involves integrating an interpolation function over the area of an element. Fortunately, as we shall see in Chapter 5, there is a convenient integration formula that allows us to write immediately

$$R_i^{(e)} = \iint_{A^{(e)}} G N_i \, dx\, dy = G \iint_{A^{(e)}} N_i \, dx\, dy$$

$$\qquad\qquad (i)$$

$$\qquad = \frac{G\,\Delta_{(e)}}{3}$$

Equations (h) and (i) give the terms of the element matrices for a typical element (e); hence we have found the element properties. Using the system

Element number	Local node number	Global node number
1	$i = 1$	1
	$j = 2$	2
	$k = 3$	4
2	$i = 1$	2
	$j = 2$	3
	$k = 3$	4
3	$i = 1$	3
	$j = 2$	1
	$k = 3$	4

Figure 3.11. Correspondence between the local and global numbering schemes for the squeeze film problem.

topology illustrated in Figure 3.11, we can write the element equations in the following symbolic form.

For element (1):

$$\begin{bmatrix} k_{11} & k_{12} & k_{14} \\ k_{21} & k_{22} & k_{24} \\ k_{41} & k_{42} & k_{44} \end{bmatrix} \begin{Bmatrix} P_1 \\ P_2 \\ P_4 \end{Bmatrix} = -\frac{G}{3} \begin{Bmatrix} \Delta_1 \\ \Delta_1 \\ \Delta_1 \end{Bmatrix}$$

For element (2):

$$\begin{bmatrix} k_{22} & k_{23} & k_{24} \\ k_{32} & k_{33} & k_{44} \\ k_{42} & k_{43} & k_{44} \end{bmatrix} \begin{Bmatrix} P_2 \\ P_3 \\ P_4 \end{Bmatrix} = -\frac{G}{3} \begin{Bmatrix} \Delta_2 \\ \Delta_2 \\ \Delta_2 \end{Bmatrix}$$

For element (3):

$$\begin{bmatrix} k_{33} & k_{31} & k_{34} \\ k_{13} & k_{11} & k_{14} \\ k_{43} & k_{41} & k_{44} \end{bmatrix} \begin{Bmatrix} P_3 \\ P_1 \\ P_4 \end{Bmatrix} = -\frac{G}{3} \begin{Bmatrix} \Delta_3 \\ \Delta_3 \\ \Delta_3 \end{Bmatrix}$$

These individual element equations can be assembled according to the procedure of Chapter 2 to give the following *system* equations:

$$\begin{bmatrix} k_{11}^{(1)} + k_{11}^{(3)} & k_{12}^{(1)} & k_{13}^{(3)} & k_{14}^{(1)} + k_{14}^{(3)} \\ k_{21}^{(1)} & k_{22}^{(1)} + k_{22}^{(2)} & k_{23}^{(2)} & k_{24}^{(1)} + k_{24}^{(2)} \\ k_{31}^{(3)} & k_{32}^{(2)} & k_{33}^{(2)} + k_{33}^{(3)} & k_{34}^{(2)} + k_{34}^{(3)} \\ k_{41}^{(1)} + k_{41}^{(3)} & k_{42}^{(1)} + k_{42}^{(2)} & k_{43}^{(2)} + k_{43}^{(3)} & k_{44}^{(1)} + k_{44}^{(2)} + k_{44}^{(3)} \end{bmatrix}$$

$$\begin{Bmatrix} P_1 \\ P_2 \\ P_3 \\ P_4 \end{Bmatrix} = -\frac{G}{3} \begin{Bmatrix} \Delta_{(1)} + \Delta_{(3)} \\ \Delta_{(1)} + \Delta_{(2)} \\ \Delta_{(2)} + \Delta_{(3)} \\ \Delta_{(1)} + \Delta_{(2)} + \Delta_{(3)} \end{Bmatrix} \qquad (j)$$

Because of the boundary conditions, this set of equations contains only one unknown, namely, P_4. The equation to be solved for the unknown pressure at the interior node is

$$(k_{44}^{(1)} + k_{44}^{(2)} + k_{44}^{(3)})P_4 = -\frac{G}{3}(\Delta_{(1)} + \Delta_{(2)} + \Delta_{(3)}) = -\frac{G\sqrt{3}}{9}$$

where

$$k_{44}^{(1)} = \frac{1}{4\,\Delta_{(1)}}(b_4^2 + c_4^2)^{(1)} = \frac{1}{4\,\Delta_{(1)}}[(y_1 - y_2)^2 + (x_2 - x_1)^2] = \frac{1}{3\,\Delta_{(1)}}$$

$$k_{44}^{(2)} = \frac{1}{4\,\Delta_{(2)}}(b_4^2 + c_4^2)^{(2)} = \frac{1}{3\,\Delta_{(2)}}$$

$$k_{44}^{(3)} = \frac{1}{4\,\Delta_{(3)}}(b_4^2 + c_4^2)^{(3)} = \frac{1}{3\,\Delta_{(3)}}$$

Now

$$\Delta_{(1)} = \frac{1}{2}\begin{vmatrix} 1 & x_1 & y_1 \\ 1 & x_2 & y_2 \\ 1 & x_3 & y_3 \end{vmatrix} = \frac{y_4}{\sqrt{3}}$$

Similarly,

$$\Delta_{(2)} = \frac{1}{2\sqrt{3}}(2 - y_4 - \sqrt{3}\,x_4)$$

$$\Delta_{(3)} = \frac{1}{2\sqrt{3}}(\sqrt{3}\,x_4 - y_4)$$

Substituting these results into the equation for the pressure at node 4 and rearranging terms gives

$$P_4(x_4, y_4) = -\frac{G\sqrt{3}}{3}\,\frac{\Delta_{(1)}\Delta_{(2)}\Delta_{(3)}}{\Delta_{(2)}\Delta_{(3)} + \Delta_{(1)}\Delta_{(3)} + \Delta_{(1)}\Delta_{(2)}}$$

or, after some manipulation

$$P_4(x_4, y_4) = -G\,\frac{(y_4 - 2 + \sqrt{3}\,x_4)(y_4 - \sqrt{3}\,x_4)\,y_4}{3[2(y_4 + \sqrt{3}\,x_4) - 3(x_4^2 + y_4^2)]}$$

It is interesting to note that, if the coordinates of node 4 coincide with the coordinates of the centroid of the triangular pad, the finite element solution for this problem is exact. It can easily be shown that, if the interior node is different from the centroid, the true solution is consistently overestimated.

Now that all nodal pressures are known, we may make additional calculations to determine nodal flows. We shall postpone the development of the necessary equations for computing such flows until we discuss lubrication problems in Chapter 8.

If we choose a solution domain of another shape and include more elements and nodes, obtaining a solution by hand becomes impractical. Suppose that we select a square domain as shown in Figure 3.12 and obtain a finite element solution by digital computer (see Chapter 10 for a sample computer program). Then the number of nodes we use is limited only by the size of the computer core storage capacity.

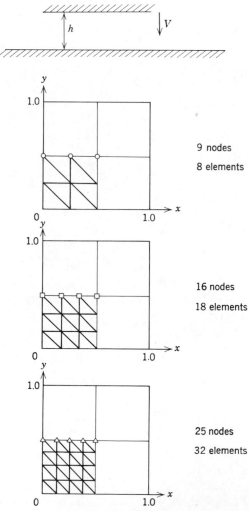

Figure 3.12.   Geometry and finite element idealization of a square pad bearing.

To illustrate the effect of mesh size or nodal spacing on the accuracy and convergence of a simple finite element solution with linear triangular elements, we consider the various meshes of Figure 3.12. Because of the symmetry in this problem, only one quadrant of the solution region need be considered. The nodal values of pressure at the indicated nodes and the exact solution for the pressure along the pad centerline are shown in Figure 3.13. In this case, we see that the finite element solution gives an underestimate of the true solution.

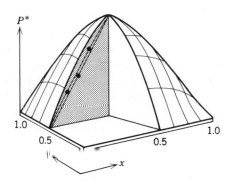

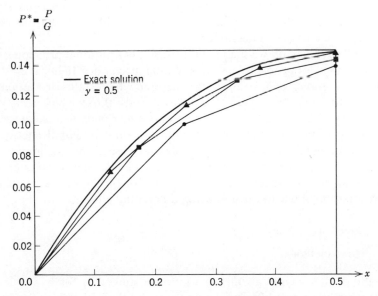

Figure 3.13. Squeeze film pressure generated by a square flat plate (comparision of FEM results and exact results).

An important point illustrated by this example is that we can develop our element equations in terms of general interpolation functions and nodal variables without specifying a particular element shape and its specific interpolation functions. The methodical procedure of first discretizing the solution domain and then picking interpolation functions is often the way whereby a practical problem is approached, but development of the element equations from a variational principle usually does not depend on these steps.

3.5 FINDING VARIATIONAL PRINCIPLES

3.5.1 Introduction

Since continuum problems are usually stated in differential form, we face the question of how to obtain the classical variational form. The functional we seek is some integrated quantity that is characteristic of the problem—a physically recognizable quantity such as energy or, perhaps, a mathematically defined entity without any physical interpretation. Variational principles yielding functionals for physical processes can be found outside as well as inside the realm of classical variational calculus. Sometimes a functional can be found from well-established theorems in physics. For example, in classical thermodynamics we know that for an isolated system at equilibrium the entropy is a maximum, whereas for a system with constant volume and temperature the Helmholtz free energy is a minimum. Consequently, to find functionals for these two cases we can express either the entropy or the Helmholtz free energy in integral form.

In Part 2 of this book we shall see that several variational principles have been established for problems in solid and fluid mechanics. In linear elasticity theory, for example, we have an energy theorem which states that among all admissible displacement fields of an elastic body those which satisfy the equilibrium conditions make the total potential energy of the system assume a stationary value. For stable equilibrium the total potential energy is a minimum. In fluid mechanics, for the special case of an incompressible, inviscid fluid experiencing irrotational flow, the kinetic energy is a minimum. Theorems such as these allow us to write the functional for the problem at once without reference to the governing differential equation and boundary conditions.

3.5.2 Three methods

If no physical principle giving a variational statement for a problem is apparent, and only the differential equations and their boundary conditions are known, we can use any one of three different approaches to decide whether a *classical variational principle* exists for the problem. In the order of their generality, from least to most general, these approaches are as follows: (1) attempt to derive a variational principle by using mathematical manipulation, (2) consult mathematics text books on variational methods and try to find the problem in question and its corresponding variational principle, or (3) use Frechet derivatives to test for the existence of the variational principle. Each of these approaches is now discussed.

1. Mathematical manipulation

Given the differential equation 3.1 and its associated boundary conditions, we seek a functional $I(\phi)$ such that its first variation with respect to ϕ vanishes. To try to find such a functional by mathematical manipulation alone, we multiply equation 3.1 by the first variation of ϕ, $\delta\phi$, and integrate over the domain D, that is,

$$\int_D \delta\phi[\mathscr{L}(\phi) - f] \, dD = 0 \tag{3.18}$$

The objective of this procedure is to manipulate the resulting expression in equation 3.18 into a form that allows the variational operator, δ, to be moved outside the integral sign. When this is accomplished, a functional has been obtained because an integrated quantity has been found whose first variation is zero. Thus, through manipulation of equation 3.18, we obtain

$$\delta \int_D [\mathscr{L}^*(\phi) - f\phi] \, dD = 0 \tag{3.19}$$

where $\mathscr{L}^*$ is the new operator resulting from the changed form of $\mathscr{L}(\phi)$. Thus the desired functional is

$$I(\phi) = \int_D [\mathscr{L}^*(\phi) - f\phi] \, dD$$

since $\delta I(\phi) = 0$.

Instructions given to indicate how one goes from equation 3.18 to equation 3.19 are vague because the procedure implied by the word "manipulation" is inexplicit and nebulous. Unfortunately, no inexorable set of rules can be given for the general case; however, it is always the situation that somewhere in the manipulation procedure Green's theorem (integration by parts) must be applied to invoke the boundary conditions. Consider a simple example illustrating the kinds of manipulations involved. Suppose that the behavior of ϕ in domain D is described by Laplace's equation, and that the value of ϕ on the boundary of D is some given function. This boundary value problem is then concisely stated as

$$\nabla^2\phi = 0 \quad \text{in } D$$
$$\phi = \Phi \quad \text{on } \Sigma, \text{ the boundary of } D$$

This is the classical Dirichlet problem. To derive the corresponding variational form for this problem we use the procedure just described and write

$$\int_D \delta\phi \, \nabla^2\phi \, dD = 0$$

Now we recognize that $\nabla^2\phi = \nabla \cdot \nabla\phi$, and we apply a vector identity to expand the integrand. Noting that

$$\nabla \cdot (u\mathbf{v}) = \nabla u \cdot \mathbf{v} + u\nabla \cdot \mathbf{v}$$

where u is a scalar quantity and $\mathbf{v}$ is a vector, we can identify u with $\delta\phi$ and $\mathbf{v}$ with $\nabla\phi$ and obtain

$$\nabla \cdot (\delta\phi \, \nabla\phi) = \nabla(\delta\phi) \cdot \nabla\phi + \delta\phi \, \nabla \cdot \nabla\phi$$
$$= \nabla(\delta\phi) \cdot \nabla\phi + \delta\phi \, \nabla^2\phi$$

or

$$\delta\phi \, \nabla^2\phi = \nabla \cdot (\delta\phi \, \nabla\phi) - \nabla(\delta\phi) \cdot \nabla\phi$$

Substituting this equation into the integral equation gives

$$\int_D \nabla \cdot (\delta\phi \, \nabla\phi) \, dD - \int_D \nabla(\delta\phi) \cdot \nabla\phi \, dD = 0$$

To invoke the boundary condition we apply Green's theorem† to the first term on the left-hand side. Thus

$$\int_\Sigma \hat{n} \cdot \delta\phi \, \nabla\phi \, d\sigma - \int_D \nabla(\delta\phi) \cdot \nabla\phi \, dD = 0$$

where $d\sigma$ is a differential surface area, and $\hat{n}$ is a unit vector normal to Σ. Now, since $\phi = \Phi$ on Σ, $\delta\phi = 0$ on Σ, and

$$\int_\Sigma d\bar{s} \cdot \delta\phi \, \nabla\phi = 0$$

we have

$$\int_D \nabla(\delta\phi) \cdot \nabla\phi \, dD = 0$$

† Green's theorem:

$$\oint_{\text{surface}} \hat{n} \cdot \mathbf{v} \, d\sigma \equiv \int_{\text{domain}} \nabla \cdot \mathbf{v} \, dD$$

Noting that $\nabla(\delta\phi \cdot \nabla\phi = \frac{1}{2}\delta(\nabla\phi \cdot \nabla\phi)$, we can write

$$\int_D \frac{1}{2}\delta(\nabla\phi \cdot \nabla\phi)\, dD = \frac{1}{2}\delta \int_D (\nabla\phi \cdot \nabla\phi)\, dD = 0$$

Our search for a functional is now complete. We have found an integrated quantity whose first variation vanishes. Hence

$$I(\phi) \equiv \int_D \nabla\phi \cdot \nabla\phi\, dD$$

and all functions ϕ which give the integral $I(\phi)$ a maximum or minimum value, or render it stationary, satisfy the Laplace equation and its boundary condition.

2. Classical procedures

If, instead of mathematical manipulation, we try to formulate variational principles by consulting mathematics text books, we can find several useful guidelines. For example, when the differential operator in equation 3.1 is linear, many helpful theorems such as those in Mikhlin's books [16, 18] are available. As Mikhlin points out, the functionals for linear differential equations are *quadratic functionals* that can be studied by constructing appropriate Hilbert and Sobolev spaces. We cite one of Mikhlin's theorems as an example.

Theorem: If $\mathscr{L}$ is a *positive* operator, that is, if

$$\int_D \phi\mathscr{L}(\phi)\, dD \geq 0$$

and if $\mathscr{L}(\phi) - f = 0$ has a solution satisfying homogeneous boundary conditions, the solution is unique and minimizes the quadratic functional

$$I(\phi) = \int_D \phi\mathscr{L}(\phi)\, dD - 2\int_D \phi f\, dD \tag{3.20}$$

When other types of boundary conditions are given, extra terms must be added to this functional; they can be found in a systematic way, as we shall see later. Consider the linear, self-adjoint† differential equation

$$\mathscr{L}(\phi) = f \tag{3.21}$$

---

† A differential operator is said to be self-adjoint in domain D if

$$\int_D u\mathscr{L}(v)\, dD = \int_D v\mathscr{L}(u)\, dD$$

A detailed mathematical definition of an adjoint for $\mathscr{L}$ can be found in [1].

to be solved subject to the boundary conditions

$$G_1(\phi) = g_1, \qquad G_2(\phi) = g_2, \qquad \ldots, \qquad G_r(\phi) = g_r \quad on\ \Sigma \qquad (3.22)$$

where $G_1, G_2, \ldots, G_r$ are linear, self-adjoint operators and $g_1, g_2, \ldots, g_r$ are given functions.† To find the minimum principle for this problem we decompose the unknown function ϕ into two parts, u and v; we let u contribute to the forcing function, f, and satisfy homogeneous boundary conditions, while v satisfies the nonhomogeneous boundary conditions. Hence, if $\phi = u + v$, then

$$\mathscr{L}(\phi) = \mathscr{L}(u) + \mathscr{L}(v) = f \qquad (3.23)$$

and

$$\mathscr{L}(u) = f - \mathscr{L}(v) \qquad (3.24)$$

with the homogeneous boundary conditions

$$G_1(u) = 0, \qquad G_2(u) = 0, \qquad \ldots, \qquad G_r(u) = 0 \quad on\ \Sigma \qquad (3.25)$$

We assume that v satisfies the nonhomogeneous boundary conditions

$$G_1(v) = g_1, \qquad G_2(v) = g_2, \qquad \ldots, \qquad G_r(v) = g_r \quad on\ \Sigma \qquad (3.26)$$

The problem formulation given by equations 3.24 and 3.25 may now be stated in an equivalent variational form if we use the Mikhlin theorem given above. The function u that satisfies equations 3.24 and 3.25 also minimizes the functional

$$I(u) = \int_D u\mathscr{L}(u)\,dD - 2\int_D u[f - \mathscr{L}(v)]\,dD \qquad (3.27)$$

Since $u = \phi - v$, we may write

$$I(u) = \int_D (\phi - v)\mathscr{L}(\phi - v)\,dD - 2\int_D (\phi - v)[f - \mathscr{L}(v)]\,dD$$

$$= \int_D [\phi\mathscr{L}(\phi) - \phi\mathscr{L}(v) - v\mathscr{L}(\phi) + v\mathscr{L}(v)]\,dD$$

$$-2\int_D [\phi f - \phi\mathscr{L}(v) - vf + v\mathscr{L}(v)]\,dD$$

$$= \int_D \phi\mathscr{L}(\phi)\,dD - 2\int_D \phi f\,dD + \int_D \phi\mathscr{L}(v)\,dD - \int_D v\mathscr{L}(\phi)\,dD$$

$$+ 2\int_D vf\,dD - \int_D v\mathscr{L}(v)\,dD \qquad (3.28)$$

† Mikhlin [8, pp. 116–121] treats precisely this problem, and our exposition closely follows his.

Equation 3.28 may be further reduced by employing Green's theorem and transforming the integrals

$$\int_D \phi \mathscr{L}(v) \, dD - \int_D v \mathscr{L}(\phi) \, dD$$

into a surface integral, which we denote by

$$\int_\Sigma R(\phi, v) \, d\Sigma$$

The expression for $R(\phi, v)$ is different for each problem because it depends on the form of the operator $\mathscr{L}$. If we use the boundary conditions of equations 3.22 and 3.26, it is usually possible to write $R(\phi, v)$ as

$$R(\phi, v) = N(\phi, g_1, g_2, \ldots, g_1) + M(v, g_1, g_2, \ldots, g_r)$$

In this case equation 3.20 may ultimately be written as

$$I(u) - \int_D \phi \mathscr{L}(\phi) \, dD \quad 2 \int_D \phi f \, dD + \int_\Sigma N(\phi, g_1, g_2, \ldots, g_R) \, d\Sigma + P(v)$$

where $P(v)$ is *constant*, a fact which may or may not be known, depending on whether we can find the function v for our particular problem. In any event, the problem of minimizing $I(u)$ is equivalent to that of minimizing the new functional $J(\phi)$, defined as

$$J(\phi) = \int_D \phi \mathscr{L}(\phi) \, dD - 2 \int_D \phi f \, dD + \int_\Sigma N(\phi, g_1, g_2, \ldots, g_R) \, d\Sigma \quad (3.29)$$

We minimize $J(\phi)$ with respect to all functions satisfying equation 3.22. The functional $J(\phi)$ is known as a *quadratic functional*.

To show how this procedure is actually used to find a minimum principle, we return to our previous example of Laplace's equation. Referring to equation 3.29 and recognizing that for our case $\mathscr{L}(\phi) = -\nabla^2 \phi$ and $f = 0$, we can write at once †

$$J(\phi) = - \int_D \phi \nabla^2 \phi \, dD + \int_\Sigma N(\phi, \Phi) \, d\Sigma$$

Now we must find $N(\phi, \Phi)$. Employing Green's theorem, we have

$$\int_D \phi \mathscr{L}(v) \, dD - \int_D v \mathscr{L}(\phi) \, dD = \int_\Sigma \left(v \frac{\partial \phi}{\partial n} - \phi \frac{\partial v}{\partial n} \right) d\Sigma$$

where n is the outward normal to the surface Σ.
Hence

$$R(\phi, v) = \left(v \frac{\partial \phi}{\partial n} - \phi \frac{\partial v}{\partial n} \right)_\Sigma$$

† Notice that we have added a minus sign to make $\mathscr{L}$ positive definite.

But our boundary condition dictates that

$$\phi = \Phi \quad \text{on } \Sigma$$

and in our formulation we also insisted that

$$v = \Phi \quad \text{on } \Sigma$$

Thus we may write

$$R(\phi, v) = \left(\Phi \frac{\partial \phi}{\partial n} - \Phi \frac{\partial v}{\partial n} \right)_{\Sigma}$$

From this expression we immediately recognize that

$$N(\phi, \Phi) = \Phi \frac{\partial \phi}{\partial n} \qquad \text{and} \qquad M(v, \Phi) = \Phi \frac{\partial v}{\partial n}$$

Our functional now becomes

$$J(\phi) = - \int_D \phi \nabla^2 \phi \, dD + \int_\Sigma \Phi \frac{\partial \phi}{\partial n} \, d\Sigma$$

and, with Green's theorem again, we have

$$J(\phi) = \int_D (\nabla \phi)^2 \, dD - \int_\Sigma \Phi \frac{\partial \phi}{\partial n} \, d\Sigma + \int_\Sigma \Phi \frac{\partial \phi}{\partial n} \, d\Sigma$$

Thus

$$J(\phi) = \int_D (\nabla \phi)^2 \, dD$$

the same result as before.

For practice in obtaining variational principles by this method of Mikhlin, the reader may want to consider the following boundary value problem:

$$\nabla^2 \phi = \frac{\partial^2 \phi}{\partial x^2} + \frac{\partial^2 \phi}{\partial y^2} + \frac{\partial^2 \phi}{\partial z^2} = f(x, y, z) \quad \text{in } D$$

with boundary conditions

$$\phi = \Phi(x, y, z) \quad \text{on } S_1$$

and

$$\frac{\partial \phi}{\partial n} + g(\sigma)\phi = h(\sigma) \quad \text{on } S_2$$

where f, g, and h are known continuous functions and the boundary segments S_1 and S_2 together comprise the complete boundary Σ ($S_1 \cup S_2 = \Sigma$). The parameter σ is measured around the boundary in such a way that, when we walk around the boundary, the domain is to our left. The outward normal to the boundary Σ is n.

We want to find an equivalent minimum principle for this classical boundary value problem.

Answer:

$$J(\phi) = \int_D (\nabla\phi)^2 \, dD - 2 \int_D f\phi \, dD + \int_{S_2} [g(\sigma)\phi^2 - 2h(\sigma)\phi] \, d\sigma$$

3. Frechet differentials

A general way to determine whether a classical variational principle exists for a given differential equation and boundary conditions is to use Frechet differentials. This method was initially based on a theorem by Vainberg [19], but was further developed by Tonti [20] and Finlayson [21]. The formalism is quite general and can be applied to nonlinear and non-self-adjoint equations. A detailed step-by-step description of the procedure is given by Finlayson [21,22], and the interested reader may refer to these clear expositions for details. Many researchers have attempted over the years to formulate variational principles for nonlinear and non-self-adjoint differential equations. Among the notable formulations are those of Glansdorff and Prigogine, who introduced the "method of the local potential," of Rosen, who treated restricted variations, and of Biot, who dealt with "Lagrangian thermodynamics." These so-called variational principles are *not* variational principles in the classical sense, and they rarely posses the advantages usually associated with classical variational principles. For instance, before these variational integrals can be defined, it is necessary to know the governing differential equations for the problem; and when the integrals have been obtained, they are not stationary in the calculations. They are never extremum principles; consequently, they cannot provide upper and lower bounds on the value of the variational integral.

In finite element analysis variational principles that are not classical have no practical use. Since few nonlinear and non-self-adjoint problems have classical variational principles, the variational approach to the derivation of element equations should be reserved for linear and self-adjoint problems. For problems not included in this category we may derive the element equations by the weighted residuals method or the energy method described in Chapter 4. The reader may well ask, "If a variational principle for a physical process can be established from the basis of physics alone or from the governing differential equation alone, is it not possible that more than one variational formulation can be derived for the same physical process? And, if more than one variational formulation is available, which is the best one to use for the finite element method?" The answer to the first question is that sometimes there are several variational formulations governing the same physical process. In regard to the second question, there is no "best" formulation for the finite element method because the formulation used depends on the desired result. Each formulation usually yields different information about the problem. Sometimes one formulation will yield an upper-bound solution, whereas another will lead to a lower-bound solution. We will encounter numerous examples of this in the applications discussed in Part 2 of this book.

3.6 CLOSURE

In this chapter we laid the foundation for deriving element equations from variational principles. We presented (a) the nature of continuum problems, (b) the variational approach to their solution, and (c) the relation of a well-known variational method, the Ritz method, to the finite element method. Taking a variational approach to the finite element method allowed us to extend the method to the broad class of problems governed by classical variational principles. We saw that the interpolation functions defined over each element give a piecewise representation of the unknown field variable, and as examples we used one- and two-dimensional interpolation functions for two simple elements. In Chapter 5 we shall consider a wide variety of elements and interpolation functions. From the variational approach, if general requirements are satisfied, we can be sure that refining the element mesh leads to a series of approximate solutions that approach the exact solution.

Although the variational approach allows us to treat many problems of practical interest, some problems do not have equivalent variational statements and cannot be treated in this way. In the next chapter we shall develop two more general approaches to the formulation of finite element equations. These approaches can be used when the variational approach cannot, and they open the way for the application of finite element techniques in many as yet unexplored areas.

REFERENCES

1. P. M. Morise and H. Feshback, *Methods of Theoretical Physics*, Vol. 1, McGraw-Hill Book Company, New York, 1953.
2. I. G. Petrovsky, *Lectures on Partial Differential Equations*, John Wiley-Interscience, New York, 1954.
3. R. Courant and K. O. Friedrichs, *Supersonic Flow and Shock Waves*, John Wiley-Interscience, New York, 1948.

4. E. F. Beckenbach (ed.), *Modern Mathematics for the Engineer*, McGraw-Hill Book Company, New York, 1956.

5. G. E. Forsythe and W. R. Wasow, *Finite Difference Methods for Partial Differential Equations*, John Wiley and Sons, New York, 1960.

6. B. A. Finlayson and L. E. Scriven, "The Method of Weighted Residuals—A Review," *Appl. Mech. Rev.*, Vol. 19, No. 9, September 1966.

7. B. A. Finlayson, *The Method of Weighted Residuals and Variational Principles*, Academic Press, New York, 1972.

8. F. B. Hildebrand, *Methods of Applied Mathematics*, Prentice-Hall, Englewood Cliffs, N.J., 1965.

9. J. T. Oden and D. Somogyi, "Finite Element Applications in Fluid Dynamics," *Proc. ASCE, J. Eng. Mech. Div.*, Vol. 95, EM3, June 1969.

10. O. C. Zienkiewicz, The Finite Element Method in Engineering Science, McGraw-Hill Book Company, England, 1971.

11. C. A. Felippa and R. W. Clough, "The Finite Element Method in Solid Mechanics," SIAM-AMS Proceedings, Vol. 2, American Mathematical Society, Providence, R.I., 1970, pp. 210–252.

12. E. R. A. Oliveira, "Theoretical Foundations of the Finite Element Method," *Int. J. Solids Structures*, Vol. 4, 1968, pp. 929–952.

13. T. H. H. Pian and P. Tong, "Basis of Finite Element Methods for Solid Continuua," *Int. J. Numer. Methods Eng.*, Vol. 1, 1969, pp. 3–28.

14. G. P. Bazeley, Y. K. Cheung, B. M. Irons and O. C. Zienkiewicz, "Triangular Elements in Plate Bending: Conforming and Nonconforming Solution," *Proceedings of the 1st Conference on Matrix Methods in Structural Mechanics*, Wright-Patterson Air Force Base, Dayton, Ohio, 1965.

15. S. M. Rohde, Private communication, 1971.

16. S. G. Mikhlin, *The Problem of the Minimum of a Quadratic Functional*, Holden-Day, San Francisco, Calif., 1965 (English translation of the 1952 Russian edition).

17. S. G. Mikhlin, *Variational Methods in Mathematical Physics*, Macmillan Company, New York, 1964 (English translation of the 1957 edition).

18. G. S. Mikhlin and K. Smolitsky, *Approximate Methods for the Solution of Differential and Integral Equations*, Elsevier, 1967.

19. M. M. Vainberg, *Variational Methods for the Study of Nonlinear Operators*, Holden-Day, San Francisco, Calif., 1964.

20. E. Tonti, "Variational Formulation of Nonlinear Differential Equations, I, II," *Bull. Acad. Roy. Belg. (Classe Sci.)* (5), Vol. 55, 1969, pp. 137–165, 262–278.

21. B. A. Finlayson, "Existence of Variational Principles for the Navier-Stokes Equation," *Phys. Fluids*, Vol. 15, No. 6, June 1972, pp. 963–967.

22. B. A. Finlayson and L. E. Scriven, "On the Search for Variational Principles," *J. Heat Mass Transfer*, Vol. 10, 1967, pp. 799–821.

4

THE MATHEMATICAL APPROACH:
A GENERALIZED INTERPRETATION

4.1 INTRODUCTION

In Chapter 3 we discussed the relation between the well-known Ritz method and the finite element method. This relation enabled us to view the finite element discretization procedure as simply another means for finding approximate solutions to variational problems. In fact, we saw how the finite element equations are derived by requiring that a given functional be stationary. This broad variational interpretation is the one most widely used to derive element equations, and it is the most convenient approach whenever a classical variational statement exists for a given problem.

However, applied scientists and engineers encounter practical problems for which classical variational principles are unknown. In these cases finite element techniques are still applicable, but more generalized procedures must be used to derive the element equations.

In this chapter we shall introduce these generalizations and show how finite element equations can be derived directly from the governing differential equations of the problem without reliance on any classical, quasi-variational, or restricted variational "principles." Two procedures are available to accomplish this; one employs the method of weighted residuals, while the other stems from a global energy balance concept. These two procedures allow us to apply the finite element method to almost all practical problems of mathematical physics.

4.2 DERIVING FINITE ELEMENT EQUATIONS FROM THE METHOD OF WEIGHTED RESIDUALS (GALERKIN'S METHOD)

The method of weighted residuals is a technique for obtaining approximate solutions to linear and nonlinear partial differential equations. It has nothing to do with the finite element method other than offering another means with which to formulate the finite element equations. In this section we shall review the general method of weighted residuals and show how one particular technique can be used to derive finite element equations. References 6 and 7 of Chapter 3 are good sources of more detailed information on general weighted residual techniques.

Applying the method of weighted residuals involves basically two steps. The first step is to assume the general functional behavior of the dependent field variable in some way so as to approximately satisfy the given differential equation and boundary conditions. Substitution of this approximation into the original differential equation and boundary conditions then results in some error called a *residual*. This residual is required to vanish in some average sense over the entire solution domain.

The second step is to solve the equation (or equations) resulting from the first step and thereby specialize the general functional form to a particular function, which then becomes the approximate solution sought.

To be more specific we shall consider a typical problem. Suppose that we want to find an approximate functional representation for a field variable ϕ governed by the differential equation

$$\mathcal{L}(\phi) - f = 0 \tag{4.1}$$

in the domain D bounded by the surface Σ. The function f is a known function of the independent variables, and we assume that proper boundary conditions are prescribed on Σ. The method of weighted residuals is applied in two steps as follows.

First, the unknown exact solution ϕ is approximated by $\tilde{\phi}$, where either the functional behavior of $\tilde{\phi}$ is completely specified in terms of unknown parameters, or the functional dependence on all but one of the independent variables is specified while the functional dependence on the remaining independent variable is left unspecified. Thus the dependent variable is approximated by

$$\phi \approx \tilde{\phi} = \sum_{i=1}^{m} N_i C_i \tag{4.2}$$

where the N_i are the assumed functions and the C_i are either the unknown parameters or unknown functions of one of the independent variables.

The m functions N_i are usually chosen to satisfy the global boundary conditions.

When $\tilde{\phi}$ is substituted into equation 4.1, it is unlikely that the equation will be satisfied, that is,

$$\mathscr{L}(\tilde{\phi}) - f \neq 0$$

In fact,

$$\mathscr{L}(\tilde{\phi}) - f = R$$

where R is the residual or error that results from approximating ϕ by $\tilde{\phi}$. The method of weighted residuals seeks to determine the m unknowns C_i in such a way that the error R over the entire solution domain is small. This is accomplished by forming a weighted average of the error and specifying that this weighted average vanish over the solution domain. Hence we choose m linearly independent weighting functions, W_i, and then insist that if

$$\int_D [\mathscr{L}(\tilde{\phi}) - f] W_i \, dD = \int_D R W_i \, dD = 0, \quad i = 1, 2, \ldots, m \qquad (4.3)$$

then $R \approx 0$ in some sense.

The form of the error distribution principle expressed in equations 4.3 depends on our choice for the weighting functions. Once we specify the weighting functions, equations 4.3 represent a set of m equations, either algebraic equations or ordinary differential equations to be solved for the C_i. The second step then is to solve equations 4.3 for the C_i, and hence obtain an approximate representation of the unknown field variable ϕ via equation 4.2. There are many linear problems and even some nonlinear problems for which it can be shown that, as $m \to \infty$, $\tilde{\phi} \to \phi$, but, in general, studies of convergence and error bounds are scarce [1].

We have a variety of weighted residual techniques because of the broad choice of weighting functions or error distribution principles that we can use (see, for example, Collatz [2]). The error distribution principle most often used to derive finite element equations is known as the *Galerkin criterion*, or *Galerkin's method*. According to Galerkin's method, the weighting functions are chosen to be the same as the approximating functions used to represent ϕ, that is, $W_i = N_i$ for $i = 1, 2, \ldots, m$. Thus Galerkin's method requires that

$$\int_D [\mathscr{L}(\tilde{\phi}) - f] N_i \, dD = 0, \quad i = 1, 2, \ldots, m \qquad (4.4)$$

In the preceding discussion we assumed that we were dealing with the entire solution domain. However, because equation 4.1 holds for any point in the solution domain, it also holds for any collection of points defining an arbitrary subdomain or element of the whole domain. For this reason we may focus our attention on an individual element and define a local approximation analogous to equation 4.2 and valid for only one element at a time. Now the familiar finite element representations of a field variable become available. The functions N_i are recognized as the interpolation functions $N_i^{(e)}$ defined over the element, and the C_i are the undetermined parameters, which may be the nodal values of the field variable or its derivatives. Then, from Galerkin's method, we can write the equations governing the behavior of an element as

$$\int_{D^{(e)}} [\mathscr{L}(\phi^{(e)}) - f^{(e)}] N_i^{(e)} \, dD^{(e)} = 0, \quad i = 1, 2, \ldots, r \quad (4.5)$$

where, as before, the superscript (e) restricts the range to one element, and

$\phi^{(e)} = \lfloor N^{(e)} \rfloor \{\phi\}^{(e)}$,

$f^{(e)}$ = forcing function defined over element (e).

r = number of unknown parameters assigned to the element.

We have a set of equations like equations 4.5 for each element of the whole assemblage. Before we can obtain the system equations from the element equations, we require that our choice of approximating functions N_i guarantee the interelement continuity necessary for the assembly process. Recall that the assembly process does not allow any spurious contributions, if we choose interpolation functions that ensure at the element boundaries continuity of ϕ, as well as continuity of its partial derivatives up to one order less than the highest-order derivative appearing in the expression to be integrated. Since the differential operator $\mathscr{L}$ in the integrand usually has higher-order derivatives than occur in the integrand resulting from a variational principle (when one exists), we can see that the Galerkin method can lead to a more stringent choice of interpolation functions than we have previously encountered in the variational approach. As we shall see in Chapter 5, the higher the order of continuity we require of the interpolations, the narrower our choice of functions becomes. Many interpolation functions provide continuity of value, fewer provide continuity of slope, and only several can ensure continuity of curvature.

Often there is a way to escape this dilemma by changing the form of equations 4.5. By applying integration by parts† to the integral expression of equations 4.5, we can obtain expressions containing lower-order derivatives, and hence we can use approximating functions with lower-order interelement continuity. When integration by parts is possible, it also offers

† Integrations by parts in three dimensions is known as Gauss's theorem; integration in two dimensions, as Green's theorem. See Kaplan [3], Chapter 5.

a convenient way to introduce the natural boundary conditions that must be satisfied on some portion of the boundary. Although the boundary terms containing the natural boundary conditions appear in the equations for each element, in the assembly of the element equations only the boundary elements give nonvanishing contributions. After the assembly process, the fixed boundary conditions (for example, fixed displacement or specified temperature) are introduced in the manner described in Chapter 2.

Several examples will help to illustrate and clarify these general ideas.

4.2.1   Example: One-dimensional heat conduction

Consider the simple differential equation governing one-dimensional steady-state heat conduction. If the material is homogeneous with temperature-independent thermal conductivity k, Fourier's law of conduction states that

$$k\frac{d^2T}{dx^2} = 0 \tag{4.6a}$$

Suppose also that at one end of the bar, $x = 0$, the temperature T_0 is specified, while at the other end, $x = L$, temperature T_L is prescribed. These boundary conditions are expressed as

$$T = T_0 \quad \text{at } x = 0$$

and

$$T = T_L \quad \text{at } x = L \tag{4.6b}$$

Although we can write the exact solution to this simple problem at once, that is, $T(x) = (T_L - T_0)(x/L) + T_0$, we shall formulate a general solution by the finite element Galerkin method.

We first approximate the unknown exact solution T by $\tilde{T}$, which has the form

$$\tilde{T} = \sum_{i=1}^{m} N_i(x)T_i \tag{4.6c}$$

for a line element with m nodes (see Figure 4.1).

Figure 4.1.   Arbitrary line element with m nodes.

The $N_i(x)$ are the interpolation functions and the T_i are the unknown nodal temperatures at the m nodes of the one-dimensional line element. We do not consider the fixed boundary conditions at the element level; instead, these are included, as before, after the assembly process. As we shall see, the natural boundary conditions emerge automatically in the formulation of the element equations.

Applying Galerkin's criterion gives

$$\int_{x_1}^{x_m} kN_j(x)\frac{d^2\tilde{T}}{dx^2}\,dx = 0, \quad j = 1, 2, \ldots, m \qquad (4.6d)$$

where x_1 and x_m are the coordinates of the end nodes of the line element. If we apply integration by parts† to the left-hand side of this equation, we obtain

$$kN_j(x)\frac{d\tilde{T}}{dx}\bigg|_{x_1}^{x_m} - \int_{x_1}^{x_m} k\frac{dN_j(x)}{dx}\frac{d\tilde{T}}{dx}\,dx = 0 \qquad (4.6e)$$

Utilizing our definition of $\tilde{T}$, we have

$$\frac{d\tilde{T}}{dx} = \sum_{i=1}^{m}\frac{\partial N_i}{\partial x}T_i = \left\lfloor\frac{\partial N}{\partial x}\right\rfloor\{T\} \qquad (4.6f)$$

where $\{T\}$ is the column vector of nodal temperatures for the element. Hence the Galerkin criterion becomes

$$\int_{x_1}^{x_m} k\frac{dN_j}{dx}\left\lfloor\frac{\partial N}{\partial x}\right\rfloor dx\{T\} = N_mQ_m - N_1Q_1 \qquad (4.6g)$$

where $Q_1 = k(d\tilde{T}/dx)$ at node 1 and $Q_m = k(d\tilde{T}/dx)$ at node m; Q_1 and Q_2 are the heat fluxes at the exterior or boundary nodes of the line element. These are the natural boundary conditions, which are left unspecified in our example problem.

We note also that by definition $N_1 = N_m = 1$. Hence the equation at node j becomes

$$\sum_{i=1}^{m} K_{t_{ij}} T_i = \begin{cases} Q_1 & \text{if } j = 1 \\ 0 & \text{if } j = 2, \ldots, m-1 \\ Q_m & \text{if } j = m \end{cases}$$

or for all nodes of the element we can write

$$\underset{m \times m}{[K_t]^{(e)}}\underset{m \times 1}{\{T\}^{(e)}} = \underset{m \times 1}{\{Q\}} \qquad (4.6h)$$

† Integration by parts in this case takes the form $\int u\,dv = uv - \int v\,du$.

where

$$K_{t_{ij}} = \int_{x_1}^{x_m} \frac{dN_i}{dx}\frac{dN_j}{dx}\,dx = \int_{x_1}^{x_m} N_i' N_j'\,dx$$

$$\{Q\} = \begin{Bmatrix} Q_1 \\ 0 \\ \vdots \\ 0 \\ Q_m \end{Bmatrix}$$

This set of equations expresses the behavior of a line element of m nodes. Explicit evaluation of these equations is possible after we specify the interpolation functions N_i. When the element equations are assembled to form the system equations, only elements with externally applied heat fluxes contribute to the flux vector $\{Q\}$. If no external fluxes are applied, internal elements contribute nothing to $\{Q\}$, and in the system equations $\{Q\}$ has only two nonzero components, representing the heat flow in and out of the ends of the bar.

In this example we see that integration by parts played two important roles. First, it brought into effect the exterior or boundary information; and, second, it lowered to one the order of the highest-order derivative appearing in the integrands of the thermal coefficient matrix $[K_t]^{(e)}$; thus the approximating functions need perserve continuity only of value and not of slope.

Another important observation can also be made. If we compare the thermal coefficient matrix $[K_t]^{(e)}$ derived from Galerkin's method with the same coefficient matrix derived from a variational principle,† we find that their forms are identical. If, in the present example, we consider only linear interpolation functions over two-node line elements, then $[K_t]^{(e)}$ becomes identical to the matrix previously derived for the same interpolation functions.

This special result follows from a far more general relationship. It can be shown that, for linear self-adjoint operators $\mathscr{L}$, application of a classical variation principle or of Galerkin's method leads to identical calculations. A thorough discussion of this equivalence can be found in Finlayson and Scriven [1] and in the references cited there.

The equivalence of the classical variational approach and the Galerkin approach to deriving finite element equations has an advantageous interpretation: by means of the Galerkin approach, an analyst can derive finite

---

† In the example problem of Section 3.4.5 we formulated finite element equations for the relation $d^2\phi/dx^2 = -f(x)$, using a classical variational principle. When $f(x) = 0$, that example problem reduces to the present case.

element equations for a given problem from the governing equations and boundary conditions, regardless of whether or not he is familar with variational calculus. When a classical variational principle exists for a given problem, he obtains the same results. When a classical variational principle is unknown or does not exist, he may still proceed to derive a finite element model.

4.2.2   Example: Two-dimensional heat conduction

As a further illustration of the use of Galerkin's method in deriving element equations, we shall consider the problem of steady two-dimensional heat conduction. The governing differential equation may be expressed in the general form as

$$\frac{\partial}{\partial x}\left(k_x \frac{\partial T}{\partial x}\right) + \frac{\partial}{\partial y}\left(k_y \frac{\partial T}{\partial y}\right) + \tilde{Q} = 0 \tag{4.7a}$$

where k_x and k_y are the thermal conductivities in the x and y directions, respectively, and $\tilde{Q}$ is the internal heat generation, these are known specified

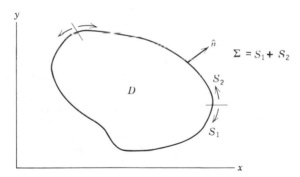

Figure 4.2.   Two-dimensional domain for the heat conduction equation.

functions of x and y. On the boundary of the domain D we have the following general boundary conditions (see Figure 4.2):

$$T = T(x, y) \quad \text{on } S_1$$

$$k_x \frac{\partial T}{\partial x} n_x + k_y \frac{\partial T}{\partial y} n_y + q + h(T - T_\infty) = 0 \quad \text{on } S_2 \tag{4.7b}$$

where $T(x, y)$ is a specified boundary temperature distribution,

n_x, n_y are the direction cosines of the outward normal vector $\hat{n}$ to the bounding curve Σ,

q is the heat loss at the boundary due to conduction, and

$h(T - T_\infty)$ is the heat loss at the boundary due to convection to ambient temperature T_∞ with convection heat transfer coefficient h.

In this example we shall compare the procedures of the variational method and the Galerkin method by deriving finite element equations in both ways.

Variational method

From the variational techniques of Section 3.5.2 we know that the function $T(x, y)$ that satisfies equations 4.7a and 4.7b also minimizes the functional

$$I(T) = \iint_D \left[\frac{k_x}{2} \left(\frac{\partial T}{\partial x} \right)^2 + \frac{k_y}{2} \left(\frac{\partial T}{\partial y} \right)^2 - \tilde{Q} T \right] dx \, dy$$

$$+ \int_{S_2} (qT + \tfrac{1}{2}hT^2 - hTT_\infty) \, ds \tag{4.7c}$$

Suppose that the domain D is divided into polygonal elements and that the distribution of T within each element is assumed to be

$$T^{(e)}(x, y) = \sum_{i=1}^{r} N_i(x, y) T_i = \lfloor N \rfloor \{T\}^{(e)} \tag{4.7d}$$

where r is the number of nodes assigned to element (e) and the T_i are the discrete nodal temperatures. By substituting equation 4.7d into equation 4.7c we obtain for one element the discretized functional $I(T^{(e)})$ expressed in terms of the discrete nodal temperatures. Requiring that $I(T^{(e)})$ be a minimum is equivalent to requiring that

$$\frac{\partial I(T^{(e)})}{\partial T_i} = 0, \quad i = 1, 2, \ldots, r$$

Hence for a typical node i we have

$$\frac{\partial I(T^{(e)})}{\partial T_i} = 0 = \iint_{D^{(e)}} \left[k_x \frac{\partial T^{(e)}}{\partial x} \frac{\partial}{\partial T_i} \left(\frac{\partial T^{(e)}}{\partial x} \right) \right.$$

$$+ k_y \frac{\partial T^{(e)}}{\partial y} \frac{\partial}{\partial T_i} \left(\frac{\partial T^{(e)}}{\partial y} \right) - \tilde{Q} \frac{\partial T^{(e)}}{\partial T_i} \left. \right] dx \, dy$$

$$+ \int_{S_2^{(e)}} \left(q \frac{\partial T^{(e)}}{\partial T_i} + hT^{(e)} \frac{\partial T^{(e)}}{\partial T_i} - hT_\infty \frac{\partial T^{(e)}}{\partial T_i} \right) d\Sigma^{(e)} \tag{4.7e}$$

The surface integral in equation 4.7e has nonzero values only for elements whose boundaries are part of the domain boundary segment S_2 or for elements that have external nodal heat fluxes specified via q.

Utilizing equation 4.7d, we may rewrite equation 4.7e as

$$0 = \iint\limits_{D^{(e)}} \left[k_x \left\lfloor \frac{\partial N}{\partial x} \right\rfloor \{T\}^{(e)} \frac{\partial N_i}{\partial x} + k_y \left\lfloor \frac{\partial N}{\partial y} \right\rfloor \{T\}^{(e)} \frac{\partial N_i}{\partial y} - \tilde{Q} N_i \right] dx \, dy$$

$$+ \int_{S_2^{(e)}} [q N_i + h \lfloor N \rfloor \{T\}^{(e)} N_i - h T_\infty N_i] \, d\Sigma^{(e)}, \quad i = 1, 2, \dots, r$$

or, in matrix notation, we have for the entire element

$$[K_t]^{(e)}\{T\}^{(e)} = \{Q\}^{(e)} - \{q\}^{(e)} - [K_h]^{(e)}\{T\}^{(e)} + \{k_{T_\infty}\}^{(e)} \qquad (4.7f)$$

where

$$K_{t_{ij}} = \iint\limits_{D^{(e)}} \left[k_x \frac{\partial N_i}{\partial x} \frac{\partial N_j}{\partial x} + k_y \frac{\partial N_i}{\partial y} \frac{\partial N_j}{\partial y} \right] dx \, dy$$

$$Q_i = \int_{D^{(e)}} \tilde{Q} N_i \, dx \, dy$$

$$q_i = \int_{S_2^{(e)}} q N_i \, d\Sigma^{(e)}$$

$$K_{h_{ij}} = \int_{S_2^{(e)}} h N_i N_j \, d\Sigma^{(e)}, \qquad k_{T_{\infty_i}} = \int_{S_2^{(e)}} h T_\infty N_i \, d\Sigma^{(e)}$$

Equation 4.7f expresses the general behavior of a two-dimensional thermal element. After we select the element shape and an appropriate set of interpolation functions, we can explicitly evaluate these element equations and then routinely assemble them for an aggregate of elements representing the whole solution domain. Nodal temperatures result from the simultaneous solution of the system equations.

We note that, since the functional contains only first-order derivatives of the temperature, our approximating functions need only guarantee continuity of temperature across interelement boundaries, and they must be able to represent constant values of both first derivatives within any element.

Galerkin's method

To derive element equations for equation 4.7a and 4.7b by Galerkin's method, we first express the approximate behavior of the temperature within each element according to equation 4.7d. Then, applying Galerkin's criterion, we may write

$$\iint_{D^{(e)}} N_i \left[\frac{\partial}{\partial x}\left(k_x \frac{\partial T^{(e)}}{\partial x} \right) + \frac{\partial}{\partial y}\left(k_y \frac{\partial T^{(e)}}{\partial y} \right) + \tilde{Q} \right] dx\, dy = 0, \qquad i = 1, 2, \dots, r$$

(4.8a)

Equation 4.8a expresses the desired averaging of the error or residual within the element boundaries, but it does not admit the influence of the boundary. Since we have made no attempt to choose the N_i so as to satisfy the boundary conditions (usually a most arduous task), we must use integration by parts to introduce the influence of the natural boundary conditions.†

Focusing our attention on a typical term of equation 4.8a, we obtain through integration by parts

$$\iint_{D^{(e)}} N_i \frac{\partial}{\partial x}\left(k_x \frac{\partial T^{(e)}}{\partial x} \right) dx\, dy = \oint_{S_2^{(e)}} k_x \frac{\partial T^{(e)}}{\partial x} N_i \, dy - \iint_{D^{(e)}} k_x \frac{\partial T^{(e)}}{\partial x} \frac{\partial N_i}{\partial x} dx\, dy$$

$$= \int_{S_2^{(e)}} k_x \frac{\partial T^{(e)}}{\partial x} n_x N_i \, d\Sigma - \iint k_x \frac{\partial T^{(e)}}{\partial x} \frac{\partial N_i}{\partial x} dx\, dy$$

where n_x is the x component of the unit normal to the boundary, and $d\Sigma$ is a differential arc length along the boundary. When we treat the second term in equation 4.8a in the same way, the equation takes the form

$$- \iint_{D^{(e)}} \left(k_x \frac{\partial T^{(e)}}{\partial x} \frac{\partial N_i}{\partial x} + k_y \frac{\partial T^{(e)}}{\partial y} \frac{\partial N_i}{\partial y} \right) dx\, dy + \iint_{D^{(e)}} N_i \tilde{Q}\, dx\, dy$$

$$+ \oint_{S_2^{(e)}} \left(k_x \frac{\partial T^{(e)}}{\partial x} n_x + k_y \frac{\partial T^{(e)}}{\partial y} n_y \right) N_i \, d\Sigma^{(e)} = 0 \qquad (4.8b)$$

The surface integral (boundary residual) in equation 4.8b now enables us to introduce the natural boundary conditions of equations 4.7b because, for elements on the boundary of D,

$$k_x \frac{\partial T^{(e)}}{\partial x} n_x + k_y \frac{\partial T^{(e)}}{\partial y} n_y = -q^{(e)} - h(T^{(e)} - T_\infty) \quad \text{on} \quad S_2$$

---

† Fixed boundary conditions are introduced in the usual manner after assembly of the element equations.

Hence, noting that $T^{(e)} = [N]\{T\}^{(e)}$, we may write equation 4.8b as

$$
\iint\limits_{D^{(e)}} \left[k_x \left\lfloor \frac{\partial N}{\partial x} \right\rfloor \{T\}^{(e)} \frac{\partial N_i}{\partial x} + k_y \left\lfloor \frac{\partial N}{\partial y} \right\rfloor \{T\}^{(e)} \frac{\partial N_i}{\partial y} \right] dx\, dy
$$

$$
- \iint\limits_{D^{(e)}} N_i \tilde{Q}\, dx\, dy + \int_{S_2^{(e)}} [qN_i + h\lfloor N \rfloor\{T\}^{(e)}N_i - hT_\infty N_i]\, d\Sigma^{(e)} = 0
$$

which we recognize to be identical to equation 4.7f, derived by the variational method.

These examples have illustrated the use of Galerkin's method in deriving finite element equations from a governing differential equation and its boundary conditions. The resulting element equations are linear algebraic equations identical to those derived from a variational principle because each of the differential operators we considered was linear and self-adjoint. In each case we had to apply integration by parts to invoke the boundary conditions. It is important to note that the procedures used in these examples of linear problems apply equally well to nonlinear problems and to problems containing more than one dependent variable. When the independent variable is a vector, such as the displacement field in an elasticity problem, we treat each component of the vector in the same way as we treated a scalar function in the foregoing.

Before considering yet another method for deriving element equations, we shall outline the procedures for handling time-dependent problems where the application of Galerkin's method leads to ordinary differential equations in time.

4.2.3   Example: Time-dependent heat conduction

The steady-state heat conduction equations in the two preceding examples are elliptic differential equations. When transient heat conduction is considered, the governing differential equation becomes parabolic, and no classical variational principle can be applied.

The differential equation governing two-dimensional transient heat conduction may be expressed in the general form as

$$
\frac{\partial}{\partial x}\left(k_x \frac{\partial T}{\partial x} \right) + \frac{\partial}{\partial y}\left(k_y \frac{\partial T}{\partial y} \right) + \tilde{Q} - c \frac{\partial T}{\partial t} = 0 \qquad (4.9a)
$$

with boundary conditions

$$T = T(x, y, t) \quad \text{on} \quad S_1, t > 0$$

$$k_x \frac{\partial T}{\partial x} n_x + k_y \frac{\partial T}{\partial y} n_y + q + h(T - T_\infty) = 0 \quad \text{on} \quad S_2, t > 0 \quad (4.9b)$$

$$T = T_0(x, y) \quad \text{in} \quad D, t = 0$$

where the nomenclature of equation 4.7a applies, and T_0 is the initial condition of the system.

Since the temperature distribution is a function of both space and time, we shall assume that the distribution of T within each element has the form

$$T^{(e)}(x, y, t) = \sum_{i=1}^{r} N_i(x, y) T_i(t) \qquad (4.9c)$$

If we carry out the steps of the preceding example, first introducing Galerkin's criterion and then integrating by parts, we obtain identical matrix equations except that the new term

$$-\int_{D^{(e)}} c \frac{\partial T^{(e)}}{\partial t} \, dx \, dy$$

arises in equation 4.8b. When equation 4.9c is substituted, the resulting equation takes the form

$$[K_t]^{(e)} \{T\}^{(e)} - [K_c] \left\{ \frac{\partial T}{\partial t} \right\}^{(e)} = \{Q\}^{(e)} - \{q\}^{(e)} [K_h]^{(e)} \{T\}^{(e)} + \{K_{T_\infty}\}^{(e)} \quad (4.9d)$$

where

$$K_{c_{ij}} = \int_{D^{(e)}} c N_i N_j \, dx \, dy,$$

and the remaining terms are the same as those in equation 4.7f. Equation 4.9d, a linear first-order differential equation in time, expresses the transient behavior of a typical element. Aral et al. [4] suggest a convenient method for obtaining computer solutions for equations of this type.

As we saw in Chapter 2, the use of classical variational principles to derive finite element equations considerably enlarges the range of problems amenable to finite element techniques. Now the use of Galerkin's method extends even farther the applicability of finite element analysis. Galerkin's method offers an alternative approach, and it not only encompasses the variational approach but also goes far beyond because it can be applied to *any* well-posed system of differential equations and their boundary conditions.

A sample of the literature [5–15] indicates that applications of Galerkin's method in finite element analysis are relatively recent. But as analysts begin to recognize the usefulness and generality of this approach, we can expect that more applications will appear in many different fields. We shall see in later chapters some specific examples in solid and fluid mechanics of how finite element equations have been formulated by Galerkin's method.

4.3   DERIVING FINITE ELEMENT EQUATIONS FROM ENERGY BALANCES

In 1969 Oden introduced in a two-part paper [16,17] a broad theory of finite element methodology as applied to continuum mechanics problems. This theory gives a generalized interpretation of finite element models and shows how finite element equations can be developed from well-established global energy balances. Such approaches obviate the need for classical variational principles and sometimes provide more insight into a problem than can be gained by a perfunctory application of Galerkin's method.

As Oden points out, the formulation of finite element equations essentially centers on one basic concept. Once the solution domain has been discretized and interpolation functions have been chosen to represent the field variable, with due consideration given to continuity requirements, "all that is needed is some means to translate a relation that holds at a point (in the solution domain) into one that must hold over a finite region" [17]. The governing differential equations are the relations that hold at a point. We have seen that one way to go from point relations to regional relations is to introduce a classical variational principle when one exists. The functional of the variational principle then encompasses the entire solution domain (or the domain of one element) and includes the natural boundary conditions. Another way to obtain regional relations is to work directly from the governing differential equations and employ Galerkin's method. After we use integration by parts in the residual equations, we obtain relations for the region inclusive of its boundary.

Oden [17] suggests that for problems of continuum mechanics there are often forms of local or global energy balances that can provide the necessary regional relations without resorting to variational principles or Galerkin's method. For example, if we employ the first law of thermodynamics, we have a versatile global energy balance principle that can be specialized or generalized as the situation demands. Any type of linear or nonlinear continuum can be treated, and thermal, mechanical, electrical, and magnetic phenomena can be taken into account.

Though a detailed exposition of Oden's general theory is beyond our purpose and scope, we shall encounter in the later chapters on applications several specific uses of the energy balance approach. To illustrate the basic ideas of this approach we shall briefly discuss the procedure for deriving finite element equations for a class of fluid mechanics problems [18]. We restrict our discussion to concepts only.

If we consider an isothermal incompressible fluid motion that satisfies the equations expressing the conservation of mass and the conservation of linear momentum, we can easily formulate an equation expressing a global balance of mechanical energy. The procedure is to write x, y, and z components of the linear momentum equation and then multiply each of these equations by the corresponding velocity components; that is, we multiply the momentum equation in the x direction by the component of velocity in the x direction, and so forth. After integrating these equations over the entire domain and summing the resulting equations, we obtain a relation expressing the conservation of mechanical energy. To construct a finite element model for a typical element we represent the velocity field and the pressure field within an element in the usual manner:

$$P^{(e)} = \Sigma N_i P_i, \qquad u^{(e)} = \Sigma N_i u_i, \qquad v^{(e)} = \Sigma N_i v_i, \qquad w^{(e)} = \Sigma N_i w_i$$

Substitution of these expressions into the energy equation and the mass conservation equation then leads to a set of matrix equations relating nodal velocities and nodal pressures for an element. If inertia terms are retained in the momentum equations, the resulting matrix equations are nonlinear; and if, in addition, the fluid motion is unsteady, nonlinear ordinary differential equations result for the velocity components.

4.4   CLOSURE

In this chapter we have been concerned in general with two more alternative approaches to the derivation of finite element models for continuum problems. Including the procedures discussed in Chapters 2 and 3, we now have at hand four distinct approaches to deriving element equations. For the simplest problems the *direct method*, employing physical reasoning, can be employed and is useful for clarifying some of the concepts in finite element analysis. A far more general approach, however, is the *variational method*, which employs classical variational principles. When only the governing differential equations and their boundary conditions are available, *Galerkin's method* is convenient; this approach surpasses the variational method in generality and further broadens the range of applicability of the finite element method. Finally, the *energy balance method* suggested by Oden offers still other possibilities by allowing us to conveniently formulate element equations for thermomechanical problems.

Before considering how these methods are applied to a multitude of practical problems, we shall examine in the next chapter the important ideas pertaining to the establishment of element shapes and appropriate interpolation functions. When the element shape and the interpolation functions N_i are specified, we can explicitly evaluate the element equations and proceed with the solution in a routine manner.

REFERENCES

1. B. A. Finlayson and L. E. Scriven, "The Method of Weighted Residuals—A Review," *Appl. Mech. Rev.*, Vol. 19, No. 9, September 1966, pp. 735–748.

2. L. Collatz, *The Numerical Treatment of Differential Equations*, Springer-Verlag, Berlin, 1966.

3. W. Kaplan, *Advanced Calculus*, Addison-Wesley Publishing Company, Reading, Mass. 1952.

4. M. M. Aral, P. G. Mayer, and C. V. Smith, Jr., "Finite Element Galerkin Method Solutions to Selected Elliptic and Parabolic Differential Equations," *Proceedings of the 3rd Conference on Matrix Methods in Structural Mechanics*, Wright-Patterson Air Force Base, Dayton, Ohio, October 1971 (in press).

5. B. A. Szabo and G. C. Lee, " Stiffness Matrix for Plates by Galerkin's Method," *Proc. ASCE, J. Eng. Mech. Div.*, Vol. 95, No. EM3, June 1969.

6. B. A. Szabo and G. C. Lee, "Derivation of Stiffness Matrices for Problems on Plane Elasticity by Galerkin Method," *Int. J. Numer. Methods Eng.*, Vol. 1, No. 3, July 1969.

7. J. W. Leonard, "Linearized Compressible Flow by the Finite Element Method," *Bell Aerospace Co. Res. Rept.* 9500-920156, December 1969.

8. J. W. Leonard and T. T. Bramlette, " Finite Element Solutions to Differential Equations," *Proc. ASCE, J. Eng. Mech. Div.*, Vol. 96, EM6, December 1970.

9. O. C. Zienkiewicz and C. J. Parekh, "Transient Field Problems: Two Dimensional and Three Dimensional Analysis by Isoparametric Finite Elements," *Int. J. Numer. Methods Eng.*, Vol. 2, No. 1, January 1970.

10. Y. Tada and G. C. Lee, "Finite Element Solution to an Elastic Problem of Beams," *Int. J. Numer. Methods Eng.*, Vol. 2, No. 2, April 1970.

11. J. W. Leonard, "Galerkin Finite Element Formulation for Incompressible Flow," *Bell Aerospace Co. Res. Rept.* 9500-920181, April 1970.

12. A. J. Baker, "Finite Element Theory for Viscous Fluid Dynamics," *Bell Aerospace Co. Res. Report* 9500-920189, August 1970.

13. O. C. Zienkiewicz and C. Taylor, "Weighted Residual Processes in F. E. M. with Particular Reference to Some Coupled and Transient Problems," Lecture for NATO Advanced Study Institute on Finite Element Methods in Continuum Mechanics, Lisbon, 1971.

14. A. J. Baker, "Finite Element Theory for the Mechanics and Thermodynamics of a Viscous, Compressible Multi-Specie Fluid," *Bell Aerospace Co. Res. Rept.* 9500-920200, June 1971.

15. A. J. Baker, "Finite Element Computational Theory for Three Dimensional Boundary Layer Flow," *AIAA Reprint* 72-108, presented at the AIAA 10th Aerospace Science Meeting, San Diego, Calif., 1972.

16. J. T. Oden, "A General Theory of Finite Elements. I: Topological Considerations," *Int. J. Numer. Methods Eng.*, Vol. 1, No. 3, 1969.

17. J. T. Oden, "A General Theory of Finite Elements. II: Applications," *Int. J. Numer. Methods Eng.*, Vol. 1, No. 2, 1969.

18. J. T. Oden, "Finite Element Analogue of the Navier Stokes Equations," *Proc. ASCE, J. Eng. Mech. Div.*, Vol. 96, No. EM4, August 1970.

5

ELEMENTS AND INTERPOLATION
FUNCTIONS

5.1 INTRODUCTION

A subject of utmost importance in finite element analysis is the selection of particular finite elements and the definition of the appropriate approximating functions within each element. As we saw in Chapters 3 and 4, after the unknown field variable has been expressed in each element in terms of appropriate nodal parameters and interpolation functions, the derivation of the element equations according to one of the three mathematical approaches follows a well-established routine.

Except for two special cases in Chapter 3 where we used the two-node line element and the three-node triangle element, our derivations of the element equations were left in terms of general interpolation function N_i and general element types. Before these general element equations and others similarly derived can be used for actual problem solving, we must specify the particular type of element and the particular interpolation functions; that is, we must choose the functions N_i in the expression

$$\phi = \Sigma \, N_i \phi_i = \lfloor N \rfloor \{\phi\}$$

In Chapter 3 we saw that the interpolation functions cannot be chosen arbitrarily; rather, certain continuity requirements must be met to ensure that the convergence criteria, which are usually different from one problem to another, are satisfied.

It is helpful at this point to introduce a standard definition and notation to express the degree of continuity of a field variable at element interfaces. If the field variable is continuous at element interfaces, we say that we have C^0 continuity; if, in addition, first derivatives are continuous, we have C^1 continuity; if second derivatives are also continuous, we have C^2 continuity;

and so on. With this definition we may now restate the compatibility and completeness requirements for the interpolation functions representing the behavior of a field variable.

Suppose that the functions appearing under the integrals in the element equations contain derivatives up to the $(r + 1)$th order. Then, to have rigorous assurance of convergence as element size decreases, we must satisfy the following requirements.

Compatibility requirement: At element interfaces we must have C^r continuity.

Completeness requirement: Within an element we must have C^{r+1} continuity.

These requirements hold regardless of whether the element equations (integral expressions) were derived using the variational method, the Galerkin method, the energy balance method, or some other method yet to be devised.

We will see in this chapter that constructing elements and interpolation functions to achieve C^0 continuity is not especially difficult, but that the difficulty increases rapidly when higher-order continuity is desired. For problems requiring C^0 continuity we can construct an infinite number of suitable elements, but from this wide variety we usually use only the simplest types in order to avoid excessive computational labor.

Construction of suitable finite elements to give a specified continuity of order $C^0, C^1, C^2, \ldots$, and so on requires skill and ingenuity stemming from much experience. Fortunately, stress analysts and, more recently, mathematicians have developed a variety of elements applicable to many different types of problems; and we may turn to the results of their work for help in formulating new elements, or we may borrow directly from their existing catalog of elements.

Each type of element in the catalog is characterized by several features. When one practitioner of the finite element method asks another what type of element he is using in his problem, the questioner is really asking for four distinct pieces of information: the shape of the element, the number and type of nodes, the type of nodal variables, and the type of interpolation function. If any one of these characterizing features is lacking, the description of an element is incomplete.

5.2 BASIC ELEMENT SHAPES

Since the fundamental premise of the finite element method is that a continuum or solution domain of arbitrary shape can be accurately modeled by an assemblage of simple shapes, most finite elements are geometrically simple.

For one-dimensional problems, that is, problems with only one independent variable, the elements are line segments (Figure 5.1). The number of nodes assigned to a particular element, as we shall see, depends on the type of nodal variables, the type of interpolation function, and the degree of continuity required.† Often there is no reason to apply finite element methods to solve one-dimensional problems because these are governed by linear or nonlinear *ordinary* differential equations that can often be solved by other standard analytical and numerical techniques [1,2]. However, for some one-dimensional problems the finite element method is the most rational approach. For example, when we are dealing with one-dimensional domains

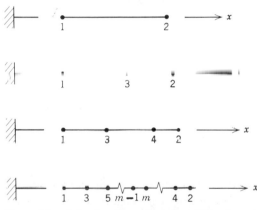

Figure 5.1.   A family of one-dimensional line elements.

that have abrupt or step changes in properties such as occur in the simple heat conduction problem in Section 2.2.2, each portion of the domain with continuously varying properties can be defined as an element. Frame analyses in solid mechanics and flow network analyses in fluid mechanics offer additional examples in which one-dimensional elements are employed. In elasticity problems where spars are used as stiffeners, one-dimensional elements can often represent the spars while being connected to other two- or three-dimensional elements that represent the rest of the elastic solid.

Common two-dimensional element shapes are shown in Figure 5.2. The three-node flat triangle element (Figure 5.2*a*) is the simplest two-dimensional

† This statement holds regardless of the dimension of the element.

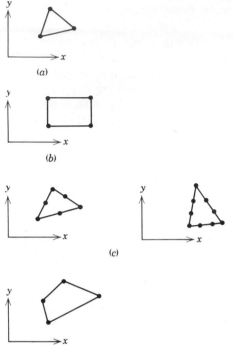

Figure 5.2. Examples of two-dimensional elements. (*a*) Three-node triangle. (*b*) Rectangle. (*c*) Six- and ten-node triangles. (*d*) General quadrilateral.

element, and it enjoys the distinction of being the first used and now the most often used basic finite element. The reason for this is that an assemblage of triangles can always represent a two-dimensional domain of any shape. A simple but less useful two-dimensional element is the four-node rectangle (Figure 5.2*b*) whose sides are parallel to the global coordinate system. This type of element is easy to construct automatically by computer because of its regular shape, but it is not well suited for approximating curved boundaries.

In addition to the simple triangle and the rectangle, other common two-dimensional elements are the six-node triangle (Figure 5.2*c*) and the general quadrilateral (Figure 5.2*d*). Quadrilateral elements may be formed directly, or they may be developed by combining two or four basic triangle elements as shown in Figure 5.3.

Other types of elements which are actually three-dimensional but are described by only one or two independent variables are axisymmetric or ring-type elements (Figure 5.4). These elements are useful when treating problems that possess axial symmetry in cylindrical coordinates. Many

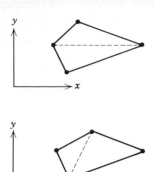

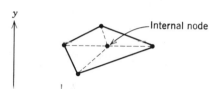

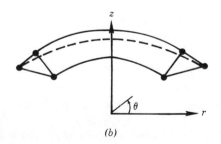

Internal node

Figure 5.3. The quadrilateral element formed by combining triangles.

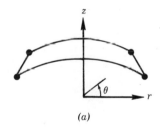

(a)

(b)

Figure 5.4. Examples of axisymmetric ring elements. (a) One-dimensional ring element. (b) Two-dimensional triangular ring element.

practical engineering problems are axisymmetric. Solids such as storage tanks, pistons, valves, shafts, rocket nozzles, and re-entry vehicle heat shields fall into this category. Before we can classify a problem as axisymmetric and define elements for it, we must be sure not only that the solution domain is axisymmetric, but also that the forcing functions are axisymmetric. We have axisymmetry only when all the parameters in the problem are the same in any plane passing through the symmetry axis of the solution domain. Since we may construct axisymmetric elements from any one- or two-dimensional element, the variety of these types of elements is virtually unlimited. We shall not discuss ring elements further in this chapter, but we shall consider them again in the following chapters on applications.

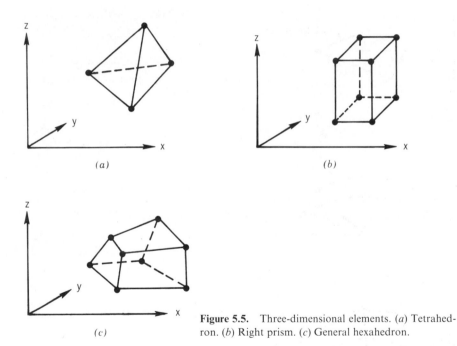

Figure 5.5. Three-dimensional elements. (*a*) Tetrahedron. (*b*) Right prism. (*c*) General hexahedron.

The four-node tetrahedron element in three dimensions (Figure 5.5*a*), is the three-dimensional counterpart of the three-node triangle element in two dimensions. Another simple three-dimensional element is the right prism shown in Figure 5.5*b*. A general hexahedron (Figure 5.5*c*), analogous to the general quadrilateral in two dimensions, may be constructed from five tetrahedra. Figure 5.6 is a photograph of a wire model formed by

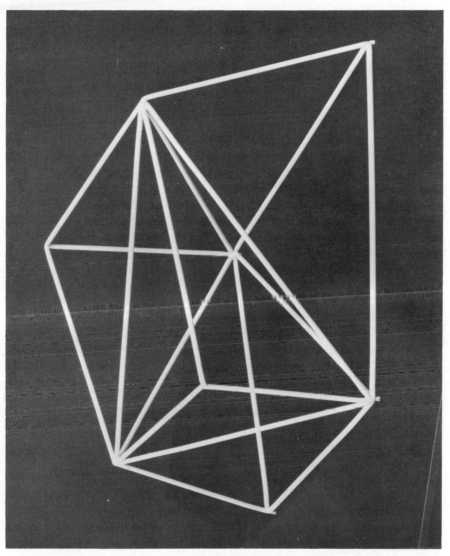

Figure 5.6. Wire model of a general hexahedron constructed from five tetrahedra.

assembling five tetrahedra. The reader may want to convince himself that there are only two possible ways to perform this assembly.

In addition to the endless variety of straight-edged elements that can be constructed, it is also possible, as we shall see, to construct elements with curved boundaries—the so-called.isoparametric element families and others. These elements, some examples of which are illustrated in Figure 5.7, are most useful when it is desirable to approximate curved boundaries with only a few elements. They have been especially helpful in the solution of three-dimensional problems, where it is often necessary to reduce the cost of computation by using fewer elements.

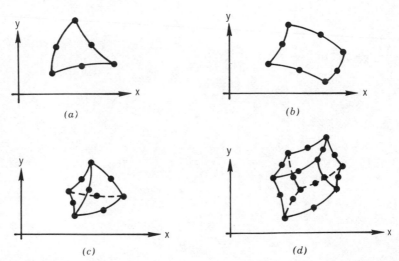

Figure 5.7.   Common isoparametric elements. (*a*) Triangle. (*b*) Quadrilateral. (*c*) Tetrahedron. (*d*) Hexahedron.

5.3   TERMINOLOGY AND PRELIMINARY CONSIDERATIONS

5.3.1   Types of nodes

In Figures 5.1–5.5 we illustrated a number of basic element shapes and typical locations of nodes assigned to these elements. We alluded to exterior and interior nodes, but now we can be more specific. Nodes are classified as either *exterior* or *interior* depending on their location relative to the geometry of an element. *Exterior nodes* lie on the boundary of an element, and they represent the points of connection between broadering elements. Nodes positioned at the corners of elements, along the edges, or on the surfaces are all exterior nodes. For one-dimensional elements such as that in Figure 5.1 there are only two exterior nodes because only the ends of the element connect to other one-dimensional elements. If one-dimensional elements are connected to two-dimensional elements along a side, all the nodes of a one-dimensional element can be exterior nodes. In contrast to exterior nodes, *interior nodes* are those that do not connect with neighboring elements. The ten-node triangle in Figure 5.2*c* has one such interior node.

5.3.2 Degrees of freedom

Two other features, in addition to shape, characterize a particular element type: (1) the number of nodes assigned to the element, and (2) the number and type of nodal variables chosen for it. Often the nodal variables or the parameters assigned to an element are called the *degrees of freedom* of the element. This terminology, which we shall adopt, is a spinoff from the solid mechanics field, where the nodal variables are usually nodal displacements and sometimes derivatives of displacements. Nodal degrees of freedom can be interior or exterior in relation to element boundaries, depending on whether they are assigned to interior or exterior nodes.

5.3.3 Interpolation functions—polynomials

In the finite element literature, the functions used to represent the behavior of a field variable within an element are called *interpolation functions, shape functions,* or *approximating functions.* We have used and will continue to use only the first terminology in this text. Although it is conceivable that many types of functions could serve as interpolation functions, only polynomials have received widespread use. The reason is that polynomials are relatively easy to manipulate mathematically—in other words, they can be integrated or differentiated without difficulty. Trigonometric functions also possess this property, but they are seldom used.† Here we shall employ only polynomials of various types and orders to generate suitable interpolation functions. The polynomials we shall consider are the following.

One independent variable

In one dimension a general complete *n*th-order polynomial may be written as

$$P_n(x) = \sum_{i=0}^{T_n^{(1)}} \alpha_i x^i \tag{5.1}$$

where the number of terms in the polynomial is $T_n^{(1)} = n + 1$.

---

† Krahula and Polhemus [3] and Chakrabarti [4] offer examples of the use of trigonometric functions in finite element analyses.

For $n = 1$, $T_1^{(1)} = 2$ and $P_1(x) = \alpha_0 + \alpha_1 x$; for $n = 2$, $T_2^{(1)} = 3$ and $P_2(x) = \alpha_0 + \alpha_1 x + \alpha_2 x^2$; and so on.

Two independent variables

In two dimensions a complete nth-order polynomial may be written as

$$P_n(x, y) = \sum_{k=1}^{T_n^{(2)}} \alpha_k x^i y^j, \quad i + j \leq n \tag{5.2}$$

where the number of terms in the polynomial is $T_n^{(2)} = (n + 1)(n + 2)/2$. For $n = 1$, $T_1^{(2)} = 3$ and $P_1(x, y) = \alpha_1 + \alpha_2 x + \alpha_3 y$; for $n = 2$, $T_2^{(2)} = 6$ and $P_2(x, y) = \alpha_1 + \alpha_2 x + \alpha_3 y + \alpha_4 xy + \alpha_5 x^2 + \alpha_6 y^2$; and so on.

Gallagher [5] suggested a convenient way to illustrate the terms in a complete two-dimensional polynomial. If the terms are placed in a triangular array in ascending order, we obtain an arrangement similar to the Pascal triangle (Figure 5.8). We note that the sum of the exponents of any term

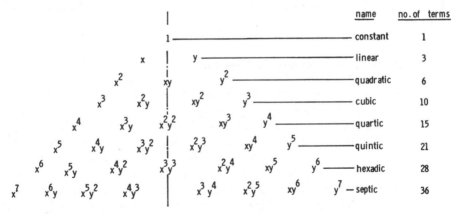

	name	no. of terms
1	constant	1
$x \quad y$	linear	3
$x^2 \quad xy \quad y^2$	quadratic	6
$x^3 \quad x^2 y \quad xy^2 \quad y^3$	cubic	10
$x^4 \quad x^3 y \quad x^2 y^2 \quad xy^3 \quad y^4$	quartic	15
$x^5 \quad x^4 y \quad x^3 y^2 \quad x^2 y^3 \quad xy^4 \quad y^5$	quintic	21
$x^6 \quad x^5 y \quad x^4 y^2 \quad x^3 y^3 \quad x^2 y^4 \quad xy^5 \quad y^6$	hexadic	28
$x^7 \quad x^6 y \quad x^5 y^2 \quad x^4 y^3 \quad x^3 y^4 \quad x^2 y^5 \quad xy^6 \quad y^7$	septic	36

Figure 5.8. Array of terms in a complete polynomial in two dimensions.

in this triangular array is the corresponding number in the well-known Pascal triangle of binomial coefficients.

Three independent variables

In three dimensions a complete nth-order polynomial may be written as

$$P_n(x, y, z) = \sum_{l=1}^{T_n^{(3)}} \alpha_l x^i y^j z^k, \quad i + j + k \leq n \tag{5.3}$$

where the number of terms in the polynomial is

$$T_n^{(3)} = \frac{(n+1)(n+2)(n+3)}{6}$$

For $n = 1$, $T_1^{(3)} = 4$ and $P_1(x, y, z) = \alpha_1 + \alpha_2 x + \alpha_3 y + \alpha_4 z$; for $n = 2$, $T_2^{(3)} = 10$ and $P_2(x, y, z) = \alpha_1 + \alpha_2 x + \alpha_3 y + \alpha_4 z + \alpha_5 xy + \alpha_6 xz + \alpha_7 yz + \alpha_8 x^2 + \alpha_9 y^2 + \alpha_{10} z^2$; and so on.

The terms in a complete three-dimensional polynomial may also be arrayed in a manner analogous to the triangular array in two dimensions. The array becomes a tetrahedron with the various terms placed at different planar levels as in Figure 5.9.

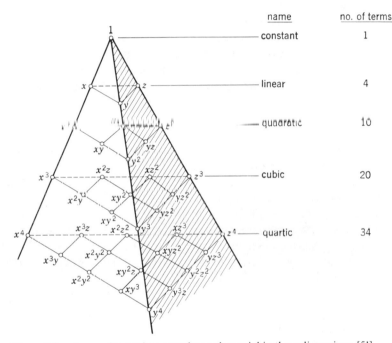

	name	no. of terms
	constant	1
	linear	4
	quadratic	10
	cubic	20
	quartic	34

Figure 5.9. Array of terms in a complete polynomial in three dimensions [51].

5.4 GENERALIZED COORDINATES AND THE ORDER OF THE POLYNOMIAL

5.4.1 Generalized coordinates

Ignoring for the moment any interelement continuity considerations, we can say that the order of the polynomial we use to represent the field variable within an element depends on the number of degrees of freedom we assign to the element. In other words, the number of coefficients in the polynomial should equal the number of nodal variables available to evaluate these coefficients. Recalling our experience in Section 3.4.3, where we obtained a piecewise linear representation for a field variable over a triangle, we wrote the linear polynomial series as

$$\phi(x, y) = \alpha_1 + \alpha_2 x + \alpha_3 y$$

The three nodal values of ϕ were then sufficient to find the three coefficients α_1, α_2, and α_3; that is, we evaluated the expression at the three nodes and thereby obtained three equations to be solved for the α_i in terms of the coordinates of the nodes and the nodal values of ϕ. If we had selected a complete or six-term polynomial to represent ϕ, we would have needed six nodal variables (six degrees of freedom) to find the six α_i.

The coefficients α_i in a polynomial series representation of a field variable are called the *generalized coordinates* of the element. The generalized coordinates are independent parameters that specify or fix the magnitude of the prescribed distribution for ϕ, while the shape of the prescribed distribution is determined by the order of the polynomial we select. As illustrated in equation 3.6, the generalized coordinates (denoted as β_i in this case) have no direct physical interpretation but rather are linear combinations of the physical nodal degrees of freedom. Also, they are not identified with particular nodes. Though the usual procedure is to use the same number of generalized coordinates and degrees of freedom, it is possible in some formulations of elasticity problems to employ an excess of generalized coordinates and associate these with internal nodes. The surplus generalized coordinates can then be determined using energy considerations. We shall not consider these types of element formulation, but the interested reader can find a discussion of the techniques in Pian [8].

5.4.2 Geometric isotropy

When we choose polynomial expansions as interpolation functions for an element, we usually try to satisfy the compatibility and completeness requirements discussed in Section 3.4.5. As we have seen, these requirements are important because they ensure (1) continuity of the field variable, and (2) convergence to the correct solution as the element mesh size is made smaller and smaller. In addition to satisfying these stipulations, we require that the field variable representation within an element, and hence the polynomial expansion for the element, remain unchanged under a linear transformation from one Cartesian coordinate system to another [6]. Polynomials that exhibit this invariance property are said to possess *geometric isotropy.* Clearly, we could not expect a good approximation to reality if our field variable representation changed with a change in origin or in the orientation of the coordinate system. Hence the need to ensure geometric isotropy in our polynomial interpolation functions is apparent.

Fortunately, we have two simple guidelines that allow us to construct polynomial series with the desired property:

1.   Polynomials of order n that are complete (contain all their terms) have geometric isotropy.

2.   Polynomials of order n that are incomplete, yet contain the appropriate terms to preserve "symmetry," have geometric isotropy.

According to the first guideline we know that the complete polynomials†

$$P_n(x, y) = \sum_{k=1}^{T_n^{(2)}} \alpha_k x^i y^j, \quad i + j \le n$$

and

$$P_n(x, y, z) = \sum_{l=1}^{T_n^{(3)}} \alpha_l x^i y^j z^l, \quad i + j + k \le n$$

when used as interpolation functions will remain invariant under linear coordinate transformations. We shall not be concerned with other types of transformations.

It is easy to illustrate what we mean by "symmetry" in the second guideline by considering a specific example. Suppose that we wish to construct a cubic polynomial expansion for an element that has eight nodal variables assigned to it. In this situation, since a complete cubic polynomial contains ten terms and hence ten undetermined coefficients, we may desire to truncate the polynomial by dropping two terms. But then the question arises, "How should the polynomial series be truncated so that geometric isotropy is preserved?" The answer is that we may drop only terms that occur in symmetric pairs, that is, (x^3, y^3) or $(x^2 y, xy^2)$. Thus, acceptable eight-term cubic polynomial interpolation functions exhibiting geometric isotropy would be

$$P(x, y) = \alpha_1 + \alpha_2 x + \alpha_3 y + \alpha_4 x^2 + \alpha_5 xy + \alpha_6 y^2 + \alpha_7 x^3 + \alpha_{10} y^3$$

or

$$P(x, y) = \alpha_1 + \alpha_2 x + \alpha_3 y + \alpha_4 x^2 + \alpha_5 xy + \alpha_6 y^2 + \alpha_8 x^2 y + \alpha_9 xy^2$$

† Polynomials in one independent variable inherently possess geometric isotropy regardless of whether they are complete.

Extending this idea to construct other incomplete polynomials is straght-forward if we refer to the arrays of terms in Figures 5.8 and 5.9. As Dunne [6] points out, meeting the criterion of geometric isotropy assures us that, when we evaluate the interpolation function along any straight edge of the element, it becomes a complete polynomial in the linear coordinate along that edge, and its order is the same as the parent two- or three-dimensional polynomial for the element.

Taylor [7] examines the completeness property of a number of commonly used interpolation functions for C^0 problems and presents guidelines for constructing computationally efficient functions for rectangular elements.

5.4.3 Deriving interpolation functions

Thus far, we have seen how a field variable can be represented within an element as a polynomial series whose coefficients are the generalized co-ordinates. We shall see in this section how the interpolation functions for the physical degrees of freedom are derived. These interpolation functions emerge from the basic procedure for expressing the generalized coordinates in terms of the nodal degrees of freedom. In Section 5.4.1 we alluded to this procedure, but now we shall consider the details.

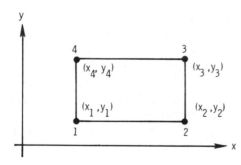

Figure 5.10. A rectangular element with sides parallel to the axes of the global coordinate system.

The basic ideas can be illustrated by a simple example in two dimensions. Suppose that we wish to construct a rectangular element with nodes position-ed at the corners of the element (Figure 5.10). If we assign one value of ϕ to each node, the element then has four degrees of freedom, and we may select as an interpolation model a four-term polynomial such as

$$\phi(x, y) = \alpha_1 + \alpha_2 x + \alpha_3 y + \alpha_4 xy \qquad (5.4)$$

The generalized coordinates may now be found by evaluating this model at each of the nodes, and then inverting the resulting set of simultaneous equations. Thus we may write

$$\phi_1 = \alpha_1 + \alpha_2 x_1 + \alpha_3 y_1 + \alpha_4 x_1 y_1$$
$$\phi_2 = \alpha_1 + \alpha_2 x_2 + \alpha_3 y_2 + \alpha_4 x_2 y_2$$
$$\phi_3 = \alpha_1 + \alpha_2 x_3 + \alpha_3 y_3 + \alpha_4 x_3 y_3$$
$$\phi_4 = \alpha_1 + \alpha_2 x_4 + \alpha_3 y_4 + \alpha_4 x_4 y_4$$

or, in matrix notation,

$$\{\phi\} = [G]\{\alpha\} \tag{5.5}$$

where

$$\{\phi\} = \begin{Bmatrix} \phi_1 \\ \phi_2 \\ \phi_3 \\ \phi_4 \end{Bmatrix}$$

$$[G] = \begin{bmatrix} 1 & x_1 & y_1 & x_1 y_1 \\ 1 & x_2 & y_2 & x_2 y_2 \\ 1 & x_3 & y_3 & x_3 y_3 \\ 1 & x_4 & y_4 & x_4 y_4 \end{bmatrix}$$

and

$$\{\alpha\} = \begin{Bmatrix} \alpha_1 \\ \alpha_2 \\ \alpha_3 \\ \alpha_4 \end{Bmatrix}$$

In principle, then, we can express the generalized coordinates as the solution of equations 5.5 for $\{\alpha\}$, that is,

$$\{\alpha\} = [G]^{-1}\{\phi\} \tag{5.6}$$

Expressing the terms of the interpolation polynomial equation 5.4 as a product of a row vector and a column vector, we can write

$$\phi = \lfloor P \rfloor \{\alpha\} \tag{5.7}$$

where

$$\lfloor P \rfloor = \lfloor 1 \quad x \quad y \quad xy \rfloor$$

Thus, by substituting equations 5.6 into equation 5.7, we have

$$\phi = \lfloor P \rfloor [G]^{-1}\{\phi\} = \lfloor N \rfloor\{\phi\} \tag{5.8a}$$

with

$$\lfloor N \rfloor = \lfloor P \rfloor [G]^{-1} \tag{5.8b}$$

Equations 5.6, 5.7, and 5.8, though obtained for one case, are generally applicable to all straight-sided elements. The original interpolation polynomial $\lfloor P \rfloor\{\alpha\}$ should not be confused with the interpolation functions N_i associated with the nodal degrees of freedom. The distinction to note here is that $\lfloor P \rfloor\{\alpha\}$ is an interpolation function that applies to the whole element and expresses the field variable behavior in terms of the generalized coordinates, whereas the interpolation functions N_i refer to individual nodes and individual nodal degrees of freedom; collectively, they represent the field variable behavior. It is easy to see from equation 5.8 that the function N_i referring to node i takes on unit value at node i and zero value at all the other nodes of the element.

This procedure for expressing the generalized coordinates in terms of the nodal degrees of freedom is actually the method commonly used to derive the nodal interpolation functions N_i. The procedure is straightforward and may be carried out perfunctorily, but sometimes difficulties are encountered. For some types of element models the inverse of the matrix $[G]$ may not exist for all orientations of the element in the global coordinate system [9]. If an explicit expression for $[G]^{-1}$ is obtained algebraically (most often an arduous task), it may be possible to see under what conditions $[G]^{-1}$ does not exist and then try to avoid these circumstances when constructing the element mesh. Such an approach, however, is seldom recommended. Another disadvantage stems from the computational effort required to obtain $[G]^{-1}$ when it exists. For a large number of elements with many degrees of freedom the cost of computation can be prohibitive.

These reasons have motivated many researchers to try to obtain the nodal interpolation functions N_i by inspection. Indeed, this has been done, and later we shall examine some results and the rationale used to obtain them. Deriving the functions N_i by inspection often relies on the use of special coordinate systems called *natural coordinates*. In the following section we shall derive various natural coordinate systems in one, two, and three dimensions.

5.5   NATURAL COORDINATES

A local coordinate system that relies on the element geometry for its definition and whose coordinates range between zero and unity within the element is known as a *natural coordinate system.* Such systems have the property that one particular coordinate has unit value at one node of the element and zero value at the other node(s); its variation between nodes is linear. We may construct natural coordinate systems for two-node line elements, three-node triangular elements, four-node quadrilateral elements, four-node tetrahedral elements, and so on into *n*-dimensional hyperspace. In the parlance of topology, natural coordinates in *n* dimensions are called barycentric coordinates [10]. Since only the simplest coordinates up to three dimensions are commonly used, we shall not consider those in higher dimensions.

The use of natural coordinates in deriving interpolation functions is particularly advantageous because special closed-form integration formulas can often be used to evaluate the integrals in the element equations. Natural coordinates also play a crucial role in the development of curve-sided elements, which we shall discuss later in this chapter.

The basic purpose of a natural coordinate system is to describe the location of a point inside an element in terms of coordinates associated with the nodes of the element. We denote the natural coordinates as L_i $(i = 1, 2, ..., n)$, where n is the number of external nodes of the element. One coordinate is associated with node i and has unit value there. It will become evident that the natural coordinates are functions of the global Cartesian coordinate system in which the element is defined.

5.5.1 Natural coordinates in one dimension

Figure 5.11 shows a line element in which we desire to define a natural coordinate system. If we select L_1 and L_2 as the natural coordinates, the location of the point x_p may be expressed as a linear combination of the nodal coordinates x_1 and x_2, that is,

$$x_p = L_1 x_1 + L_2 x_2 \tag{5.9}$$

Since x_p can be any point on the line element, we can drop the subscript p for convenience. The coordinates L_1 and L_2 may be interpreted as weighting functions relating the coordinates of the end nodes to the coordinate of any

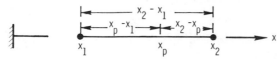

Figure 5.11.   Two-node line element with the global coordinate x_p defining some point within the element.

interior point. Clearly, the weighting functions are not independent since we must have

$$L_1 + L_2 = 1 \qquad (5.10)$$

Equations 5.9 and 5.10 may be solved simultaneously for the functions L_1 and L_2 with the following result:

$$L_1(x) = \frac{x - x_1}{x_2 - x_1}, \qquad L_2(x) = \frac{x_2 - x}{x_2 - x_1} \qquad (5.11)$$

The functions L_1 and L_2 are seen to be simply ratios of lengths and are often called *length coordinates*. The variation of L_1 is shown in Figure 5.12. We

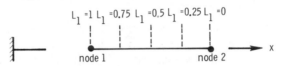

Figure 5.12.   Variation of a length coordinate within an element.

recognize that the natural coordinates for the line element are precisely the same as the linear interpolation functions we used in the example of Section 3.4.5. Hence the linear interpolation used for the field variable ϕ in that example can be written directly as

$$\phi(x) = \phi_1 L_1 + \phi_2 L_2$$

If ϕ is taken to be a function of L_1 and L_2, differentiation of ϕ follows the chain rule formula

$$\frac{d\phi}{dx} = \frac{\partial \phi}{\partial L_1}\frac{\partial L_1}{\partial x} + \frac{\partial \phi}{\partial L_2}\frac{\partial L_2}{\partial x} \qquad (5.12)$$

where

$$\frac{\partial L_1}{\partial x} = \frac{1}{x_2 - x_1}, \qquad \frac{\partial L_2}{\partial x} = \frac{-1}{x_2 - x_1} \qquad (5.13)$$

Integration of length coordinates over the length of an element is simple with the aid of the following convenient formula:

$$\int_{x_1}^{x_2} L_1{}^{\alpha} L_2{}^{\beta} \, dx = \frac{\alpha!\,\beta!\,(x_2 - x_1)}{(\alpha + \beta + 1)!} \qquad (5.14)$$

where α and β are integer exponents and, for instance, $\alpha!$ is the factorial of α. Table 5.1 gives the values of equation 5.14 for various integers α and β.

Table 5.1.   Integrals of length co-ordinates

$$l_{\alpha\beta} = \frac{1}{x_2 - x_1} \int_{x_1}^{x_2} L_1{}^{\alpha} L_2{}^{\beta}\, dx = \frac{A}{B}$$

$\alpha + \beta$	α	β	A	B
0	0	0	1	1
1	1	0	1	2
2	2	0	2	6
2	1	1	1	6
3	3	0	3	12
3	2	1	1	12
4	4	0	12	60
4	3	1	3	60
4	2	2	2	60
5	5	0	10	60
5	4	1	2	60
5	3	2	1	60
6	6	0	60	420
6	5	1	10	420
6	4	2	4	420
6	3	3	3	420

5.5.2   Natural coordinates in two dimensions

The development of natural coordinates for triangular elements follows the same procedure we used for the one-dimensional case. Again, the goal is to choose coordinates L_1, L_2, and L_3 to describe the location of any point x_p within the element or on its boundary (Figure 5.13). The original Cartesian coordinates of a point in the element should be linearly related to the new coordinates by the following equations:

$$x = L_1 x_1 + L_2 x_2 + L_3 x_3$$
$$y = L_1 y_1 + L_2 y_2 + L_3 y_3 \tag{5.15}$$

in addition to these equations we impose a third condition requiring that the weighting functions sum to unity, that is,

$$L_1 + L_2 + L_3 = 1 \tag{5.16}$$

From equation 5.16 it is clear that only two of the natural coordinates can be independent, just as in the original coordinate system, where there are only two independent coordinates.

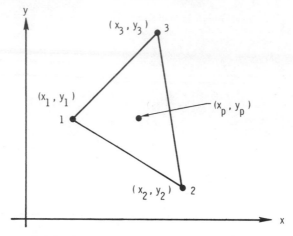

Figure 5.13. Three-node triangle element with global coordinates (x_p, y_p) defining some point within the element.

Inversion of equations 5.15 and 5.16 gives the natural coordinates in terms of the Cartesian coordinates. Thus

$$L_1(x, y) = \frac{1}{2\Delta}(a_1 + b_1 x + c_1 y)$$

$$L_2(x, y) = \frac{1}{2\Delta}(a_2 + b_2 x + c_2 y) \tag{5.17}$$

$$L_3(x, y) = \frac{1}{2\Delta}(a_3 + b_3 x + c_3 y)$$

where

$$2\Delta = \begin{vmatrix} 1 & x_1 & y_1 \\ 1 & x_2 & y_2 \\ 1 & x_3 & y_3 \end{vmatrix} = 2(\text{area of triangle 1-2-3}) \tag{5.18}$$

$$a_1 = x_2 y_3 - x_3 y_2, \qquad b_1 = y_2 - y_3, \qquad c_1 = x_3 - x_2 \tag{5.19}$$

The other coefficients are obtained by cyclically permuting the subscripts.

Recalling our example of piecewise linear interpolation in Section 3.4.3, we see that the natural coordinates L_1, L_2, and L_3 are precisely the interpolation functions for linear interpolation over a triangle, that is, $N_i = L_i$ for the linear triangle. A little algebraic manipulation will reveal that the natural coordinates for a triangle have an interpretation analogous to that of length coordinates for a line. Just as $L_1(x)$ for the line element is a ratio of lengths, $L_1(x, y)$ for the triangle element is a ratio of areas. Figure 5.14 shows how the natural coordinates, often called *area coordinates*, are related to areas. As shown in Figure 5.14, when the point (x_p, y_p) is located on the boundary of the element, one of the area segments vanishes and hence the

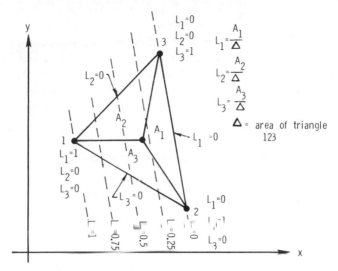

Figure 5.14.   Area coordinates for a triangle.

appropriate area coordinate along that boundary is identically zero. For example, if (x_p, y_p) is on line 1-3, then

$$L_2 = \frac{A_2}{\Delta} = 0 \quad \text{since} \quad A_2 = 0$$

If we interpret the field variable ϕ as a function of L_1, L_2, and L_3 instead of x, y, differentiation becomes

$$\frac{d\phi}{dx} = \frac{\partial\phi}{\partial L_1}\frac{\partial L_1}{\partial x} + \frac{\partial\phi}{\partial L_2}\frac{\partial L_2}{\partial x} + \frac{\partial\phi}{\partial L_3}\frac{\partial L_3}{\partial x}$$

$$\frac{d\phi}{dy} = \frac{\partial\phi}{\partial L_1}\frac{\partial L_1}{\partial y} + \frac{\partial\phi}{\partial L_2}\frac{\partial L_2}{\partial y} + \frac{\partial\phi}{\partial L_3}\frac{\partial L_3}{\partial y}$$

(5.20)

where

$$\frac{\partial L_i}{\partial x} = \frac{b_i}{2\Delta}, \quad \frac{\partial L_i}{\partial y} = \frac{c_i}{2\Delta}, \quad i = 1, 2, 3$$

(5.21)

There is also a convenient formula for integrating area coordinates over the area of a triangular element. In the example of Section 3.4.7, we alluded to the following formula:

$$\int_{A^{(e)}} L_1{}^\alpha L_2{}^\beta L_3{}^\gamma \, dA^{(e)} = \frac{\alpha!\beta!\gamma!}{(\alpha + \beta + \gamma + 2)!} 2\Delta \qquad (5.22)$$

Table 5.2 gives the values of equation 5.22 for various integers α, β, and γ.

Another type of natural coordinate system can be established for a four-node quadrilateral element in two dimensions. Figure 5.15 shows a general quadrilateral element in the global Cartesian coordinate system and a local natural coordinate system. In the natural coordinate system whose

Table 5.2.   Integrals of area coordinates

$$l_{\alpha\beta\gamma} = \frac{1}{\Delta} \int_{A^{(e)}} L_1{}^\alpha L_2{}^\beta L_3{}^\gamma \, dA^{(e)} = \frac{A}{B}$$

$\alpha + \beta + \gamma$	α	β	γ	A	B
0	0	0	0	1	1
1	1	0	0	1	3
2	2	0	0	2	12
2	1	1	0	1	12
3	3	0	0	6	60
3	2	1	0	2	60
3	1	1	1	1	60
4	4	0	0	12	180
4	3	1	0	3	180
4	2	2	0	2	180
4	2	1	1	1	180
5	5	0	0	60	1,260
5	4	1	0	12	1,260
5	3	2	0	6	1,260
5	3	1	1	3	1,260
5	2	2	1	2	1,260
6	6	0	0	180	5,040
6	5	1	0	30	5,040
6	4	2	0	12	5,040
6	4	1	1	6	5,040
6	3	3	0	9	5,040
6	3	2	1	3	5,040
6	2	2	2	1	5,040

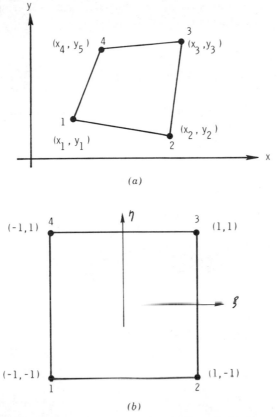

Figure 5.15. Natural coordinates for a general quadrilateral. (*a*) Cartesian coordinates. (*b*) Natural coordinates.

origin is at the centroid the quadrilateral element is a square with sides extending to $\xi = \pm 1$, $\eta = \pm 1$. The local and global coordinates are related by the following equations:

$$x = \tfrac{1}{4}[(1 - \xi)(1 - \eta)x_1 + (1 + \xi)(1 - \eta)x_2 + (1 + \xi)(1 + \eta)x_3$$
$$+ (1 - \xi)(1 + \eta)x_4]$$
$$y = \tfrac{1}{4}[(1 - \xi)(1 - \eta)y_1 + (1 + \xi)(1 - \eta)y_2 + (1 + \xi)(1 + \eta)y_3$$
$$+ (1 - \xi)(1 + \eta)y_4]$$

(5.23)

Rather than trying to solve equations 5.23 for ξ and η in terms of x, y, and the nodal coordinates and proceed as before, we use numerical methods to carry out differentiation and integration [11]. We shall postpone discussion of numerical procedures until we consider the isoparametric element concepts later in this chapter.

5.5.3   Natural coordinates in three dimensions

We can define natural coordinates for the four-node tetrahedron in a manner analogous to the procedures for the three-node triangle. The result, as the reader may expect, is a set of *volume coordinates* whose physical interpretation turns out to be a ratio of volumes in the tetrahedron. Figure 5.16 shows a typical element and defines the node-numbering scheme.

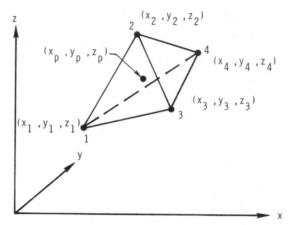

Figure 5.16.   A tetrahedral element whose nodes are numbered according to the right hand rule. The point (x_p, y_p, z_p) is some point *within* the element.

The global Cartesian coordinates and the local natural coordinates are related by

$$
\begin{aligned}
x &= L_1 x_1 + L_2 x_2 + L_3 x_3 + L_4 x_4 \\
y &= L_1 y_1 + L_2 y_2 + L_3 y_3 + L_4 y_4 \\
z &= L_1 z_1 + L_2 z_2 + L_3 z_3 + L_4 z_4 \\
&\quad L_1 + L_2 + L_3 + L_4 = 1
\end{aligned}
\tag{5.24}
$$

Equations 5.24 can be inverted to give

$$
L_i = \frac{1}{6V}(a_i + b_i x + c_i y + d_i z), \quad i = 1, 2, 3, 4
\tag{5.25}
$$

where

$$6V = \begin{vmatrix} 1 & x_1 & y_1 & z_1 \\ 1 & x_2 & y_2 & z_2 \\ 1 & x_3 & y_3 & z_3 \\ 1 & x_4 & y_4 & z_4 \end{vmatrix} = 6 \begin{pmatrix} \text{volume of the tetrahedron} \\ \text{defined by nodes 1, 2, 3, 4} \end{pmatrix} \quad (5.26)$$

and, for instance,

$$a_1 = \begin{vmatrix} x_2 & y_2 & z_2 \\ x_3 & y_3 & z_3 \\ x_4 & y_4 & z_4 \end{vmatrix}, \qquad c_1 = - \begin{vmatrix} x_2 & 1 & z_2 \\ x_3 & 1 & z_3 \\ x_4 & 1 & z_4 \end{vmatrix}$$

$$b_1 = - \begin{vmatrix} 1 & y_2 & z_2 \\ 1 & y_3 & z_3 \\ 1 & y_4 & z_4 \end{vmatrix}, \qquad d_1 = - \begin{vmatrix} x_2 & y_2 & 1 \\ x_3 & y_3 & 1 \\ x_4 & y_4 & 1 \end{vmatrix} \quad (5.27)$$

The other constants are obtained through a cyclic permutation of subscripts 1, 2, 3, and 4. Since the constants are the cofactors of the determinant in equation 5.26, attention must be given to the appropriate sign. If the tetrahedron is defined in a right-handed Cartesian coordinate system, equations 5.26 and 5.27 are valid only when the nodes are numbered so that nodes 1, 2, and 3 are ordered counterclockwise when viewed from node 4.

Figure 5.17 illustrates the physical interpretation of natural coordinates for a tetrahedron.

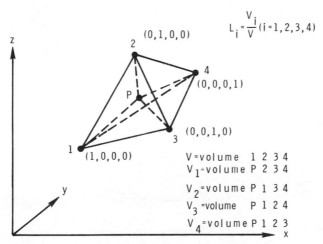

Figure 5.17.   Volume coordinates.

Table 5.3. Integrals of volume coordinates

$$l_{\alpha\beta\gamma\delta} = \frac{1}{V} \int_{V^{(e)}} L_1{}^\alpha L_2{}^\beta L_3{}^\gamma L_4{}^\delta \, dV^{(e)} = \frac{A}{B}$$

$\alpha + \beta + \gamma + \delta$	α	β	γ	δ	A	B
0	0	0	0	0	1	1
1	1	0	0	0	1	4
2	2	0	0	0	2	20
2	1	1	0	0	1	20
3	3	0	0	0	6	120
3	2	1	0	0	2	120
3	1	1	1	0	1	120
4	4	0	0	0	24	840
4	3	1	0	0	6	840
4	2	2	0	0	4	840
4	2	1	1	0	2	840
4	1	1	1	1	1	840
5	5	0	0	0	60	3,360
5	4	1	0	0	12	3,360
5	3	2	0	0	6	3,360
5	3	1	1	0	3	3,360
5	2	2	1	0	2	3,360
5	2	1	1	1	1	3,360
6	6	0	0	0	360	30,240
6	5	1	0	0	60	30,240
6	4	2	0	0	24	30,240
6	4	1	1	0	12	30,240
6	3	3	0	0	18	30,240
6	3	2	1	0	6	30,240
6	3	1	1	1	3	30,240
6	2	2	2	0	4	30,240
6	2	2	1	1	2	30,240

The appropriate differentiation formulas are as follows:

$$\frac{\partial \phi}{\partial x} = \sum_{i=1}^{4} \frac{\partial \phi}{\partial L_i} \frac{\partial L_i}{\partial x}$$

$$\frac{\partial \phi}{\partial y} = \sum_{i=1}^{4} \frac{\partial \phi}{\partial L_i} \frac{\partial L_i}{\partial y} \qquad (5.28)$$

$$\frac{\partial \phi}{\partial z} = \sum_{i=1}^{4} \frac{\partial \phi}{\partial L_i} \frac{\partial L_i}{\partial z}$$

where

$$\frac{\partial L_i}{\partial x} = \frac{b_i}{6V}, \qquad \frac{\partial L_i}{\partial y} = \frac{c_i}{6V}, \qquad \frac{\partial L_i}{\partial z} = \frac{d_i}{6V} \qquad (5.29)$$

And the integration formula is

$$\int_{V^{(e)}} L_1{}^\alpha L_2{}^\beta L_3{}^\gamma L_4{}^\delta = \frac{\alpha!\,\beta!\,\gamma!\,\delta!}{(\alpha + \beta + \gamma + 3)!}\, 6V \qquad (5.30)$$

which is evaluated in Table 5.3 for several integer combinations of α, β, γ, and δ.

Natural coordinates can also be established for general hexahedral elements in three dimensions [11,12]. For the eight-node hexahedron shown in Figure 5.18 the equations relating the Cartesian coordinates and the

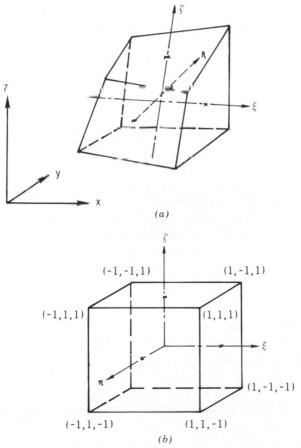

(a)

(b)

Figure 5.18. Hexahedral coordinates [12]. (a) Cartesian coordinates. (b) Natural coordinates.

natural coordinates are

$$x = \sum_{i=1}^{8} x_i L_i$$

$$y = \sum_{i=1}^{8} y_i L_i \tag{5.31}$$

$$z = \sum_{i=1}^{8} z_i L_i$$

where

$$L_i = \tfrac{1}{8}(1 + \xi\xi_i)(1 + \eta\eta_i)(1 + \xi\xi_i), \quad i = 1, 2, \ldots, 8 \tag{5.32}$$

Again, inversion of equations 5.31 and 5.32 is not possible, so we must use numerical methods to carry out differentiation and integration. This will become clearer in later examples.

5.6   INTERPOLATION CONCEPTS IN ONE DIMENSION

Constructing nodal interpolation functions according to the procedure discussed in Section 5.4.3 involves computing and inverting the matrix $[G]$ for each element of the assemblage. Even for a particular formulation in which $[G]^{-1}$ always exists, the process can consume much computer time. For a number of different elements it is possible to write the nodal interpolation functions directly by employing natural coordinates *or* classical interpolation polynomials. In this section, we review the features of two classical interpolation functions that we shall use later to construct the functions N_i for several types of elements.

5.6.1   Lagrange polynomials

One type of useful interpolation function is the Lagrange polynomial, defined as

$$L_k(x) = \prod_{\substack{m=0 \\ m \neq k}}^{n} \frac{x - x_n}{x_k - x_m} = \frac{(x - x_0) \cdots (x - x_{k-1})(x - x_{k+1}) \cdots (x - x_n)}{(x_k - x_0) \cdots (x_k - x_{k-1})(x_k - x_{k+1}) \cdots (x_k - x_n)}$$

$$\tag{5.33}$$

Since $L_k(x)$ is a product of n linear factors, it is clearly a polynomial of degree n. We note that, when $x = x_k$, the numerator and denominator of $L_k(x_k)$ are identical and the polynomial has unit value. But, when $x = x_m$ and $m \neq k$, the polynomial vanishes. This fact can be used to represent an arbitrary function $\phi(x)$ over an interval on the x axis. For example, suppose that $\phi(x)$ is given by discrete values at four points in the closed interval $[x_0, x_3]$ (Figure 5.19a). A polynomial of degree 3 passing through the four

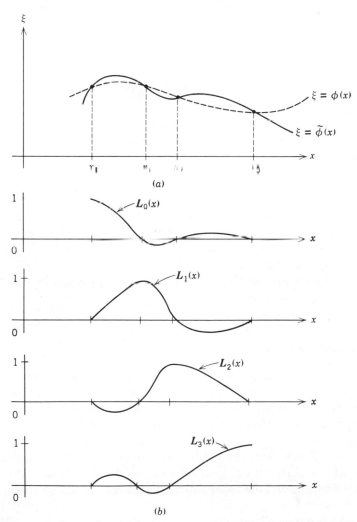

Figure 5.19.   Interpolation using Lagrange polynomials. (*a*) The given function and its approximate representation. (*b*) Lagrange interpolation coefficients.

discrete values $\phi_i = \phi(x_i)$ $(i = 0, 1, 2, 3)$ and approximating the function $\phi(x)$ in the interval may be written at once as

$$\phi(x) \simeq \tilde{\phi}(x) = \sum_{i=0}^{3} \phi_i L_i(x) = [L]\{\phi\}$$

and we recognize that $L_i(x)$ plays the role $N_i(x)$.

The Lagrange polynomials L_i in the expression for $\tilde{\phi}(x)$ are sometimes called Lagrangian interpolation coefficients. Figure 5.19b shows how these coefficients take on the value zero or unity as required. The coefficient for ϕ_2, for instance, would become

$$L_2(x) = \frac{(x - x_0)(x - x_1)(x - x_3)}{(x_2 - x_0)(x_2 - x_1)(x_2 - x_3)}$$

Since the Lagrangian coefficients possess the desired properties of the nodal interpolation functions, we may write immediately for any line element with only ϕ_i specified at the nodes (not derivatives)

$$N_i(x) = L_i(x)$$

with the order of the interpolation polynomials depending on the number of nodes we assign to the element. Since Lagrangian interpolation functions guarantee continuity of ϕ at the connecting nodes, they are suitable for elements used in problems requiring C^0 continuity.

Later in this chapter we shall see how these concepts can be generalized to two- and three-dimensional elements.

5.6.2   Hermite polynomials

Just as Lagrange polynomials enable us to write at once various orders of interpolation functions when values of the field variable ϕ are specified at the nodes, Hermite polynomials enable us to construct interpolation functions when ϕ, as well as its derivatives, is specified at the nodes. An nth-order Hermite polynomial in x is denoted as $H^n(x)$ and is a polynomial of order $2n + 1$. Thus, for example, $H^1(x)$ is a first-order Hermite polynomial which is cubic in x. Hermite polynomials are useful as interpolation functions because their value and the value of their derivatives up to order n are unity or zero at the end points of the closed interval $[0, 1]$. This property can be represented symbolically if we assign two subscripts and write $H_{mi}{}^n(x)$, where m is the order of the derivative and i designates either node 1 or node 2, the end nodes of a line element. Then we have, when $x = x_2$ or when $m \neq n$ and $x = x_1$,

$$\frac{d^m H_{mi}{}^n(x)}{dx^m} = 0 \qquad (5.34a)$$

However, when $x = x_1$ and $m = n$,

$$\frac{d^m H_{m1}{}^n(x)}{dx^m} = 1 \qquad (5.34b)$$

A similar statement can be made for node 2.

To illustrate a simple application of Hermite polynomials, suppose that we wish to construct interpolation functions for a two-node line element with ϕ and $\partial\phi/\partial x = \phi_x$ specified as degrees of freedom at each node (Figure 5.20). Since the element has four degrees of freedom (enough to determine

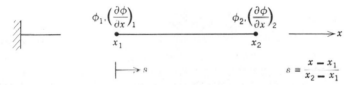

Figure 5.20. Higher-order two-node line element with four degrees of freedom.

uniquely a cubic expansion), we may select first-order or cubic Hermite polynomials as interpolation functions. These are defined as

$$H_{01}{}^1(s) = 1 - 3s^2 + 2s^3$$
$$H_{11}{}^1(s) = (x_2 - x_1)s(s - 1)^2$$
$$H_{02}{}^1(s) = s^2(3 - 2s) \qquad (5.35)$$
$$H_{12}{}^1(s) = (x_2 - x_1)s^2(s - 1)$$

with

$$s = \frac{x - x_1}{x_2 - x_1}, \qquad 0 \leq s \leq 1, \qquad \frac{\partial}{\partial x} = \frac{1}{x_2 - x_1}\frac{\partial}{\partial s}$$

Figure 5.21 shows the general behavior of a set of Hermite functions for $m = 0, 1, 2, 3$. Using the shape functions for $m = 1$, we may write the interpolation model as

$$\phi = H_{01}{}^1\phi_1 + H_{11}{}^1\phi_{x1} + H_{02}{}^1\phi_2 + H_{12}{}^1\phi_{x2} = \lfloor N \rfloor \{\phi\} \quad (5.36)$$

where

$$\{\phi\} = \begin{Bmatrix} \phi_1 \\ \phi_{x_1} \\ \phi_2 \\ \phi_{x_2} \end{Bmatrix}, \qquad \lfloor N \rfloor = \lfloor H_{01}{}^1, H_{11}{}^1, H_{02}{}^1, H_{12}{}^1 \rfloor$$

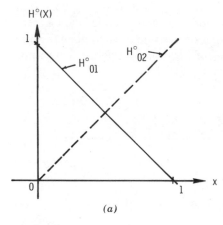

(a)

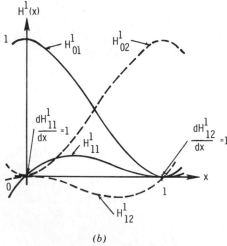

(b)

Figure 5.21. Hermitian polynomials. *(a)* Zeroth-order Hermite polynomials. *(b)* First-order Hermite polynomials.

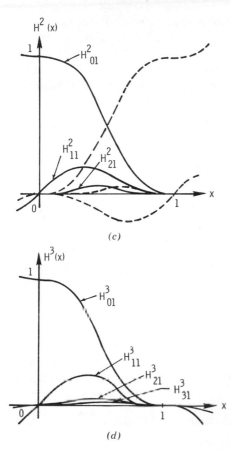

(c)

(d)

Figure 5.21 (*continued*). (*c*) Second-order Hermite polynomials. (*d*) Third-order Hermite polynomials.

We note that this interpolation model guarantees continuity of both ϕ and ϕ_x at the nodes, and hence it is suitable for one-dimensional problems requiring C^1 continuity. Later in this chapter, we shall discuss the use of Hermite polynomials for two- and three-dimensional elements and elements giving continuity of higher derivatives.

5.7 TREATING INTERNAL NODES—CONDENSATION/SUBSTRUCTURING

Before considering many different types of elements, we need to discuss the problem of how to handle elements that have internal nodes. As we saw in Section 5.3.1, these nodes, by definition, do not connect with the nodes of other elements during the assembly process, and consequently the degrees of freedom associated with internal nodes do not affect interelement continuity. Though interelement continuity is unaffected, extra degrees of freedom and internal nodes are sometimes used to improve the field variable representation *within* an element.† To reduce the overall size of the assembled system matrices and save equation-solving expense, we may eliminate internal nodal degrees of freedom at the element level before assembly. This is done by a process called *condensation* [13].

As we have seen, element equations have the standard form $[K]\{x\} = \{R\}$, where $\{x\}$ is a column vector of all the degrees of freedom of the element. If the element has internal nodal degrees of freedom, we may rearrange and partition $[K]$ as follows:

$$
\begin{bmatrix} [k_{11}] & \vdots & [k_{12}] \\ -\;-\;- & + & -\;-\;- \\ [k_{21}] & \vdots & [k_{22}] \end{bmatrix} \begin{Bmatrix} \{x_1\} \\ -\;- \\ \{x_2\} \end{Bmatrix} = \begin{Bmatrix} \{R_1\} \\ -\;- \\ \{R_2\} \end{Bmatrix} \tag{5.37a}
$$

where $\{x_2\}$ is the column vector of internal nodal degrees of freedom and $\{R_2\}$ is the associated vector of resultant nodal actions or forcing functions. In expanded form these equations become

$$
[k_{11}]\{x_1\} + [k_{12}]\{x_2\} = \{R_1\} \tag{5.37b}
$$

$$
[k_{21}]\{x_1\} + [k_{22}]\{x_2\} = \{R_2\} \tag{5.37c}
$$

When equation 5.37c is solved for $\{x_2\}$ and the result is substituted into equation 5.37b, we obtain

$$
[[k_{11}] - [k_{12}][k_{22}]^{-1}[k_{21}]]\{x_1\} = \{R_1\} - [k_{12}][k_{22}]^{-1}\{R_2\} \tag{5.38}
$$

or

$$
[\tilde{K}]\{x_1\} = \{\tilde{R}\}
$$

Equation 5.38 is the condensed form that contains only the degrees of freedom associated with external nodes as unknowns. Thus in the assembly process the matrices $[\tilde{K}]$ and $[\tilde{R}]$ are used instead of $[K]$ and $[R]$. The internal degrees of freedom, having served their purpose in the element formulation, are then eliminated from further consideration. In addition to reducing the size of the assembled system matrix, the condensation process considerably reduces its bandwidth [13].

In practice, the procedure symbolized by equations 5.37 and 5.38 is actually carried out by performing a symmetric backward Gaussian elimination in equation 5.37a [13]. This procedure, which is described in detail

† Pian [8] discusses these techniques for improving the element stiffness matrices for certain types of elasticity problems.

and programed in FORTRAN by Desai and Abel [14], leads to far more efficient digital computation.

The condensation process we have applied to one element can also be applied to groups of elements to eliminate nodal degrees of freedom not lying on the boundary of the assembled group. One simple example of a case where this would be desirable is shown in Figure 5.3c. The internal node resulting when four triangular elements are assembled to form a quadrilateral element would usually be eliminated. By this same procedure it is possible to devise very complex elements from assemblies of simple elements. After the individual element equations are assembled, the internal degrees of freedom are eliminated, leaving only boundary nodes and their degrees of freedom.

A procedure called *substructuring* in structural mechanics problems offers a physical interpretation of the process of eliminating internal nodal variables. Substructuring is a method whereby a large complex structure such as a bridge, aircraft frame, ship, or automobile body is viewed as an assemblage of a small number of very complex finite elements. There are several reasons for employing substructuring. Sometimes we may encounter complex structures requiring hundreds or thousands of elements, nodes, and degrees of freedom. If available computers cannot solve the resulting large-order matrix equations, we must divide the structure into smaller parts that can be handled. For example, in the case of an aircraft structure, we could choose to analyze separately the wings, fuselage, and tail sections. Each section would be analyzed in the usual way by discretizing it with simple finite elements; then the individual element equations would be assembled to form the coefficient matrix $[K]$ for the whole section. Since only the external nodes of the section interconnect with the external nodes of the other sections, we can use the process of condensation to eliminate the internal nodal degrees of freedom and to form a matrix $[\tilde{K}]$ for each section.

Figure 5.22 shows how substructuring was used on a practical structural design problem—the analysis of portions of the 747 aircraft [15]. Four substructures were chosen for detailed discretization, as shown in Figure 5.23. Substructuring was necessary for this problem because the number of nodes and total number of degrees of freedom posed a problem far too large for existing computers. As Table 5.4 indicates, even the substructures themselves contain many degrees of freedom. But after the stiffness matrices for the substructures had been assembled, the internal degrees of freedom not interconnecting with the other substructures were eliminated by condensation and the problem became more tractable.

The substructures themselves may be viewed as very complex elements with many boundary nodes. Such elements and their matrix equations may then be stored for later use in similar problems. It is important to note that

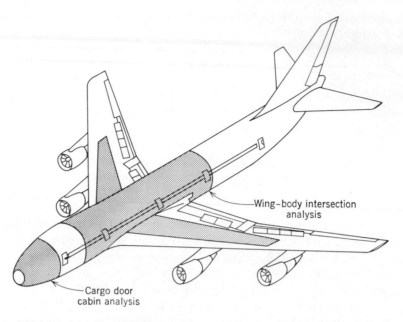

Figure 5.22. Portions of an aircraft structure treated by substructuring [15] (a) Section of the 747 analyzed by the finite element method.

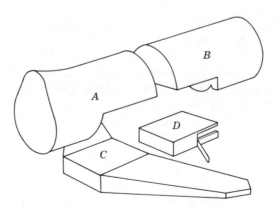

Figure 5.22 (*continued*). (*b*) Schematic of the individual substructures (complex elements).

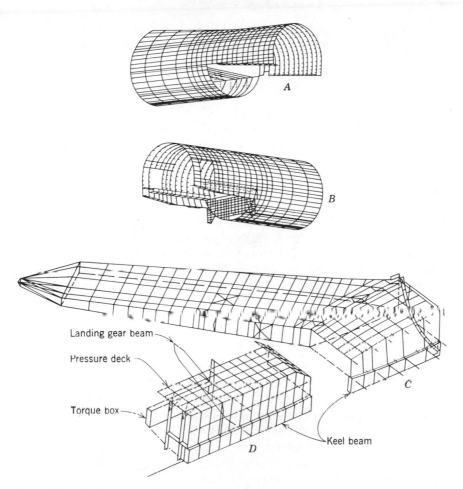

Figure 5.23. Finite element idealizations of the individual substructures indicated in Figure 5.22b.

Table 5.4. Substructure statistics for the 747 problem [15]

Substructure	Nodes	Elements	Simultaneous equations
A	1009	2897	1003
B	1014	2728	1017
C	1060	3546	~6000
D	894	2526	~5000

the concept of substructuring and the formulation of complex elements, though originally devised for structural problems, can be used in the finite element analysis of any continuum problem.

5.8   TWO-DIMENSIONAL ELEMENTS

5.8.1   Elements for C^0 problems

In this section we shall discuss some families of two-dimensional elements that can be used for problems requiring only continuity of the field variable ϕ at element interfaces. For such problems we usually choose the nodal values of ϕ to be the degrees of freedom of the element. To ensure interelement continuity we require that the number of nodes along a side of the element and hence the number of nodal values of ϕ be sufficient to determine uniquely the variation of ϕ along that side. For example, if ϕ is assumed to have a quadratic variation within the element and retains its quadratic behavior along the element sides, then three values of ϕ or three external nodes must lie along each side.

Although we shall consider only triangular and rectangular element families, the number of elements capable of satisfying C^0 continuity is infinite—infinite because we can continue to add nodes and degrees of freedom to the elements to form ever-increasing higher-order elements. In general, as the complexity of the elements is increased by adding more nodes and more degrees of freedom and using high-order polynomials, the number of elements and total number of degrees of freedom needed to achieve a given accuracy in a given problem are less than would be required if simpler elements were used. But this does not suggest that we should always favor higher-order over lower-order elements. The computational effort saved by having fewer degrees of freedom for the assembled matrices may be overshadowed by the increased effort required to formulate and compute the individual element equations.

Unfortunately, no general guidelines for choosing the optimum element for a given problem can be presented because the type of element that yields good accuracy with low computing time is problem dependent. For C^0 problems, elements that require polynomials of order greater than 3 are rarely used because little additional accuracy is gained for the extra effort expended. Also, if we need to model a complicated boundary, it is more advantageous to use a large number of simple elements than a few complex ones.

Triangular elements

The original three-node triangle formulated by Turner et al. [16] is only the first of an infinite series of triangular elements that can be specified. Figure 5.24 shows a portion of the family of higher-order elements obtained by assigning additional external and interior nodes to the triangles. Note that each element in this series has a sufficient number of nodes (degrees of freedom in this case) to specify uniquely a complete polynomial of the order necessary to give C^0 continuity. Hence the compatibility, completeness, and geometric isotropy requirements are satisfied. For example, in Figure 5.24c the element contains ten nodes and each side has four degrees of freedom. This is enough to specify a cubic variation of ϕ within the element and along the element boundaries. In general, for triangles with nodes arrayed as in Figure 5.24, a complete nth-order polynomial requires $\frac{1}{2}(n+1)(n+2)$ nodes for its specification.

For the three-node triangle the linear variation of ϕ is written as

$$\phi(x,y) = \alpha_1 + \alpha_2 x + \alpha_3 y = \lfloor 1\ x\ y \rfloor \{\alpha\} = \lfloor P \rfloor \{\alpha\}$$

and by evaluating this expression at each node we find, as before,

$$\{\phi\} = [G]\{\alpha\}$$

According to the procedure of Section 5.4.3, we may write

$$\phi = \lfloor P \rfloor [G]^{-1}\{\phi\} = \lfloor N \rfloor \{\phi\}$$

Hence

$$\lfloor N \rfloor = \lfloor P \rfloor [G]^{-1}$$

where the $N_i = L_i$, the area coordinates for the triangle.

For the second element of the series, the six-node triangular element introduced by Veubeke [17], the quadratic variation of ϕ is written as

$$\phi(x,y) = \alpha_1 + \alpha_2 x + \alpha_3 y + \alpha_4 x^2 + \alpha_5 xy + \alpha_6 y^2$$
$$= \lfloor 1\ x\ y\ x^2\ xy\ y^2 \rfloor \{\alpha\} = \lfloor P \rfloor \{\alpha\}$$

And by the same procedure the interpolation functions are

$$\lfloor N \rfloor = \lfloor P \rfloor [G]^{-1}$$

For the ten-node triangle, the cubic variation of ϕ is written as

$$\phi(x,y) = \alpha_1 + \alpha_2 x + \alpha_3 y + \alpha_4 x^2 + \alpha_5 xy + \alpha_6 y^2$$
$$+ \alpha_7 xy^2 + \alpha_8 yx^2 + \alpha_9 x^3 + \alpha_{10} y^3$$

Again,

$$\lfloor N \rfloor = \lfloor P \rfloor [G]^{-1}$$

The calculations represented symbolically here may be carried out for all the triangular elements of this series because it can be shown that for every element $[G]^{-1}$ always exists [9]. But for this family of elements it is possible to establish interpolation functions directly, thereby avoiding evaluation of $[G]^{-1}$ for each element. Irons et al. [9] demonstrated by using geometric reasoning that, if the interpolation functions for one element of

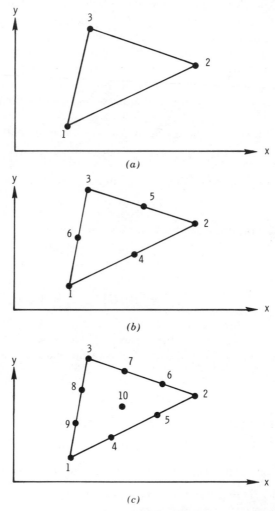

Figure 5.24. Linear and higher-order triangular elements with ϕ specified at the nodes. (The nodes along any line are equally spaced.) (*a*) Linear (3 nodes). (*b*) Quadratic (6 nodes). (*c*) Cubic (10 nodes).

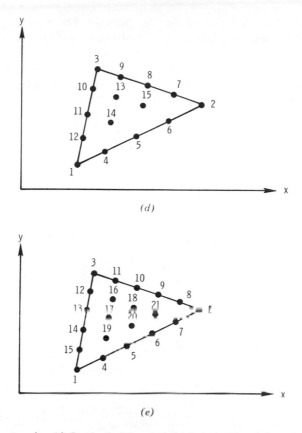

Figure 5.24 (*continued*). (*d*) Quadratic (15 nodes). (*e*) Quintic (21 nodes).

the series are known, it is possible to establish a recurrence relation to find the interpolation functions for the next higher-order element. This means that we can begin with the interpolation functions of the linear triangle, and then derive the interpolation functions for *all* the higher-order triangles. Although the process is conceptually appealing, it requires some judicious scaling and manipulating. For these reasons we shall discuss in detail a more straightforward procedure advanced by Silvester [18].

Seeking an orderly method for designating nodes within higher-order triangles, Silvester introduced a triple-index numbering scheme. After we discuss this numbering scheme, we may use it to express the interpolation functions for any order of triangle.

The nodes of the elements in Figure 5.24 can be given the three-digit label $\alpha\beta\gamma$, where α, β, and γ are integers satisfying $\alpha + \beta + \gamma = n$, and n is the order of the interpolation polynomial for the triangle. These integers designate constant coordinate lines in the area coordinate system and indicate the number of steps or levels by which a particular node is located from a side of the triangle. Figure 5.25a shows this designation for a typical node within a triangle, while Figure 5.25b shows the complete specification of the nodes for a ten-node cubic triangle (the interior node, for instance, has the designation 111). We may use the same digit notation for the interpolation functions for the element. Employing a triple subscript, we may write $N_{\alpha\beta\gamma}(L_1, L_2, L_3)$ to denote the interpolation function for node $\alpha\beta\gamma$ as a function of the area coordinates L_1, L_2, and L_3.

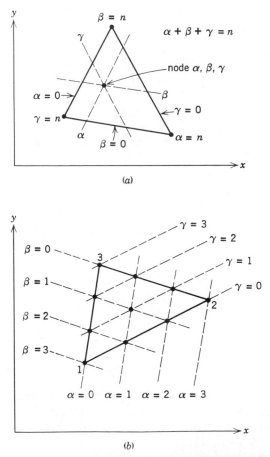

Figure 5.25. Node identification with digits. (a) Three-digit identification of a node within a triangle. (b) Example of digit node identification for the 10-node cubic triangle.

Silvester [18] has shown that the interpolation functions for an nth-order triangular element may be expressed by the following simple and convenient formula:

$$N_{\alpha\beta\gamma}(L_1, L_2, L_3) = N_\alpha(L_1)N_\beta(L_2)N_\gamma(L_3)$$

where

$$N_\alpha(L_1) = \prod_{i=1}^{\alpha}\left(\frac{nL_1 - i + 1}{i}\right), \quad \alpha \geq 1 \qquad (5.39)$$

$$= 1, \qquad\qquad\qquad \alpha = 0 \qquad (5.40)$$

For $N_\beta(L_2)$ and $N_\gamma(L_3)$ the formula has the same form. The symbol π signifies, as before, the product of all the terms. For example,

$$\prod_{i=1}^{4}(i^2 + 1) = (1^1 + 1)(2^2 + 1)(3^2 + 1)(4^2 + 1)$$

$$= \quad 2 \ \times \ 5 \ \times \ 10 \ \times \ 17 \quad = 1700$$

Equations 5.39 and 5.40 now provide the means for constructing the interpolation functions for all higher-order elements of this series. Suppose, for example, that we want to develop the interpolation functions for the six-node triangular element. Since the functions are quadratic, $n - 2$, and with the three-digit node notation the interpolation functions we seek are N_{200}, N_{020}, N_{002} and N_{101}, N_{110}, N_{011} for the three corner nodes and the three side nodes, respectively. Consider N_{020}, that is, $\alpha - 0$, $\beta = 2$, $\gamma = 0$. From equation 5.40

$$N_\alpha = N_0 = 1$$

$$N_\beta = N_2 = \prod_{i=1}^{2}\left(\frac{2L_2 - i + 1}{i}\right)$$

$$= \left(\frac{2L_2 - 1 + 1}{1}\right)\left(\frac{2L_2 - 2 + 1}{2}\right)$$

$$= L_2(2L_2 - 1)$$

$$N_\gamma = N_0 = 1$$

Hence

$$N_{020} = N_0(L_1)N_2(L_2)N_0(L_3)$$
$$= L_2(2L_2 - 1)$$

$$(5.41a)$$

Similarly, it is easy to show that

$$
\begin{aligned}
N_{200} &= L_1(2L_1 - 1) \\
N_{020} &= L_3(2L_3 - 1) \\
N_{101} &= 4L_3L_1 \\
N_{110} &= 4L_1L_2 \\
N_{011} &= 4L_2L_3
\end{aligned}
\qquad (5.41b)
$$

As an exercise the reader may want to show that for the ten-node cubic triangle the interpolation functions are

$$
\begin{aligned}
N_{100} &= \tfrac{1}{2}(3L_1 - 1)(3L_1 - 2), \text{ etc.,} && \text{for side nodes} \\
N_{120} &= \tfrac{9}{4}L_1L_2(3L_1 - 1), \text{ etc.,} && \text{for corner nodes} \\
N_{111} &= 27L_1L_2L_3 && \text{for the interior node}
\end{aligned}
\qquad (5.42)
$$

When we use these interpolation functions, the element equations will contain derivatives and integrals of the area coordinates L_1, L_2, L_3. These may be easily evaluated by referring to equations 5.29 and 5.30.

An alternative way to obtain a cubic variation of ϕ over a triangle is to specify ϕ, $\partial\phi/\partial x$, and $\partial\phi/\partial y$ at the three corner nodes and ϕ alone at an interior node, as shown in Figure 5.26. For this case ten conditions are available for uniquely specifying a cubic polynomial. Thus, $\phi = \lfloor N \rfloor \{\phi\}$, but the vector of nodal values of degrees of freedom now becomes

$$
\{\tilde{\phi}\}^T = \lfloor \phi_1 \quad \phi_{x_1} \quad \phi_{y_1}, \phi_2 \quad \phi_{x_2} \quad \phi_{y_2}, \phi_3 \quad \phi_{x_3} \quad \phi_{y_3}, \quad \phi_4 \rfloor
\qquad (5.43)
$$

By writing the complete cubic expansion for ϕ and evaluating it at each of the nodes for each degree of freedom, we find

$$
\begin{Bmatrix} \phi_1 \\ \phi_{x_1} \\ \phi_{y_1} \\ \phi_2 \\ \phi_{x_2} \\ \phi_{y_2} \\ \phi_3 \\ \phi_{x_3} \\ \phi_{y_3} \\ \phi_4 \end{Bmatrix}
=
\begin{bmatrix}
1 & x_1 & y_1 & x_1y_1 & x_1^2 & y_1^2 & y_1x_1^2 & x_1y_1^2 & x_1^3 & y_1^3 \\
0 & 1 & 0 & y_1 & 2x_1 & 0 & 2y_1x_1 & y_1^2 & 3x_1^2 & 0 \\
0 & 0 & 1 & x_1 & 0 & 2y_1 & x_1^2 & 2x_1y_1 & 0 & 3y_1^2 \\
\cdot & \cdot & \cdot & \cdot & \cdot & \cdot & \cdot & \cdot & \cdot & \cdot \\
\cdot & \cdot & \cdot & \cdot & \cdot & \cdot & \cdot & \cdot & \cdot & \cdot \\
\cdot & \cdot & \cdot & \cdot & \cdot & \cdot & \cdot & \cdot & \cdot & \cdot \\
\cdot & \cdot & \cdot & \cdot & \cdot & \cdot & \cdot & \cdot & \cdot & \cdot \\
\cdot & \cdot & \cdot & \cdot & \cdot & \cdot & \cdot & \cdot & \cdot & \cdot \\
\cdot & \cdot & \cdot & \cdot & \cdot & \cdot & \cdot & \cdot & \cdot & \cdot \\
1 & x_4 & y_4 & x_4y_4 & x_4^2 & y_4^2 & y_4x_4^2 & x_4y_4^2 & x_4^3 & y_4^3
\end{bmatrix}
\begin{Bmatrix} \alpha_1 \\ \alpha_2 \\ \alpha_3 \\ \alpha_4 \\ \alpha_5 \\ \alpha_6 \\ \alpha_7 \\ \alpha_8 \\ \alpha_9 \\ \alpha_{10} \end{Bmatrix}
$$

or

$$
\{\tilde{\phi}\} = [G]\{\alpha\}
$$

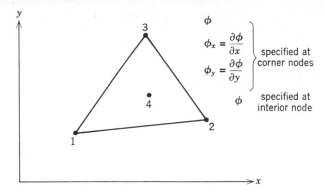

Figure 5.26. An alternative form of cubic interpolation over a triangle.

By the usual procedure,

$$\lfloor N \rfloor = \lfloor P \rfloor [G]^{-1} = \lfloor 1 \quad x \quad y \quad xy \quad x^2 \quad y^2 \quad yx^2 \quad xy^2 \quad x^3 \quad y^3 \rfloor [G]^{-1}$$

From Felippa and Clough [13]

$$\{N\} = \begin{Bmatrix} L_1^2(3 - 2L_1) - 7L_1 L_2 L_3 \\ L_1^2(x_{21}L_2 - x_{13}L_3) + (x_{13} - x_{21})L_1 L_2 L_3 \\ L_1^2(y_{21}L_2 - y_{13}L_3) + (y_{13} - y_{21})L_1 L_2 L_3 \\ \cdots\cdots\cdots\cdots\cdots\cdots\cdots\cdots\cdots\cdots\cdots\cdots\cdots \\ \cdots\cdots\cdots\cdots\cdots\cdots\cdots\cdots\cdots \\ 27L_1 L_2 L_3 \end{Bmatrix} \qquad (5.44)$$

where

$$x_{ij} = x_i - x_j \qquad \text{and} \qquad y_{ij} = y_i - y_j$$

In some applications this type of cubic formulation is more convenient than the previous one, where only ϕ is specified at the nodes [13].

Rectangular elements

Interpolation functions for rectangular elements with sides parallel to the global axes are easily developed using the Lagrangian interpolation concepts discussed in Sections 5.6.1. Although we discussed Lagrange interpolation in only one dimension, we may generalize these concepts to two or more dimensions simply by forming products of the functions which hold for the individual one-dimensional coordinate directions.

Suppose, for instance, that we wish to derive the four interpolation functions for the four-node rectangle in Figure 5.27. We may proceed as in

Elements and Interpolation Functions

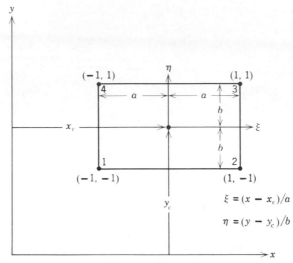

Figure 5.27. A rectangular element showing the relation between local and global coordinates. This local coordinate system is convenient for developing Lagrangian interpolation functions for rectangular elements.

Section 5.4.3, but an easier way is to use Lagrangian interpolation functions and write them directly. After we define the local coordinates as shown in Figure 5.27, we may write

$$\phi(\xi, \eta) = N_1(\xi, \eta)\phi_1 + N_2(\xi, \eta)\phi_2 + N_3(\xi, \eta)\phi_3 + N_4(\xi, \eta)\phi_4$$

where

$$N_1(\xi, \eta) = L_1(\xi)L_1(\eta), \qquad N_2(\xi, \eta) = L_2(\xi)L_2(\eta), \qquad \text{etc.}$$

and the L_i are the Lagrange polynomials defined in equation 5.33.

Interpolation functions formed as products in this way satisfy the requirements of possessing unit value at the node for which they are defined and zero value at the other nodes. For instance, $L_1(\xi_1) = 1$ and $L_1(\xi_2) = L_1(\xi_3) = L_1(\xi_4) = 0$. Also, L_1 varies linearly between nodes 1 and 2, and hence preserves C^0 continuity along edge 1–2. Similar comments hold for the other first-order Lagrange polynomials for the other nodes. Referring to equation 5.33, we can write the explicit expressions at once:

$$N_1(\xi, \eta) = L_1(\xi)L_1(\eta) = \frac{\xi - \xi_2}{\xi_1 - \xi_2} \times \frac{\eta - \eta_4}{\eta_1 - \eta_4} = \frac{\xi - 1}{-1 - 1} \times \frac{\eta - 1}{-1 - 1} \qquad (5.45a)$$

$$N_2(\xi, \eta) = L_2(\xi)L_2(\eta) = \frac{\xi - \xi_1}{\xi_2 - \xi_1} \times \frac{\eta - \eta_3}{\eta_2 - \eta_4} = \frac{\xi + 1}{1 - 1(-1)} \times \frac{\eta - 1}{-1 - 1}$$

$$(5.45b)$$

$$N_3(\xi, \eta) = L_3(\xi)L_3(\eta) = \frac{\xi - \xi_3}{\xi_3 - \xi_4} \times \frac{\eta - \eta_3}{\eta_3 - \eta_2} = \frac{\xi - 1}{1 - (-1)} \times \frac{\eta - 1}{1 - (-1)}$$

(5.45c)

$$N_4(\xi, \eta) = L_4(\xi)L_4(\eta) = \frac{\xi - \xi_4}{\xi_4 - \xi_3} \times \frac{\eta - \eta_4}{\eta_4 - \eta_1} = \frac{\xi + 1}{-1 - 1} \times \frac{\eta - 1}{1 - (-1)}$$

(5.45d)

These interpolation functions are called *bilinear* for obvious reasons.

Other higher-order rectangular elements can be formulated in precisely the same way [19,20]. Figure 5.28 shows just a sample of the infinite number of elements that can be constructed. Regardless of the number of nodes in the ξ and η directions, we can write the interpolation function for node k as

$$N_k(\xi, \eta) = L_k(\xi)L_k(\eta)$$ (5.46)

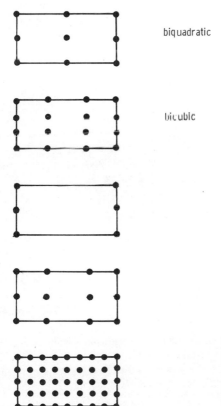

biquadratic

bicubic

Figure 5.28. A sample of elements from the infinite series of Lagrange rectangles.

and the variation of the field variable within the element as

$$\phi(\xi, \eta) = \sum_{k=1}^{n} L_k(\xi)L_k(\eta)\phi_k$$

where n is the total number of nodes assigned to the element. The functions given by equation 5.46 are expressed in terms of local coordinates, but a simple substitution recovers their form in the global coordinate system. Note that the order of the polynomials given by equation 5.46 is always that needed to ensure C^0 continuity along element boundaries.

The usefulness of Lagrangian elements is limited, however, because the higher-order elements contain a large number of interior nodes. These may be eliminated by the condensation process of Section 5.7 when desired, but extra manipulation is required. Another important aspect of Lagrangian elements is that the interpolation functions are never complete polynomials and they possess geometric isotropy only when equal numbers of nodes are used in the x and y directions.

A more useful set of rectangular elments is that known as the "serendipity family,"† devised by Ergatoudis et al. [21] and shown in Figure 5.29. These elements contain only exterior nodes, and their interpolation

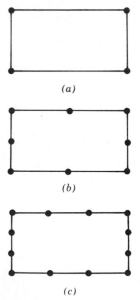

(a)

(b)

(c)

Figure 5.29. Useful elements of the "serendipity family" [21]. (a) Linear. (b) Quadratic. (c) Cubic.

† This terminology, coined in ref. 21, stems from a fairy tale of old Ceylon (once called Serendip), where there were once three princes "who in their travels were always discovering, by chance or by sagacity, [agreeable] things they did not seek."

functions were derived by inspection. In terms of the natural coordinate system defined in Figure 5.27, the interpolation functions are, from ref. 21, as follows.

1. Linear element:

$$N_i(\xi, \eta) = \tfrac{1}{4}(1 + \xi\xi_i)(1 + \eta\eta_i) \tag{5.47}$$

2. Quadratic element:

$$N_i(\xi, \eta) = \tfrac{1}{4}(1 + \xi\xi_i)(1 + \eta\eta_i)(\xi\xi_i + \eta\eta_i - 1) \quad \text{for nodes at} \xi = \pm 1, \eta = \pm 1$$
$$N_i(\xi, \eta) = \tfrac{1}{2}(1 - \xi^2)(1 + \eta\eta_i) \quad\quad\quad \text{for nodes at } \xi = 0, \eta = \pm 1$$
$$N_i(\xi, \eta) = \tfrac{1}{2}(1 + \xi\xi_i)(1 - \eta^2) \quad\quad\quad \text{for nodes at } \xi = \pm 1, \eta = 0 \tag{5.48}$$

3. Cubic element:

$$N_i(\xi, \eta) = \tfrac{1}{32}(1 + \xi\xi_i)(1 + \eta\eta_i)[9(\xi^2 + \eta^2) - 10]$$
$$\text{for nodes at } \xi = \pm 1, \eta = \pm 1$$
$$N_i(\xi, \eta) = \tfrac{9}{32}(1 + \xi\xi_i)(1 - \eta^2)(1 + 9\eta\eta_i) \quad \text{for nodes at } \xi = \pm 1, \eta = \pm \tfrac{1}{3} \tag{5.49}$$

and similarly for the other side nodes.

Each of these elements preserves C^0 continuity along the element boundaries because the interpolation functions are complete polynomials in the linear coordinates along the boundaries.

Rectangular elements, in general, are of limited use because they are not well suited for representing curved boundaries. However, an assemblage of rectangular and triangular elements, with triangular elements near the boundary, can be very effective. The appeal of rectangular elements stems from the ease with which interpolation functions can be found for them.

Arbitrary quadrilateral elements

Interpolation functions for arbitrary quadrilateral elements, such as those in Figure 5.15a, can be easily derived by introducing a special set of four homogeneous coordinates (only two of which are independent) first used by Taig and Kerr [22]. The coordinates are similar to those of Figure 5.15b in that they take on the values ± 1 along the sides of the element, but they are not orthogonal as before. For the details of this special coordinate system and the procedure for obtaining interpolation functions, the reader is referred to Zienkiewicz and Cheung [23]. Examples can be found in Felippa and Clough [13].

5.8.2   Elements for C^1 problems

Constructing two-dimensional elements that can be used for problems
requiring continuity of the field variable ϕ as well as its normal derivative
$\partial\phi/\partial n$ along element boundaries is far more difficult than constructing
elements for C^0 continuity alone. To preserve C^1 continuity, we must be
sure that ϕ and $\partial\phi/\partial n$ are uniquely specified along the element boundaries
by the degrees of freedom assigned to the nodes along a particular boundary.
As Felippa and Clough [13] point out, the difficulties arise from the following
principles:

1. The interpolation functions must contain at least some cubic terms
because the three nodal values ϕ, $\partial\phi/\partial x$, and $\partial\phi/\partial y$ must be specified at
each corner of the element.
2. For nonrectangular elements, C^1 continuity requires the specification
of at least the six nodal values ϕ, $\partial\phi/\partial x$, $\partial\phi/\partial y$, $\partial^2\phi/\partial x\,\partial y$, $\partial^2\phi/\partial x^2$, and
$\partial^2\phi/\partial y^2$ at the corner nodes. For a rectangular element with sides parallel
to the global axes, we need to specify at the corners nodes only ϕ, $\partial\phi/\partial x$,
$\partial\phi/\partial y$, and $\partial^2\phi/\partial x\,\partial y$.

It is sometimes very convenient to specify only ϕ, $\partial\phi/\partial x$, and $\partial\phi/\partial y$ at
corners; but when this is done, it is impossible to have continuous second
derivatives at the corner nodes. In general, the cross derivative $\partial^2\phi/\partial x\,\partial y$
will be directionally dependent and hence nonunique at intersections of the
sides of the element.

Analysts first began to encounter difficulties in formulating elements for
C^1 problems when they attempted to apply finite element techniques to
plate-bending problems. For such problems, the displacement of the mid-
plane of the plate for Kirchhoff plate-bending theory is the field variable
in each element, and interelement continuity of the displacement and its
slope is a desirable *physical* requirement. Also, since the functional for plate
bending involves second-order derivatives, continuity of slope at element
interfaces is a mathematical requirement because it ensures convergence as
element size is reduced. For these reasons, analysts have labored to find
elements giving continuity of slope and value, and we shall now discuss
several that they have found.

Rectangular elements

Whereas triangles are the simplest element shapes to establish C^0 continuity
in two dimensions, rectangles with sides parallel to the global axes are the
simplest element shapes of C^1 continuity in two dimensions. The reason is

that the element boundaries meet at right angles, and imposing continuity of the cross derivatives $\partial^2\phi/\partial x\,\partial y$ at the corners guarantees continuity of the derivatives that otherwise might be nonunique.

Figure 5.30 shows a four-node rectangle with ϕ, $\partial\phi/\partial x$, $\partial\phi/\partial y$, and $\partial^2\phi/\partial x\,\partial y$ specified at the corner nodes. This is a 16-degree-of-freedom element, and it would be possible to write

$$\overset{1\times16\ \ 16\times1}{\phi(x,y) = \lfloor P \rfloor \ \{\alpha\}}\tag{5.50}$$

Since a quintic polynomial contains 21 terms, 5 would have to be eliminated from $\lfloor P \rfloor$.

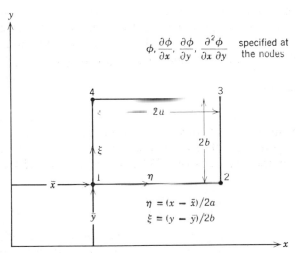

Figure 5.30.   First-order Hermite rectangle ensuring C^1 continuity with 16 degrees of freedom.

Evaluating equation 5.50 at each node, we obtain as usual

$$\overset{16\times16}{\{\tilde{\phi}\} = [G]\,\{\alpha\}}$$

But for some choices of the polynomials $\lfloor P \rfloor\{\alpha\}$, $[G]^{-1}$ may not exist.

To avoid these difficulties, we can make use of the Hermite interpolation formulas discussed in Section 5.62. Just as we have done with Lagrange formulas, we may form products of one-dimensional Hermite polynomials and obtain interpolation functions ensuring continuity of ϕ and its normal derivatives $\partial\phi/\partial n$ along element boundaries. Bogner et al. [24] have shown that the appropriate cubic interpolation polynomials in the expansion for ϕ, that is,

$$\phi(\xi, \eta) = \sum_{j=1}^{4} \left[{}_1N_j\phi_j + {}_2N_j\left(\frac{\partial\phi}{\partial x}\right)_j + {}_3N_j\left(\frac{\partial\phi}{\partial y}\right)_j + {}_4N_j\left(\frac{\partial^2\phi}{\partial x\,\partial y}\right)_j \right] \quad (5.51a)$$

are

$$\begin{aligned}
{}_1N_j(\xi, \eta) &= H_{0j}^{(1)}(\xi)H_{0j}^{(1)}(\eta) \\
{}_2N_j(\xi, \eta) &= H_{0j}^{(1)}(\xi)H_{1j}^{(1)}(\eta) \\
{}_3N_j(\xi, \eta) &= H_{1j}^{(1)}(\xi)H_{0j}^{(1)}(\eta) \\
{}_4N_j(\xi, \eta) &= H_{1j}^{(1)}(\xi)H_{1j}^{(1)}(\eta)
\end{aligned} \qquad (5.51b)$$

where ξ and η are the normalized coordinates defined in Figure 5.30. Actually, any number of nodes and any order of derivatives may be used by simply extending this procedure as indicated by Smith and Duncan [25] and Birkhoff et al. [26]. Smith [27] shows the results when Hermitian rectangles of various orders are applied to plate-bending problems.

Triangular elements

A number of investigators [28–30] have simultaneously devised complex triangular elments for C^1 continuity. By assigning 21 degrees of freedom to the element (Figure 5.31), they were able to use a complete quintic polynomial to represent the field variable ϕ. When ϕ and all first and second deriva-

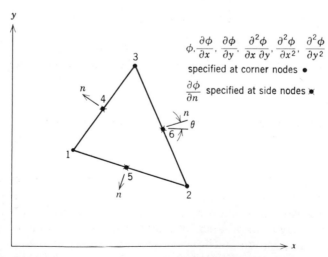

Figure 5.31. Twenty-one-degree of freedom triangle for C^1 continuity.

tives are specified at the corner nodes,† there are only 18 degrees of freedom, so 3 more are needed to specify the 21-term quintic polynomial. The 3 extra degrees of freedom are obtained by specifying the normal derivatives $\partial\phi/\partial n$ at the midside nodes. Hence the expression for ϕ becomes

$$\phi(x, y) = \alpha_1 + \alpha_2 x + \alpha_3 y + \alpha_4 xy + \alpha_5 x^2 + \cdots + \alpha_{21} y^5$$
$$= \lfloor P \rfloor \{\alpha\} \tag{5.52}$$

This element guarantees continuity of ϕ along element boundaries because, along a boundary where s is the linear coordinate, ϕ varies in s as a quintic function which is uniquely determined by six nodal values, namely, ϕ, $\partial\phi/\partial s$, and $\partial^2\phi/\partial s^2$ at each end node.

Slope continuity is also assured because the normal slope along each edge varies as a quartic function which is uniquely determined by five nodal variables, namely, $\partial\phi/\partial n$ and $\partial^2\phi/\partial n^2$ at each end node plus $\partial\phi/\partial n$ at the midside node.

To find the interpolation functions we evaluate equation 5.52 for each of the nodal variables at the nodes as usual and obtain

$$\overset{21\times21\ \ 21\times1}{\{\tilde{\phi}\} = \lfloor G \rfloor \{\alpha\}} \tag{5.53}$$

The three equations in equation 5.53 corresponding to the normal slopes at the midside nodes are found from the expression

$$\frac{\partial\phi}{\partial n} = \frac{\partial\phi}{\partial x}\frac{\partial x}{\partial n} + \frac{\partial\phi}{\partial y}\frac{\partial y}{\partial n}$$

in which we recognize that

$$\frac{\partial x}{\partial n} = -\sin\theta, \qquad \frac{\partial y}{\partial n} = \cos\theta$$

where θ is the angle of the normal vector to the x axis. Then

$$\phi(x, y) = \lfloor P \rfloor [G]^{-1}\{\tilde{\phi}\}$$

Finding $[G]^{-1}$ involves a tremendous amount of algebra; hence, in practice, $[G]^{-1}$ is evaluated numerically and the interpolation functions are never explicitly written.

Care must be taken when assembling these elements because the midside nodes have only 1 degree of freedom, whereas the corner nodes have 6 degrees

---

† The use of elements that impose continuity of second derivatives at the nodes is limited to situations in which such continuity is indeed physically possible. In some cases, such as in the analysis of plate bending when inhomogeneous materials are present, the plate properties may vary abruptly from one element to another and the imposition of continuous second derivatives or, equivalently, continuous moments normal to the plate is not permissible.

of freedom. The presence of such midside nodes is undesirable because they require special "bookkeeping" in the coding process, and they increase the bandwidth of the final matrix.

Fortunately, several investigators [28,29,31] have devised a way to eliminate the midside nodal degrees of freedom and to convert the 21-degree-of-freedom element into a more useful and convenient 18-degree-of-freedom element. The procedure is to express the normal slopes $\partial\phi/\partial n$ in terms of the 18 degrees of freedom at the corner nodes while constraining $\partial\phi/\partial n$ to have a cubic variation along the sides. The reader is referred to Brebbia and Connor [32] for the details.

Other, rather complex elements, of which several are shown in Figure 5.32, have been formulated to give C^1 continuity. We shall briefly discuss how these elements were constructed, but omit their interpolation functions as they are exceedingly complex by comparison with the others we have discussed.

The element in Figure 5.32a is a six-node triangle with only 12 degrees of freedom. It is formed by assembling three triangles [13]. The procedure is to assume a ten-term cubic polynomial over each subtriangle and then perform a C^1 matching between the elements. Such triangular elements can be further combined by constraining the variation of ϕ along one side to be linear. This eliminates node 0 and allows a quadrilateral element (Figure 5.32b) to be formed with 19 degrees of freedom. The internal degrees of freedom can then be condensed out. De Veubeke [33] started with four triangles, assumed a complete cubic variation in each one, and then performed a similar C^1 matching to form the 16-degree-of-freedom element shown in Figure 5.32c.

This listing of various elements for C^1 problems is by no means exhaustive. Instead, it is intended to indicate the difficulties in finding suitable elements and to cite some references where more information and details can be found.

Because full slope continuity is difficult to achieve, many investigators have experimented with elements that guarantee slope continuity at the nodes and satisfy the other requirements, but violate slope continuity along the element boundaries. These are the so-called *incompatible* or *nonconforming* elements. Incompatible elements have often been used with surprising success in plate-bending problems. Although the interpolation functions for plate-bending problems must satisfy the C^1 continuity and completeness requirements of Section 3.4.5 to ensure convergence, it is sometimes possible to achieve convergence without preserving C^1 continuity [34–37,40–42]. This is vividly illustrated by a number of studies that compare the performance of various compatible and incompatible elements [13,38,39]. Apparently, C^1 continuity is not always a necessary condition for convergence in C^1 problems. Experience has indicated that convergence is more dependent on the completeness than on the compatibility property of the element. Some investi-

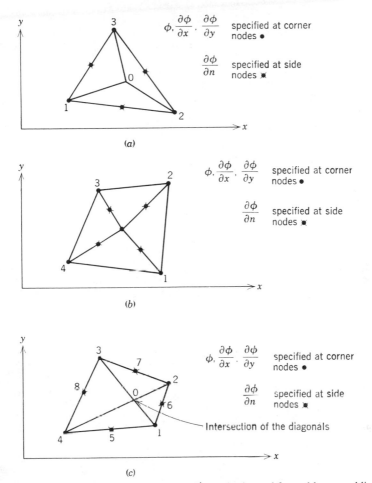

Figure 5.32.   Several complex elements giving C^1 continuity and formed by assembling other elements. (*a*) Triangular element with 12 degrees of freedom. (*b*) Quadrilateral element with 19 degrees of freedom. (*c*) Quadrilateral devised by De Vebeuke (16 degrees of freedom).

gators have achieved convergence with incompatible but complete elements [23], when no convergence was obtained with incomplete but compatible elements [43].

Table 5.5 shows a sample of incompatible elements, and Holand [44] lists several more. Any of these elements can be used in the solution of continuum problems involving functionals containing up to (and including) second-order derivatives. Confronted by this fact, the analyst may ask, "Which element should I use to solve my problem?" Unfortunately, no general answer can be given because the answer is problem dependent.

Table 5.5. Some incompatible elements for C^1 problems

Element	Nodal Variables	Order of Polynomial	Degrees of freedom per element	References	Comments
	ϕ specified at ● $\dfrac{\partial \phi}{\partial n}$ specified at *	Complete quadratic	6	39	Simplest possible plate-bending element. Gives convergent answers comparable to those for more complex triangular elements.
	$\phi, \dfrac{\partial \phi}{\partial x}, \dfrac{\partial \phi}{\partial y}$ specified at ●	Incomplete cubic: either xy^2 or x^2y term omitted	9	38	Geometric isotropy is not preserved. For certain orientations of the element $[G]^{-1}$ may not exist. Area coordinates can be used to express the interpolation functions and thus avoid $[G]^{-1}$ problem. Gives poor results
	$\phi, \dfrac{\partial \phi}{\partial x}, \dfrac{\partial \phi}{\partial y}$ specified at ●	Incomplete quartic: x^4, x^2y^2, and y^4 terms omitted	12	34, 36	Geometric isotropy is preserved. $[G]^{-1}$ given explicitly in ref. 34. Gives satisfactory results when the rectangular elements can fit the given geometry.
	$\phi, \dfrac{\partial \phi}{\partial x}, \dfrac{\partial \phi}{\partial y}$ specified at ●	Incomplete quartic	12	42	Sometimes a more convenient element for plate bending.

5.9   THREE-DIMENSIONAL ELEMENTS

5.9.1   Elements for C^0 problems

Constructing three-dimensional elements to give C^0 continuity at element interfaces follows immediately from a natural extension of the corresponding elements in two dimensions. Instead of requiring continuity of the field variable along the *edges* of the element, we require continuity on the *faces* of the elements. We must now ensure that the nodes on element interfaces uniquely define the variation of the function on the interfaces.

The three-dimensional counterparts of the triangle and the quadrilateral are the tetrahedron and the hexahedron, respectively. We shall indicate the development of interpolation functions for these elements in a fashion directly analogous to that used for the two-dimensional elements.

Tetrahedral elements

The simplest element in three dimensions is the four-node tetrahedron (Figure 5.5a), suggested first by Martin [45] in 1961 and independently by Gallagher et al. [46] in 1962. By adding more nodes to this basic tetrahedron in the manner indicated in Figure 5.33, we may form an infinite family of higher-order three-dimensional elements. If we take the nodal value of the field variable ϕ as the only degree of freedom at the nodes, each layer of nodes contains just enough degrees of freedom to specify uniquely a complete polynomial of the next highest order. To have a complete polynomial of nth order, the tetrahedron must have $\frac{1}{6}(n + 1)(n + 2)(n + 3)$ nodes when ϕ is the only nodal variable. Again, this family of tetrahedral elements satisfies the compatibility, completeness, and isotropy requirements.

A linear variation of ϕ within the element would be expressed as

$$\phi = \alpha_1 + \alpha_2 x + \alpha_3 y + \alpha_4 z$$

whereas a quadratic variation is given by

$$\phi = \alpha_1 + \alpha_2 x + \alpha_3 y + \alpha_4 z + \alpha_5 x^2 + \alpha_6 y^2$$
$$+ \alpha_7 z^2 + \alpha_8 xy + \alpha_9 xz + \alpha_{10} yz$$

and so on for the higher-order elements of this series.

Interpolation functions for this family of elements can be obtained by the usual procedure involving the calculation of $[G]^{-1}$, but Silvester [47] has introduced a greatly simplified approach by providing convenient

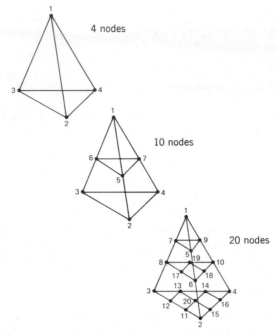

Figure 5.33.   First three elements of the tetrahedral family.

formulas analogous to those for the triangular element family. Using a four-digit subscript notation† for the interpolation functions, we may write

$$N_{\alpha\beta\gamma\delta}(L_1, L_2, L_3, L_4) = N_\alpha(L_1)N_\beta(L_2)N_\gamma(L_3)N_\delta(L_4) \qquad (5.54a)$$

where

$$N_\alpha(L_1) = \prod_{i=1}^{\alpha}\left(\frac{nL_1 - i + 1}{i}\right) \quad \text{for } \alpha \geq 1 \qquad (5.54b)$$

$$= 1 \qquad\qquad\qquad \text{for } \alpha = 0$$

and n is the order of the polynomial. The formulas for N_β, N_γ, and N_δ are similar. Silvester notes that these interpolation functions when evaluated on any face of the tetrahedron degenerate into precisely the interpolation functions defined previously for the corresponding triangular elements.

Other tetrahedral elements can be constructed in a manner similar to that for the plane triangular elements. Figure 5.34 shows a sample of two such

---

† Since this notation is an obvious extension of that used for the triangular elements, we shall not elaborate on it further.

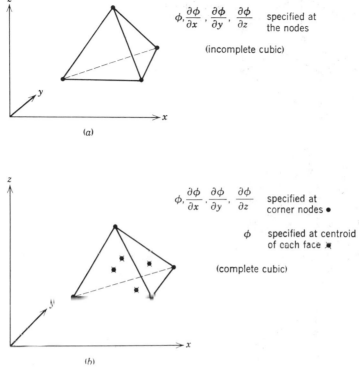

$$\phi, \frac{\partial \phi}{\partial x}, \frac{\partial \phi}{\partial y}, \frac{\partial \phi}{\partial z} \quad \text{specified at the nodes}$$

(incomplete cubic)

(a)

$$\phi, \frac{\partial \phi}{\partial x}, \frac{\partial \phi}{\partial y}, \frac{\partial \phi}{\partial z} \quad \text{specified at corner nodes} \bullet$$

$$\phi \quad \text{specified at centroid of each face} \ast$$

(complete cubic)

(b)

Figure 5.34. Examples of some tetrahedral elements for problems requiring C^0 continuity. (a) Tetrahedral element with 16 degrees of freedom. (b) Tetrahedral element with 20 degrees of freedom.

elements that several investigators [48,49] have used in three-dimensional stress analyses. Fjeld [50] discusses a number of other popular three-dimensional elements and compares their computational efficiencies.

Hexahedral elements

The concepts of Lagrange and Hermite interpolation for two-dimensional elements extend also to hexahedral elements in three dimensions. The first three members of the Lagrange hexahedral family (right prisms) are shown in Figure 5.35. As before, interpolation functions for this family of elements may be written as the product of the Lagrange polynomials in all of the orthogonal coordinate directions ξ, η, ζ (origin at the centroid of the element) [20]. Hence for node k we have

$$N_k(\xi, \eta, \zeta) = \boldsymbol{L}_k(\xi) \, \boldsymbol{L}_k(\eta) \, \boldsymbol{L}_k(\zeta) \tag{5.55}$$

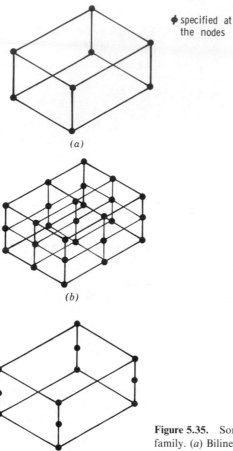

ϕ specified at
the nodes

(a)

(b)

Figure 5.35. Some hexahedral elements of the Lagrange family. (*a*) Bilinear (8 degrees of freedom). (*b*) Biquadratic (27 degrees of freedom). (*c*) Mixed linear and quadratic (12 degrees of freedom).

(c)

where it is understood that each function L_k is properly formed to account for the number of subdivisions (nodes) in the particular coordinate direction. As in the case of the Lagrange rectangular element, the Lagrange hexahedron contains undesirable interior nodes when higher-order elements are formed (quadratic, cubic, etc.). These interior nodes may be "condensed out" by the standard procedure (Section 5.7), but an alternative is to construct element shape functions directly, using only exterior nodes. Zienkiewicz et al. [51] generated the series of such elements shown in Figure 5.36. The interpolation functions for these so-called serendipity elements are incomplete polynomials and were derived by inspection. We quote the results in the following with notation analogous to that used in Section 5.8.1.

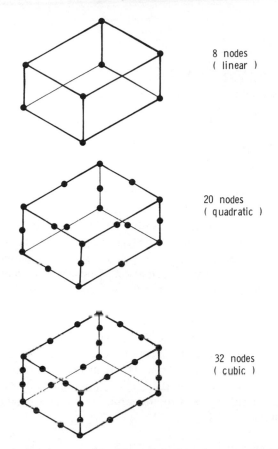

8 nodes
(linear)

20 nodes
(quadratic)

32 nodes
(cubic)

Figure 5.36.   Hexahedral elements of the "Serendipity" family containing only exterior nodes.

Linear element

$$N_i = \tfrac{1}{8}(1 + \xi\xi_i)(1 + \eta\eta_i)(1 + \zeta\zeta_i) \tag{5.56}$$

Quadratic element

Corner nodes:

$$N_i = \tfrac{1}{8}(1 + \xi\xi_i)(1 + \eta\eta_i)(1 + \zeta\zeta_i)(\xi\xi_i + \eta\eta_i + \zeta\zeta_i - 2)$$

Typical midside node, $\xi_i = 0, \eta_i = \pm 1, \zeta\zeta_i = \pm 1$:

$$N_i = \tfrac{1}{4}(1 - \xi^2)(1 + \eta\eta_i)(1 + \zeta\zeta_i)$$

$$\left.\begin{array}{c}\\ \\ \\ \\ \\ \end{array}\right\} \tag{5.57}$$

Cubic element

Corner nodes:

$$N_i = \tfrac{1}{64}(1 + \xi\xi_i)(1 + \eta\eta_i)(1 + \zeta\zeta_i)[9(\xi^2 + \eta^2 + \zeta^2) - 19]$$

Typical midside node, $\xi_i = \pm\tfrac{1}{3}, \eta_i = \pm 1, \zeta_i = \pm 1$: $\qquad$ (5.58)

$$N_i = \tfrac{9}{64}(1 - \xi^2)(1 + 9\xi\xi_i)(1 + \eta\eta_i)(1 + \zeta\zeta_i)$$

Higher-order elements of this family are seldom considered because interior nodes must be introduced to continue the construction of the interpolation functions.

The hexahedral elements we have discussed thus far have only the nodal values of the field variable defined as degrees of freedom. We may also construct hexahedral elements with ϕ and its derivatives defined at the nodes. The procedure employs products of Hermite polynomials and follows the scheme illustrated in Section 5.8.1 for two-dimenional rectangular elements.

We have discussed the direct formation of hexahedral elements, but it is also possible to form them by assembling groups of tetrahedral elements. In this connection, Felippa and Clough [13] make the useful observation that such assemblies for two- and three-dimensional elements lead to a practical difference; that is, experience has shown that hexahedral elements directly formed in the manner we have discussed seem to give more accurate results than the corresponding elements† formed by assembling tetrahedral elements. In two dimensions, on the other hand, quadrilaterals formed directly and quadrilaterals formed by assembling triangles produce essentially the same results.

Triangular prisms

Modeling complex-shaped, three-dimensional solution domains with hexahedral elements may pose some difficulties because these "brick"-shaped elements may not fit the boundary well. Rather than using a large number of small "bricks," it is advantageous to mix hexahedra and triangular prisms to obtain a good fit.

Interpolation functions for a family of triangular prism elements can be easily generated by forming products of the interpolation functions for the triangular cross sections with Lagrange functions or "serendipity" functions in the length dimension. The resulting triangular prism elements may then be assembled with Lagrange hexahedra or serendipity hexahedra, respectively. These two types of element families are shown in Figure 5.37.

---

† In this context "corresponding elements" is taken to mean elements that have the same number of external degrees of freedom.

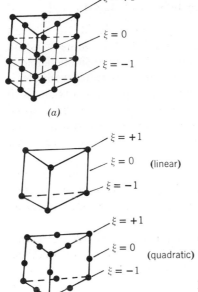

$\xi = +1$
$\xi = 0$
$\xi = -1$

(a)

$\xi = +1$
$\xi = 0$ (linear)
$\xi = -1$

$\xi = +1$
$\xi = 0$ (quadratic)
$\xi = -1$

(b)

Figure 5.37. Families of triangular prism elements. (a) One member of the Lagrange family. (b) Two members of the "serendipity" family.

As an example of the form of the interpolation functions, we quote general results from Zienkiewicz et al. [51] for the quadratic prism of the serendipity type (Figure 5.37b).

$$N_i = \tfrac{1}{2}L_i(2L_i - 1)(1 + \xi) - \tfrac{1}{2}L_i(1 - \xi^2)$$

Midsides of triangles:

$$N_i = 2L_i L_j(1 + \xi)$$

(5.59)

Midsides of rectangles:

$$N_i = L_i(1 - \xi^2)$$

The reader may wish to verify, as an exercise, that these functions are indeed suitable.

5.9.2   Elements for C^1 problems

We shall not discuss the formulation of three-dimensional elements for problems requiring C^1 continuity because no practical elements have yet been developed for this case. The excessive number of degrees of freedom required for such elements precludes their usefulness.

5.10　ELEMENTS WITH CURVED SIDES FOR C^0 PROBLEMS

Fitting a curved boundary with straight-sided elements (like those dealt with thus far in this chapter) often leads to a satisfactory representation of the boundary, but better fitting would be possible if curve-sided elements could be formulated for the task. If curve-sided elements were available, it would be permissible to use a smaller number of larger elements and still achieve a close boundary representation. Also, in practical three-dimensional problems where the great number of degrees of freedom can overburden even the largest computers, it is sometimes essential to have a means for reducing problem size by using fewer elements.

This reasoning encouraged the development of curve-sided elements. Apparently, Taig [52] was the first to introduce curved quadrilateral elements; later his ideas were generalized by Irons [53] and Ergatoudis et al. [54] to other element configurations.

The essential idea underlying the development of curved-sided elements centers on mapping or transforming simple geometric shapes in some local coordinate system into distorted shapes in the global Cartesian coordinate system and then evaluating the element equations for the curve-sided elements that result. An example will help to clarify the concepts. For the purpose of discussion we shall restrict our example to two dimensions, but all the concepts extend immediately to one dimension or three dimensions.

Suppose that we wish to represent a solution domain in x, y Cartesian coordinates by a network of curved-sided quadrilateral elements, and furthermore we desire the field variable ϕ to have a quadratic variation within each element. According to our previous discussion, if we choose the nodal values of ϕ as degrees of freedom, three nodes must be associated with each side of the element. The solution domain and the desired finite element model might appear as shown in Figure 5.38. To construct one typical element of this assemblage we focus our attention on the simpler "parent" element in the ξ, η local coordinate system shown in Figure 5.39. We know from Section 5.8.1. that this element is the second member of the serendipity family of rectangular elements, and the quadratic variation of ϕ within the element may be expressed as

$$\phi(\xi, \eta) = \sum_{i=1}^{8} N_i(\xi, \eta)\phi_i \tag{5.60}$$

where N_i are the serendipity functions given in equations 5.48. The nodes in the ξ-η plane may be mapped into corresponding nodes in the x-y plane by defining the relations

$$x = \sum_{i=1}^{8} F_i(\xi,\eta)x_i, \qquad y = \sum_{i=1}^{8} F_i(\xi, \eta)y_i \tag{5.61}$$

The extension of this mapping procedure to elements with a different number of nodes and elements in other dimensions is obvious.

For this example, the mapping functions F_i must be quadratic since the curved boundaries of the element in the x-y plane need three points for their unique specification, and the F_i should take on the proper values of unity and zero when evaluated at the nodes in the ξ-η plane. Functions meeting these requirements are precisely the quadratic serendipity interpolation functions presented in Section 5.8.1. Hence we can write

$$x = \sum_{i=1}^{8} N_i(\xi, \eta)x_i, \qquad y = \sum_{i=1}^{8} N_i(\xi, \eta)y_i \qquad (5.62)$$

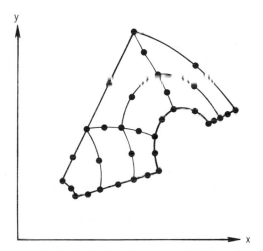

Figure 5.38. A two-dimensional solution domain represented by curved quadrilateral elements.

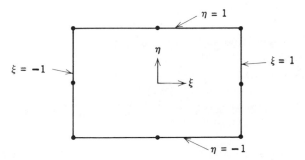

Figure 5.39. Parent rectangular element in local coordinates.

where the N_i are given by equations 5.48. The mapping defined in equations 5.62 results in a curve-sided quadratic element of the type shown in Figure 5.40. For this particular element, the functional representation of the field variable and the functional representation of its curved boundaries are expressed by interpolation functions of the same order. Curve-sided elements formulated in this way are called *isoparametric* elements. Different terminology is used to describe curve-sided elements whose geometry and field variable representations are described by polynomials of different order.†

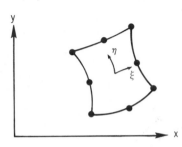

Figure 5.40. A curve-sided quadrilateral element resulting from mapping the rectangular parent element.

In contrast to isoparametric elements, we define *subparametric* elements as elements whose geometry is described by a lower-order polynomial than that used for the field variable, and *superparametric* elements as those whose geometry is described by a higher-order polynomial. Of the three categories of curve-sided elements, isoparametric elements are the most commonly used, but some forms of subparameteric and superparametric elements are employed in the stress analysis of shell structures. In this discussion we shall confine our attention to isoparametric elements.

When writing equations like equations 5.62, we assume that indeed the transformation between the local ξ, η coordinates and the global x, y coordinates is unique; that is, we assume that each point in one system has a corresponding point in the other system. If the transformation is nonunique, we can expect violent and undesirable distortions in the x, y system that may fold the curved element back upon itself. A method for checking for nonuniqueness and violent distortion will be pointed out later when we discuss the derivation of the curved-element properties.

An important consideration in the construction of curved elements is preservation of the continuity conditions in the global coordinate system.

† The number of nodes used to define a curved element may be different from the number at which the element degrees of freedom are specified.

In this regard, Zienkiewicz et al. [51] has advanced three useful guidelines:

1. If two adjacent curved elements are generated from parent elements whose interpolation functions satisfy interelement continuity, these curved elements will be continuous.

2. If the interpolation functions are given in the local coordinate system and they ensure continuity of ϕ in the parent element, ϕ will also be continuous in the curved element.

3. The isoparametric element formulation offers the attractive feature that, if the completeness criterion (Section 3.45) is satisfied in the parent element, it is automatically satisfied in the curved element.

Having described the element shape via equations of the form of equation 5.61, we face the task of evaluating the element equations by carrying out the usual integrations appearing in them. In general, the terms in the element equations will contain integrals of the form

$$\int_{A^{(e)}} f\left(\phi, \frac{\partial \phi}{\partial x}, \frac{\partial \phi}{\partial y}\right) dx \, dy \qquad (5.63)$$

where $A^{(e)}$ is the area of the curved element. Since ϕ is expressed as a function of the local coordinates ξ and η as in equation 5.60, it is necessary to express $\partial \phi / \partial x$, $\partial \phi / \partial y$, and $dx \, dy$ in terms of ξ and η also. This can be done as follows. From equation 5.60

$$\frac{\partial \phi}{\partial x} = \sum_{i=1}^{8} \frac{\partial N_i}{\partial x} \phi_i, \qquad \frac{\partial \phi}{\partial y} = \sum_{i=1}^{8} \frac{\partial N_i}{\partial y} \phi_i \qquad (5.64a)$$

Hence we must express $\partial N_i / \partial x$ and $\partial N_i / \partial y$ in terms of ξ and η. Because of the inverse form of equations 5.62 we write, by the chain rule of differentiation,

$$\frac{\partial N_i}{\partial \xi} = \frac{\partial N_i}{\partial x} \frac{\partial x}{\partial \xi} + \frac{\partial N_i}{\partial y} \frac{\partial y}{\partial \xi}$$

$$\frac{\partial N_i}{\partial y} = \frac{\partial N_i}{\partial x} \frac{\partial x}{\partial \eta} + \frac{\partial N_i}{\partial y} \frac{\partial y}{\partial \eta}$$

or

$$\begin{Bmatrix} \dfrac{\partial N_i}{\partial \xi} \\[2mm] \dfrac{\partial N_i}{\partial \eta} \end{Bmatrix} = \begin{bmatrix} \dfrac{\partial x}{\partial \xi} & \dfrac{\partial y}{\partial \xi} \\[2mm] \dfrac{\partial x}{\partial \eta} & \dfrac{\partial y}{\partial \eta} \end{bmatrix} \begin{Bmatrix} \dfrac{\partial N_i}{\partial x} \\[2mm] \dfrac{\partial N_i}{\partial y} \end{Bmatrix} = [J] \begin{Bmatrix} \dfrac{\partial N_i}{\partial x} \\[2mm] \dfrac{\partial N_i}{\partial y} \end{Bmatrix} \qquad (5.64b)$$

where $[J]$ is defined as a Jacobian matrix evaluated from equation 5.62 for each element. To find the desired derivatives, we must invert equation 5.64, and this involves finding the inverse of the Jacobian matrix as indicated:

$$
\left\{ \begin{array}{c} \dfrac{\partial N_i}{\partial x} \\[2mm] \dfrac{\partial N_i}{\partial y} \end{array} \right\} = [J]^{-1} \left\{ \begin{array}{c} \dfrac{\partial N_i}{\partial \xi} \\[2mm] \dfrac{\partial N_i}{\partial \eta} \end{array} \right\} \tag{5.65}
$$

Now with equation 5.65 expressions for $\partial\phi/\partial x$ and $\partial\phi/\partial y$ can be found directly. Similarly, it is easy to show that [56]

$$
dx\, dy = |J|\, d\xi\, d\eta \tag{5.66}
$$

where $|J|$ is the determinant of $[J]$. The operations indicated in equations 5.65 and 5.66 depend on the existence of $[J]^{-1}$ for each element of the assemblage, and the coordinate mapping described by equations 5.62 is unique only if $[J]^{-1}$ exists. We can test for uniqueness and acceptable distortion by evaluating $[J]$ and checking its sign. If the sign of $[J]$ does not change in the solution domain, we can be assured of an acceptable mapping.

With these transformations integrals such as expression 5.63 reduce to the form

$$
\int_{-1}^{1} \int_{-1}^{1} f'(\xi, \eta)\, d\xi\, d\eta \tag{5.67}
$$

where f' is the transformed function f.

Although the integration limits are now those of the simple parent element, the transformed integrand f' is not a simple function that permits closed-form integration. For this reason, it is necessary to resort to numerical integration, but this poses no particular difficulty. Always associated with the procedure of numerical integration is the question of how accurately the integration needs to be done to ensure convergence and to guard against making the resulting system matrix singular. Irons [53–55] provides the following guideline for isoparametric element formulations: Convergence of the finite element process should occur if the numerical integration is accurate enough to evaluate the area or volume of the curved element exactly.

The sample isoparametric element we have considered is just one of many possibilities. Actually, we can start with basic elements in any dimension, elements which may be described by local coordinates ξ, η, ζ or natural coordinates L_1, L_2, L_3, and so on, and transform these into curved elements. When the parent element such as a triangle or a tetrahedron is expressed in terms of natural coordinates, it is necessary to express one of the L's in terms of the others since not all the natural coordinates are independent. Also in this case the limits of integration must be changed to correspond to the boundaries of the element. In any event numerical integration is still necessary.

The reader can find a rather extensive state-of-the-art review of the developments and progress in formulating curve-sided elements in Zienkiewicz [57]. Later chapters on applications will illustrate the usefulness of such elements.

5.11 CLOSURE

Many different types of finite elements suitable for a wide variety of problems were presented in this chapter. Beginning with the basic ideas of polynomial interpolation functions and the selection of the order of the polynomial, we discussed the general procedure for constructing finite elements. From linear interpolation concepts we saw how local, natural coordinate systems can be established for line, triangle, and tetrahedron elements. With these natural coordinates we saw how it is possible to write down interpolations functions at once for any members of the higher-order elements in the triangular and tetrahedral families. The interpolation concepts associated with Lagrange and Hermite polynomials also permitted direct construction of elements. Though the list of elements we compiled for C^0 and C^1 problems is by no means exhaustive, it is sufficiently extensive to make possible the finite element solution of many problems of continuum mechanics. If the reader happens to encounter a problem for which special-purpose elements are needed, he should be able to devise his own elements by following the procedures we have discussed.

Now that the physical and mathematical foundations of the finite element method have been presented, we consider in the following chapters some applications of the method in the realms of solid and fluid mechanics. In these chapters on applications we shall discuss the special techniques employed for each class of problem.

REFERENCES

1. M. Golomb and M. Shanks, *Elements of Ordinary Differential Equations*, McGraw-Hill Book Company, New York, 1950.

2. B. Carnahan, H. A. Luther, and J. O. Wilkes, *Applied Numerical Methods*, John Wiley and Sons, New York, 1969.

3. J. Krahula and J. Polhemus, "Use of Fourier Series in the Finite Element Method," *AIAA J.*, Vol. 6, No. 4, April 1968.

4. S. Chakrabarti, "Trigonometric Function Representations for Rectangular Plate Bending Elements," *Int. J. Numer. Methods Eng.*, Vol. 3, No. 2,

5. R. H. Gallagher, "Analysis of Plate and Shell Structures," *Proceedings of Symposium on Application of Finite Element Methods in Civil Engineering*, ASCE-Vanderbilt University, Nashville, Tenn., November 1969.

6. P. Dunne, "Complete Polynomial Displacement Fields for the Finite Element Method," *Aeronaut. J.*, Vol. 72, March 1968.

7. R. L. Taylor, "On Completeness of Shape Functions for Finite Element Analysis," *Int. J. Numer. Methods in Eng.*, Vol. 4, No. 1, 1972, pp. 17–22.

8. T. H. H. Pian, "Derivation of Element Stiffness Matrices," *AIAA J.*, Vol. 2, No. 3, pp. 576–577.

9. B. M. Irons, J. G. Ergatoudis, and O. C. Zienkiewicz, Discussion of the Paper by P. Dunne, *Aeronaut. J.*, Vol. 72, March 1968.

10. I. Fried, "Some Aspects of the Natural Coordinate System in the Finite Element Method," *AIAA J.*, Vol. 7, No. 7, 1969, pp. 1366–1368.

11. B. M. Irons, "Numerical Integration Applied to Finite Element Methods," Conference on Use of Computers in Structural Engineering, University of Newcastle, England, 1966.

12. R. W. Clough, "Comparison of Three-Dimensional Finite Elements," in *Finite Element Methods in Stress Analysis*, I. Holland, and K. Bell (eds.), Tapir Press, Trondhein, Norway, 1969.

13. C. A. Felippa and R. W. Clough, "The Finite Element Method in Solid Mechanics," *Numerical Solution of Field Problems in Continuum Physics*, SIAM-AMS Proceedings, Vol. 2, American Mathematical Society, Providence, R.I., 1970, pp. 210–252.

14. C. S. Desai and J. F. Abel, *Introduction to the Finite Element Method*, Van Nostrand-Reinhold, New York, 1972, pp. 125–126.

15. S. D. Hansen, G. L. Anderton, N. E. Connacher, and C. S. Dougherty, "Analysis of the 747 Aircraft Wing-Body Intersection," *Proceedings of 2nd Conference on Matrix Methods in Structural Mechanics*, Wright-Patterson Air Force Base, Dayton, Ohio, October 15–17, 1968.

16. W. J. Turner, R. W. Clough, H. C. Martin, and L. S. Topp, "Stiffness and Deflection Analysis of Complex Structures," *J. Aeronaut. Sci.*, Vol. 23, 1956, pp. 805–823.

17. B. Fraeijs de Veubeke, "Displacement and Equilibrium Models in the Finite Element Method," *Stress Analysis*, John Wiley and Sons, New York, 1965, Chapter 9.

18. P. Silvester, "Higher-Order Polynomial Triangular Finite Elements for Potential Problems," *Int. J. Eng. Sci.*, Vol. 7, No. 8, pp. 849–861.

19. J. G. Ergatoudis, "Quadrilateral Elements in Plane Analysis and Introduction to Solid Analysis," M.S. Thesis, University of Wales, Swansea, U.K., 1966.

20. J. H. Argyris, K. E. Buck, I. Fried, G. Mareczek, and D. W. Scharpf, "Some New Elements for the Matrix Displacement Method," *Proceedings of 2nd Conference on Matrix Methods in Structural Mechanics*, Wright-Patterson Air Force Base, Dayton, Ohio, 1968.

21. J. G. Ergatoudis, B. M. Irons, and O. C. Zienkiewicz, "Curved Isoparametric Quadrilateral Elements for Finite Element Analysis," *Int. J. Solids Struct.*, Vol. 4, 1968, pp. 31–42.

22. I. C. Taig and R. I. Kerr, "Some Problems in the Discrete Element Representation of Aircraft Structures," in *AGARD-ograph 72*, B. F. de Veubeke (ed.), Pergamon Press, New York, 1964.

23. O. C. Zienkiewicz and Y. K. Cheung, *The Finite Element Method in Structural and Continuum Mechanics*, McGraw-Hill Book Company, New York, 1967, pp. 67–70.

24. F. K. Bogner, R. L. Fox, and L. A. Schmit, "The Generation of Inter-element—Compatible Stiffness and Mass Matrices by the Use of Interpolation Formulae," *Proceedings of 1st Conference on Matrix Methods in Structural Mechanics*, Wright-Patterson Air Force Base, Dayton, Ohio, 1968.

25. I. M. Smith and W. Duncan, "The Effectiveness of Excessive Nodal Continuities in Finite Element Analysis of Thin Rectangular and Skew Plates in Bending," *Int. J. Numer. Methods Eng.*, Vol. 2, 1970, pp. 253-258.

26. G. Birkhoff, M. H. Schultz, and R. S. Varga, "Piecewise Hermite Interpolation in One and Two Variables with Applications to Partial Differential Equations," *Numer. Math.*, Vol. II, 1968, pp. 232–256.

27. I. M. Smith, "A Finite Element Analysis for Moderately Thick Rectangular Plates in Bending," *Int. J. Mech. Sci.*, Vol. 10, 1968, pp. 563–570.

28. G. A. Butlin and R. Ford, "A Compatible Triangular Plate Bending Finite Element," *Rept.* 68-15, Eng. Dept., University of Leicester, England, October 1968.

29. J. H. Argyris, I. Fried, and D. W. Scharpf, "The TUBA Family of Plate Elements for the Matrix Displacement Method," Technical Note, *J. Roy. Aeronaut. Soc.*, Vol. 72, August 1968.

30. K. Bell, "A Refined Triangular Plate Bending Finite Element," *Int. J. Numer. Methods Eng.*, Vol. 1, No. 1, January 1969, pp. 101–122.

31. G. R. Cowper, E. Kosko, G. M. Lindberg, and M. D. Olson, "Formulation of a New Triangular Plate Bending Element," *Trans. Can. Aeronaut. Inst.*, Vol. 1, 1968, pp. 86-90.

32. C. Brebbia and J. Connor, "Plate Bending," Chapter 4 in *Finite Element Techniques in Structural Mechanics*, H. Tottenham and C. Brebbia (eds.), Stress Analysis Publishers, Southampton, England, 1971, pp. 112-114.

33. B. Fraeijs de Veubeke, "Bending and Stretching of Plates—Special Models for Upper and Lower Bounds," *Proceedings of 1st Conference on Matrix Methods in Structural Mechanics*, Wright-Patterson Air Force Base, Dayton, Ohio, 1965.

34. R. J. Melosh, "Basis for Derivation of Matrices for the Direct Stiffness Method," *AIAA J.*, Vol. 1, 1963, pp. 1631–1637.

35. R. J. Melosh, "A Stiffness Matrix for the Analysis of Thin Plates in Bending," *J. Aerospace Sci.*, Vol. 28, 1961, p. 34.

36. A. Adini and R. W. Clough, "Analysis of Plate Bending by the Finite Element Method," Report to the National Science Foundation, Grant G7337, 1961.

37. L. S. D. Morley, "A Triangular Equilibrium Element with Linearly Varying Bending Moments for Plate Bending Problems," *J. Roy. Aeronaut. Soc.*, Vol. 71, 1967, p. 715.

38. G. P. Bazeley, Y. K. Cheung, B. M. Irons, and O. C. Zienkiewicz, "Triangular Elements in Bending—Conforming and Non-conforming Solutions," *Proceedings of 1st Conference on Matrix Methods in Structural Mechanics*, Wright-Patterson Air Force Base, Dayton, Ohio, October 1965.

39. R. W. Clough and C. A. Felippa, "A Refined Quadrilateral Element for Analysis of Plate Bending," *Proceedings of 2nd Conference on Matrix Methods in Structural Mechanics*, Wright-Patterson Air Force Base, Dayton, Ohio, October 1968.

40. J. H. Argyris, "Continua and Discontinua," *Proceedings of 1st Conference on Matrix Methods in Structural Mechanics*, Wright-Patterson Air Force Base, Dayton, Ohio, October 1965.

41. L. S. D. Morley, "On the Constant Moment Plate Bending Element," *J. Strain Anal.*, Vol. 6, No. 1, 1971, pp. 20–24.

42. D. J. Dawe, "Parallelogramic Elements in the Solution of Rhombic Cantilever Plate Problems," *J. Strain Anal.*, Vol. 1, No. 3, 1966.

43. R. W. Clough and J. L. Tocher, "Finite Element Stiffness Matrices for the Analysis of Plate Bending," *Proceedings of 1st Conference on Matrix Methods in Structural Mechanics*, Wright-Patterson Air Force Base, Dayton, Ohio, October 1965.

44. I. Holand, "Stiffness Matrices for Plate Bending Elements," in *Finite Element Methods in Stress Analysis*, I. Holand and K. Bell (eds.), Tapir Press, Trondheim, Norway, 1969, p. 160.

45. H. C. Martin, "Plane Elasticity Problems and the Direct Stiffness Method," *Trend Eng. Univ. Wash.*, Vol. 13, 1961.

46. R. H. Gallagher, J. Padlog, and P. P. Bijlaar, "Stress Analysis of Heated Complex Shapes," *J. Aerospace Sci.*, 1962, p. 700.

47. P. Silvester, "Tetrahedral Polynomial Finite Elements for the Helmholtz Equation," *Int. J. Numer. Methods Eng.*, Vol. 4, No. 3, 1972, pp. 405–413.

48. J. H. Argyris, I. Fried, and D. W. Scharpf, "The TET20 and the TEA 8 Elements for the Matrix Displacement Method," *Aeronaut. J.*, Vol. 72, No. 691, July 1968, pp. 618–623.

49. Y. R. Rashid, P. D. Smith, and N. Prince, "On Further Application of the Finite Element Method to Three-Dimensional Elastic Analysis," *Proceedings of IVTAM Symposium on High Speed Computing of Elastic Structures*, University of Liege, Liege, Belgium, August 1970.

50. S. A. Fjeld, "Three-Dimensional Theory of Elasticity," in *Finite Element Methods in Stress Analysis*, I. Holand and K. Bell (eds.), Tapir Press, Trondheim, Norway, 1969, pp. 333–363.

51. O. C. Zienkiewicz, B. M. Irons, J. Ergatoudis, S. Ahmad, and F. C. Scott, "Iso-Parameter and Associated Element Families for Two- and Three-Dimensional Analysis," in *Finite Elements Methods in Stress Analysis*, I. Holand and K. Bell (eds.), Tapir Press, Trondheim, Norway, 1969, pp. 383–432.

52. I. C. Taig, "Structural Analysis by the Matrix Displacement Method," *English Electric Aviation Rept.* SO-17, 1961.

53. B. M. Irons, "Engineering Application of Numerical Integration in Stiffness Method," *AIAA J.*, Vol. 14, 1966, pp. 2035–2037.

54. J. Ergatoudus, B. Irons, and O. C. Zienkiewicz, "Curved Isoparametric Quadrilateral Elements for Finite Element Analysis," *Int. J. Solids Struct.*, Vol. 4, 1968, pp. 31–42.

55. B. M. Irons, Discussion of "Stiffness Matrices for Sector Element" by I. R. Raju and A. K. Rao, *AIAA J.*, Vol. 7, 1969, pp. 156–157.

56. W. Kaplan, *Advanced Calculus*, Addison-Wesley Publishing Company, Reading, Mass., 1952, p. 93.

57. O. C. Zienkiewicz, "Isoparametric and Allied Numerically Integrated Elements—A Review," *Proceedings of Symposium on Numerical and Computer Methods in Structural Mechanics*, University of Illinois, Urbana, Ill., September 1971.

PART TWO

6

ELASTICITY PROBLEMS

6.1 INTRODUCTION

Most applications of the finite element method to solid mechanics problems rely on the use of a variational principle to derive the necessary element properties or equations. The three most commonly used variational principles are the principle of minimum potential energy, the principle of complementary energy, and the Reissner principle. In Appendix C the form of each of these principles is presented for comparison.

The particular variational principle that we decide to use dictates the unknown field variable represented in the elements. When we invoke the potential energy principle, we must assume the form of the displacement field within each element. This is sometimes called the *displacement method* or the *compatibility method* in finite element analysis. When we use the complementary energy principle, we assume the form of the stress field, and this is commonly called the *force method* or the *equilibrium method*. To use the Riessner principle we must assume some displacements as well as some stresses. Hence this approach is known as the *mixed method*. Pian and Tong [1] have tabulated (Table 6.1) these and other variational bases of the finite element method in solid mechanics. For particular problems one principle may be more suitable than another, but for a large class of problems the displacement method is the simplest to apply. Consequently, it remains the most widely used.

The introductory nature of our exposition here precludes a discussion of the advanced approaches beyond the displacement method. Hence our derivations of element equations in this chapter rely solely on the minimum potential energy principle. Readers interested in the other methods may consult the references given in Table 6.1.

This chapter begins with the development of the general element equations for two-dimensional elasticity problems. Then these equations are specialized

197

Table 6.1 Classification of finite element methods in elasticity
(after Pian and Tong [1])

Model	Variational principle	Inside each element	Along Interelement boundary	Unknown in final equations	references
Compatible	Minimum potential energy	Continuous displacements	Displacement compatibility	Nodal displacements	2
Equilibrium	Minimum complementary energy	Continuous and equilibrating stresses	Equilibrium boundary tractions	Stress parameters / Generalized nodal displacements	3, 4 / 5
Hybrid 1	Modified complementary energy	Continuous and equilibrating stresses	Assumed compatible displacements	Nodal displacements	6
Hybrid 2	Modified potential energy	Continuous displacements	Assumed equilibrating boundary tractions	Displacement parameters and boundary forces	7
Hybrid 3	Modified potential energy	Continuous displacements	Assumed boundary tractions for each element and assumed boundary displacements	Nodal displacements	8
Mixed (Plate-bending problems)	Reissner's principle	Continuous stresses and displacements	Combinations of boundary displacements and tractions	Combination of boundary displacements and tractions	9

198

for the particular cases of plane stress, plane strain, axisymmetric solids, and plate bending. After discussing these two-dimensional problems, we consider the generalizations necessary to treat three-dimensional problems. Examples are given along the way to indicate the variety of solid mechanics problems amenable to finite element analysis.

6.2   GENERAL FORMULATION FOR TWO-DIMENSIONAL PROBLEMS

6.2.1   The variational principle

The potential energy of a two-dimensional elastic body acted upon by surface and body forces and in equilibrium can be written from equation C.22 (Appendix C) as

$$\Pi(u, v) = \frac{1}{2} \iint_A \left[\lfloor \tilde{\delta} \rfloor [B]^T [C][B]\{\tilde{\delta}\} - 2\lfloor \tilde{\delta} \rfloor [B]^T [C]\{\epsilon_0^*\} \right] t \, dA$$

$$- \iint_A \lfloor F^* \rfloor \{\tilde{\delta}\} t \, dA \quad \int_{C_1} [T^*]\{\tilde{\delta}\} \, d_0 \tag{0.1}$$

where $t = t(x, y)$ = thickness of the body (usually constant for plane strain problems),

$\{\delta\} = \begin{Bmatrix} u(x, y) \\ v(x, y) \end{Bmatrix}$ = column matrix of the components of the displacement field measured from some datum,

$$[B] = \begin{bmatrix} \dfrac{\partial}{\partial x} & 0 \\[2mm] 0 & \dfrac{\partial}{\partial y} \\[2mm] \dfrac{\partial}{\partial x} & \dfrac{\partial}{\partial y} \end{bmatrix} = \text{matrix relating strains and displacements,}$$

$[C]$ = material stiffness matrix, which takes different forms according to the problem considered,

$\{\epsilon_0^*\}$ = column vector of initial strains, which may be due to nonuniform temperature distributions, shrink fits, shot-peening, and so on,

$\lfloor F^* \rfloor = \lfloor X^*, Y^* \rfloor$ = body force components due to gravity, centrifugal action, and the like,

$\lfloor T^* \rfloor = \lfloor T_x^*, T_y^* \rfloor$ = boundary traction components acting on portion C of the boundary. These are defined per unit length for a unit thickness.

At equilibrium we know that the displacement field (u, v) in the body is such that the total system potential energy assumes a minimum value.

Having at hand a suitable variational principle, we may now develop general finite element equations for the elastic continuum. First, we imagine the continuum to be subdivided into elements of some shape; then we assume the form of the displacement function over each element. For this general formulation we need not specify the type of element or the particular displacement function. Instead we can develop the equations for the general case first and then specialize them for particular cases later.

6.2.2 Requirements for the displacement interpolation functions

Assume that the area A (Figure 6.1) is divided into M discrete elements. We may write the potential energy of the assemblage of elements as the sum of the potential energies of all elements provided that the interpolation functions expressing the variation of displacement within each element satisfy the compatability and completeness requirements specified in Chapter 3. In other words, to write

$$\Pi(u, v) = \sum_{e=1}^{M} \Pi^{(e)}(u, v) \tag{6.2}$$

and to be rigorously assured of convergence as element mesh size decreases, the interpolation functions must satisfy the following requirements.

Compatability: Since only first-order derivatives of displacement appear in the integrand of the functional for the potential energy, the interpolation

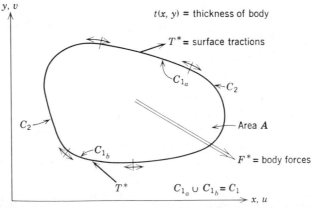

Figure 6.1. Arbitrary two-dimensional elastic body experiencing surface tractions and body forces.

functions must be such that at element interfaces the displacement is continuous. In other words, we must have C^0 continuity in displacement at element interfaces.

Completeness: The interpolation functions representing the displacement within an element must be such that rigid body displacements (uniform states of displacement) and constant strain states (uniform first-derivative states of displacement) are represented in the limit as element size is reduced; that is, we must have C^1 continuity of displacement within elements.

For plane stress and plane strain as well as three-dimensional elasticity problems, polynomial interpolation functions satisfy the compatibility and completeness requirements when the polynomials contain at least the constant and linear terms (see Figures 5.8 and 5.9).

If we intend to select interpolation functions according to the above guidelines, we may focus our attention on an isolated element. To express $\Pi^{(e)}(u, v)$, the potential energy functional for one element, in terms of discrete values of displacement components, we assume that within each element having r nodes the displacement field is approximately related to its nodal values by r interpolating functions $N_i(x, y)$, so that the distributed displacement field is expressed as

$$\{\delta\}^{(e)} = \left\{ \begin{matrix} u(x, y) \\ v(x, y) \end{matrix} \right\}^{(e)} = \left\{ \begin{matrix} \sum\limits_{i=1}^{r} N_i(x, y)u_i \\ \sum\limits_{i=1}^{r} N_i(x, y)v_i \end{matrix} \right\}^{(e)} = \left\{ \begin{matrix} \lfloor N \rfloor \{u\} \\ \lfloor N \rfloor \{v\} \end{matrix} \right\}^{(e)}. \tag{6.3}$$

Here we can see a principal difference between the finite element formulations of two-dimensional elasticity problems and the finite element formulations we considered in earlier chapters. Before, the unknown field variable was a scalar, and at each nodal point there was only one unknown—the nodal value of the field variable. In the present elasticity problem, however, the unknown field variable (the displacement) is a vector with two components. Thus, at each node point, there are two unknowns, the two nodal values of the components of displacement.

6.2.3   The element stiffness equations

By substituting equations 6.3 into the potential energy functional, we obtain potential energy of one element in terms of the nodal values of the displacement field for that element. Since the displacement field for the element has been expressed in terms of *known* interpolation functions and *unknown* nodal displacements, the potential energy functional will be similarly expressed.

Thus, for element (e), the discretized functional is

$$\Pi^{(e)}(\{\bar{\delta}\}^{(e)}) = \Pi^{(e)}(u_1, u_2, \ldots, u_r, v_1, v_2, \ldots, v_r) \tag{6.4}$$

or, more explicitly,

$$\Pi^{(e)}(\{\bar{\delta}\}^{(e)}) = \frac{1}{2} \iint_{A^{(e)}} [\lfloor \bar{\delta} \rfloor^{(e)} [B]^{T(e)} [C]^{(e)} [B]^{(e)} \{\bar{\delta}\}^{(e)}$$

$$- 2\lfloor \bar{\delta} \rfloor^{(e)} [B]^{T(e)} [C]^{(e)} \{\epsilon_0^*\}^{(e)}] t^{(e)} \, dA^{(e)}$$

$$- \iint_{A^{(e)}} \lfloor F^* \rfloor^{(e)} \{\bar{\delta}\}^{(e)} t^{(e)} \, dA^{(e)} - \int_{C_1^{(e)}} \lfloor T^* \rfloor^{(e)} \{\bar{\delta}\}^{(e)} \, dA^{(e)} \tag{6.5}$$

We know that at equilibrium the potential energy of the *system* assumes a minimum value. Because of the summation principle expressed in equation 6.2 we may carry out the minimization process element by element. We note that the potential energy of the discretized *system* assumes its minimum value when the first variation of the functional vanishes, that is,

$$\delta\Pi(u, v) = \sum_{e=1}^{M} \delta\Pi^{(e)}(u, v) = 0 \tag{6.6}$$

where

$$\delta\Pi^{(e)}(u, v) = \sum_{i=1}^{r} \frac{\partial\Pi^{(e)}}{\partial u_i} \delta u_i + \sum_{i=1}^{r} \frac{\partial\Pi^{(e)}}{\partial v_i} \delta v_i = 0 \tag{6.7}$$

But the δu_i and the δv_i are independent (not necessarily zero) variations; hence we must have

$$\frac{\partial\Pi^{(e)}}{\partial u_i} = \frac{\partial\Pi^{(e)}}{\partial v_i} = 0, \quad i = 1, 2, \ldots, r \tag{6.8}$$

for every element (e) of the system. Equations 6.8 express the conditions we use to find the element equations. It is most convenient when applying the operations of equations 6.8 to consider a typical node q of the element. Then from equation 6.5 we have at node q

$$\begin{Bmatrix} \dfrac{\partial\Pi^{(e)}}{\partial u_q} \\[2mm] \dfrac{\partial\Pi^{(e)}}{\partial v_q} \end{Bmatrix} = \{0\} = \iint_{A^{(e)}} [B]_q^{T(e)} [C]_q^{(e)} [B]_q^{(e)} \{\bar{\delta}\}^q t^{(e)} \, dA^{(e)}$$

$$- \iint_{A^{(e)}} [B]_q^{T(e)} [C]_q^{(e)} \{\epsilon_0^*\}_q^{(e)} t^{(e)} \, dA^{(e)}$$

$$- \iint_{A^{(e)}} N_q \{F^*\}_q^{(e)} t^{(e)} \, dA^{(e)} - \int_{C_{1_q}^{(e)}} N_q \{T^*\}_q^{(e)} \, ds_q^{(e)} \tag{6.9a}$$

where

$$\{\bar{\delta}\}^q = \begin{Bmatrix} u_q \\ v_q \end{Bmatrix} = \text{column vector of the two displacement components at node } q \tag{6.9b}$$

$$[B]_q^{(e)} = \begin{bmatrix} \dfrac{\partial N_q}{\partial x} & 0 \\[2ex] 0 & \dfrac{\partial N_q}{\partial y} \\[2ex] \dfrac{\partial N_q}{\partial y} & \dfrac{\partial N_q}{\partial x} \end{bmatrix} \tag{6.9c}$$

and the remaining quantities retain their previous definition except that they apply only for element (e), as the superscript indicates. The reader may wish to "unravel" the matrix form of equation 6.5 and then perform the differentiations with respect to the nodal displacements u_q and v_q to convince himself that indeed equation 6.9 results. Since the traction vector $\{T^*\}$ is a boundary effect, the last term of equation 6.9 applies only if element (e) lies on the boundary, where traction is specified.

We recognize that equation 6.9 is the force-displacement relation for node q. In matrix notation we have

$$\overset{2\times2}{\underset{}{\lfloor k\rfloor^q}}\overset{2\times1}{\{\delta\}^q} = \overset{2\times1}{\{F_0\}^q} + \overset{2\times1}{\{F_B\}^q} + \overset{2\times1}{\{F_T\}^q} = \overset{2\times1}{\{F\}^q} \tag{6.10}$$

where

$$\overset{2\times2}{[k]^q} = \int\!\!\int_{A^{(e)}} \overset{2\times3}{[B]_q^{T(e)}}\overset{3\times3}{[C]_q^{(e)}}\overset{3\times2}{[B]_q^{(e)}}t^{(e)}\,dA^{(e)} = \text{stiffness matrix at node } q \tag{6.11}$$

$$\overset{2\times1}{\{F_0\}^q} = \int\!\!\int_{A^{(e)}} \overset{2\times3}{[B]_q^{T(e)}}\overset{3\times3}{[C]^{(e)}}\overset{3\times1}{\{\epsilon_0^*\}_q^{(e)}}t^{(e)}\,dA^{(e)} = \begin{array}{l}\text{initial force vector}\\\text{at node } q\end{array} \tag{6.12}$$

$$\overset{2\times1}{\{F_B\}^q} = \int\!\!\int_{A^{(e)}} N_q(x,y)\overset{2\times1}{\{F^*\}_q^{(e)}}t^{(e)}\,dA^{(e)} = \text{nodal body force vector} \tag{6.13}$$

$$\overset{2\times1}{\{F_T\}^q} = \int_{C_1^{(e)}} N_q(x,y)\overset{2\times1}{\{T^*\}_q^{(e)}}\,ds_q^{(e)} = \begin{array}{l}\text{nodal force vector due to}\\\text{surface loading (present only for}\\\text{boundary elements.)}\end{array} \tag{6.14}$$

$$\overset{2\times1}{\{F\}^q} = \text{resultant external load vector at node } q \tag{6.15}$$

Equation 6.10 expresses the stiffness matrix associated with a typical node, but since each element has r nodes, the complete stiffness matrix for the element is a $2r \times 2r$ matrix of the form

$$\underset{2r \times 2r}{[K]^{(e)}} = \begin{bmatrix} [k]^1 & & & & \\ & [k]^2 & & & 0 \\ & & [k]^3 & & \\ & & & \ddots & \\ & 0 & & & \ddots \\ & & & & [k]^r \end{bmatrix} \qquad (6.16)$$

The arrangement of terms in the element stiffness matrix implies that the column matrix of discrete nodal displacements for the element has the form

$$\{\delta\}^{(e)} = \begin{Bmatrix} \{\hat{\delta}\}^1 \\ \{\hat{\delta}\}^2 \\ \vdots \\ \{\hat{\delta}\}^r \end{Bmatrix} = \begin{Bmatrix} u_1 \\ v_1 \\ u_2 \\ v_2 \\ \vdots \\ u_r \\ v_r \end{Bmatrix} \qquad (6.17)$$

Thus the force-displacement equations for the element take the standard form:

$$\underset{2r \times 2r}{[K]^{(e)}} \underset{2r \times 1}{\{\delta\}^{(e)}} = \underset{2r \times 1}{\{F\}^{(e)}} \qquad (6.18)$$

where

$$\underset{2r \times 1}{\{F\}^{(e)}} = \begin{Bmatrix} \{F\}^1 \\ \{F\}^2 \\ \vdots \\ \{F\}^r \end{Bmatrix} \qquad (6.19)$$

It is important to note that $\{\delta\}^{(e)}$, defined by equation 6.17, is the column vector of *discrete* nodal displacements for element (e), whereas $\{\hat{\delta}\}^{(e)}$, defined by equation 6.3, is the column vector of the *continuous* displacement field within the element.

6.2.4 The system equations

Equation 6.16 with its components given by equation 6.11 is the general form of the element stiffness matrix for two-dimensional elasticity problems. The complete force-displacement equations for the discretized elastic solid (the system) are assembled from the sets of element equations like equations 6.18. The assembly process follows the procedure described in Section 2.3. Again, the *system* equations have the same form as the *element* equations except that they are expanded in dimension to include all nodes. Hence, when the discretized system has m nodes, the system equations become

$$\underset{2m \times 2m}{[K]} \; \underset{2m \times 1}{\{\delta\}} = \underset{2m \times 1}{\{F\}} \tag{6.20}$$

where $\{\delta\}$ is a column vector of nodal displacement components for the entire system, and $\{F\}$ is the column vector of resultant nodal forces.

For the displacement formulation we have developed here, either force or displacement is known at *every* node of the system. If body forces and initial strains are absent, the vector $\{F\}$ has zero components except for the components corresponding to nodes where concentrated external *or* displacements are specified. To account for prescribed boundary displacements the system equation 6.20 are modified according to one of the procedures described in Section 2.4. After the displacement boundary conditions have been inserted we may solve equations 6.20 by using any of the standard techniques for solving linear simultaneous algebraic equations.

Determining the explicit element equations for a given problem involves evaluating equations 6.11–6.16, and, in general, this calls for numerical integration. When linear interpolation functions are used, evaluating the necessary integrals can be simplified by using the closed-form integration formula given in Table 5.2. For elements affected by body forces $\{F^*\}$ or boundary elements experiencing surface tractions $\{T^*\}$, we face the additional task of evaluating the integrals in equations 6.13 and 6.14. A simplification, known as the *lumped force* technique, can be used to facilitate evaluation of the boundary force integrals. The procedure is to find concentrated nodal force components which are the static equivalents of the given distributed loading. Then we simply allocate these to the boundary nodes in question and disregard the line integrals. This intuitive procedure leads to the correct values for the nodal forces when the interpolation functions N_i are linear; but when higher-order interpolation functions are used, the lump force technique leads to "inconsistent" nodal loads.

In contrast to the lumped force technique, we can use the *consistent force* technique. This means we introduce no intuitive reasoning about how the distributed loading should be assigned to the nodes. Instead, we evaluate the line integrals as they appear and allow the mathematics to dictate what the nodal loads should be. If the N_i are nonlinear, the nonlinear weighting of the distributed load is then naturally incorporated. The advantage of the consistent force technique is that the analysis is inherently more rational. The assurance of an upper bound on the potential energy is also preserved, but this is of little engineering value.

Treatment of the body force term depends on how the body force originates. If the body forces are due to dynamic action, D'Alembert's principle can be used to express them in terms of acceleration components. In this case we must consider a dynamics problem. Dynamics problems are discussed later in this chapter.

For the steady-state problem, once the system equations are solved for the nodal displacements, we may return to the basic relations between stress and strain, and strain and displacement, to find the stress at any point in any of the elements. Referring to equations C.4 and C.19, we may write

$$\{\sigma\}^{(e)} = [C]^{(e)}[B]^{(e)}\{\bar{\delta}\}^{(e)} \tag{6.21}$$

for the stresses due to displacement. The *actual* stress is obtained by accounting for the initial strains and adding the term $-[C]^{(e)}\{\epsilon_0^*\}^{(e)}$ to equation 6.21. If any initial stresses are present, these must also be added.

6.3 APPLICATION TO PLANE STRESS AND PLANE STRAIN [10, 11]

Often three-dimensional elasticity problems can be reduced to more tractable two-dimensional problems by recognizing that the essential descriptions of the geometry and the loading require only two independent coordinates. Plane stress and plane strain problems are two examples of this simplification. The stress analysis of very long solids such as concrete dams or walls whose geometry and loading are constant in the longest dimension falls into the category of *plane strain* problems. For these kinds of problems we may determine stresses and displacements by studying only a unit-thickness slice of the solid in the x-y plane (Figure 6.2). Similarly, if we are investigating very thin, flat plates, whose loading occurs *only* in the x-y plane of the plate and not transverse to the plane, we have a problem in *plane stress* (Figure 6.3).

The basic equations for plane stress and plane strain are summarized in Appendix C. In the following we shall derive the element equations for the linear three-node triangle in plane stress or plane strain. Substitution of the appropriate matrix of elastic constants $[C]$ specializes the equations for one case or the other.

Displacement model for a triangular element

A typical triangular element with forces and displacements defined at its nodes is shown in Figure 6.1. If we use the natural coordinates described in Section 5.5, we may express the linear variation of the displacement field

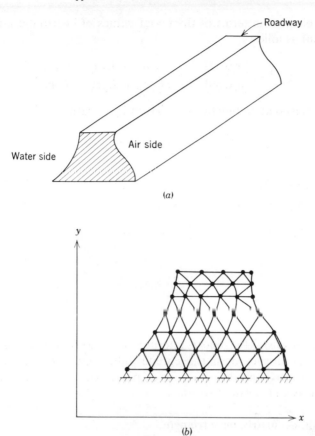

Figure 6.2.   Example of a plane strain problem. (a) Earth dam. (b) Finite element model of a
unit slice of the dam.

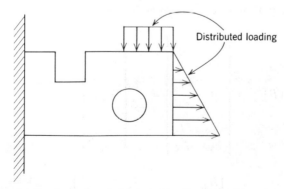

Figure 6.3.   Example of a plane stress problem; a thin, flat plate subjected to in-plane loading.

within the element in terms of the nodal values of horizontal and vertical displacement as follows:

$$\{\tilde{\delta}\}^{(e)} = \begin{Bmatrix} u(x, y) \\ v(x, y) \end{Bmatrix}^{(e)} = \begin{Bmatrix} L_1 u_1 + L_2 u_2 + L_3 u_3 \\ L_1 v_1 + L_2 v_2 + L_3 v_3 \end{Bmatrix}^{(e)} \tag{6.22}$$

where the interpolation functions are given by equation 5.17;

$$L_i = \frac{1}{2\Delta}(a_i + b_i x + c_i y); \quad i = 1, 2, 3$$

$$a_1 = x_2 y_3 - x_3 y_2, \qquad a_2 = x_3 y_1 - x_1 y_3, \qquad a_3 = x_1 y_2 - x_2 y_1$$

$$b_1 = y_2 - y_3, \qquad b_2 = y_3 - y_1, \qquad b_3 = y_1 - y_2$$

$$c_1 = x_3 - x_2, \qquad c_2 = x_1 - x_2, \qquad c_3 = x_1 - x_3$$

and

$$2\Delta = \begin{vmatrix} 1 & x_1 & y_1 \\ 1 & x_2 & y_1 \\ 1 & x_3 & y_3 \end{vmatrix}$$

As noted previously, the linear interpolation model for displacement in equation 6.22 satisfies both the compatibility and the completeness requirements. Clearly this definition for displacement results in an element of constant stress and constant strain.

Element stiffness matrix for a triangle

Having chosen the type of element and the variation of the displacement within the element, we can now determine the element equations from equation 6.10–6.19. The matrix relating strains and displacements at each node is, from equation 6.9c,

$$[B]_i^{(e)} = \begin{bmatrix} \dfrac{\partial L_i}{\partial x} & 0 \\ 0 & \dfrac{\partial L_i}{\partial y} \\ \dfrac{\partial L_i}{\partial y} & \dfrac{\partial L_i}{\partial x} \end{bmatrix} = \frac{1}{2\Delta}\begin{bmatrix} b_i & 0 \\ 0 & c_i \\ c_i & b_i \end{bmatrix}, \quad i = 1, 2, 3 \tag{6.23}$$

$$[B]_i^{T^{(e)}} = \frac{1}{2\Delta}\begin{bmatrix} b_i & 0 & c_i \\ 0 & c_i & b_i \end{bmatrix}$$

For an isotropic material in plane strain we have by definition

$$[C] = \frac{E}{(1-v)(1-2v)} \begin{bmatrix} 1-v & v & 0 \\ v & 1-v & 0 \\ 0 & 0 & \dfrac{1-2v}{2} \end{bmatrix} \qquad (C.34)$$

while for plane stress

$$[C] = \frac{E}{1-v^2} \begin{bmatrix} 1 & v & 0 \\ v & 1 & 0 \\ 0 & 0 & \dfrac{1-v}{2} \end{bmatrix} \qquad (C.41)$$

where E is Young's modulus and v is Poisson's ratio.

To be specific, let us consider the case of plane stress. Substituting equations C.41 and 6.23 into equation 6.11 gives for a typical node

$$[k]^i = \frac{1}{4\Delta^2} \frac{E}{1-v^2} \iint_{A^{(e)}} \begin{bmatrix} b_i & 0 & c_i \\ 0 & c_i & b_i \end{bmatrix} \begin{bmatrix} 1 & v & 0 \\ v & 1 & 0 \\ 0 & 0 & \dfrac{1-v}{2} \end{bmatrix} \begin{bmatrix} b_i & 0 \\ 0 & c_i \\ c_i & b_i \end{bmatrix} t^{(e)} \, dA^{(e)}$$

$$= \frac{1}{4\Delta^2} \frac{E}{1-v^2} \begin{bmatrix} b_i & 0 & c_i \\ 0 & c_i & b_i \end{bmatrix} \begin{bmatrix} 1 & v & 0 \\ v & 1 & 0 \\ 0 & 0 & \dfrac{1-v}{2} \end{bmatrix} \begin{bmatrix} b_i & 0 \\ 0 & c_i \\ c_i & b_i \end{bmatrix} \iint_{A^{(e)}} t^{(e)} \, dA^{(e)},$$

$$i = 1, 2, 3, \quad (6.24)$$

If the thickness of the element $t^{(e)}$ is constant, the integral becomes the area of the triangle, Δ, and we have

$$[k]^i = \frac{Et^{(e)}}{4\Delta(1-v^2)} \begin{bmatrix} b_i & 0 & c_i \\ 0 & c_i & b_i \end{bmatrix} \begin{bmatrix} 1 & v & 0 \\ v & 1 & 0 \\ 0 & 0 & \dfrac{1-v}{2} \end{bmatrix} \begin{bmatrix} b_i & 0 \\ 0 & c_i \\ c_i & b_i \end{bmatrix} \qquad (6.25)$$

If the thickness of the element varies, we must evaluate the integral

$$\iint_{A^{(e)}} t^{(e)} \, dA^{(e)}$$

To do this we may approximately represent the variation of $t^{(e)}$ over the element by writing

$$t^{(e)} \approx t_1{}^{(e)}L_1 + t_2{}^{(e)}L_2 + t_3{}^{(e)}L_3 \qquad (6.26)$$

that is, we can give $t^{(e)}$ linear variation over the element by expressing it in terms of the interpolation function L_i and its nodal values $t_i{}^{(e)}$. Then

$$\iint_{A^{(e)}} t^{(e)}\, dA \approx \iint_{A^{(e)}} (t_1{}^{(e)}L_1 + t_2{}^{(e)}L_2 + t_3{}^{(e)}L_3)\, dA^{(e)} \qquad (6.27a)$$

and, using the integration formula of Table 5.2, we have

$$\iint_{A^{(e)}} t^{(e)}\, dA^{(e)} \approx \frac{\Delta}{3}(t_1{}^{(e)} + t_2{}^{(e)} + t_3{}^{(e)}) \qquad (6.27b)$$

Another optional approach would be to evaluate the integral numerically using Gaussian quadrature.

Proceeding on the assumption that our triangular element has a constant thickness, we may carry out the matrix multiplications indicated in equation 6.25 and substitute the result into equation 6.16 to determine the element stiffness matrix. The result is exactly the same as obtained previously (equation 2.27) from intuitive reasoning.

Element force matrices for a triangle

The initial force vector is simply formed by substituting the prescribed initial strain vector $\{\epsilon_0^*\}$ into equation 6.12 and carrying out the integration over the element. Hence, from equation 6.12,

$$\{F_0\}^i = \iint_{A^{(e)}} \begin{bmatrix} b_i & 0 & c_i \\ 0 & c_i & b_i \end{bmatrix} \begin{bmatrix} 1 & v & 0 \\ v & 1 & 0 \\ 0 & 0 & \dfrac{1-v}{2} \end{bmatrix} \begin{Bmatrix} \epsilon_{x0}^* \\ \epsilon_{y0}^* \\ \gamma_{xy0}^* \end{Bmatrix}_i t^{(e)}\, ds^{(e)} \qquad (6.28)$$

The body force vector at node i is obtained from equation 6.13:

$$\{F_B\}^i = \iint_{A^{(e)}} L_i \begin{Bmatrix} X_i^* \\ Y_i^* \end{Bmatrix} t^{(e)}\, dA^{(e)} = \frac{t^{(e)}\Delta}{3} \begin{Bmatrix} X_i^* \\ Y_i^* \end{Bmatrix} \qquad (6.29)$$

Evaluating the formula of equation 6.14 for the surface tractions requires a note of explanation. Figure 6.4 shows several boundary elements where tractions are specified. At node i on the boundary we have from equation 6.14

$$\{F_T\}^i = \int_{C_1^{(e)}} L_i \begin{Bmatrix} X_i^* \\ Y_i^* \end{Bmatrix} ds_i = \int_0^{l_{ij}} L_i \begin{Bmatrix} X_i^* \\ Y_i^* \end{Bmatrix} ds \qquad (6.30)$$

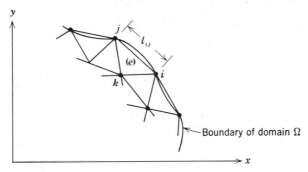

Figure 6.4.   Typical boundary elements of domain Ω.

where l_{ij} is the length of the side of the element between nodes i and j on the boundary. We shall assume that the traction components X_i^* and Y_i^* are constant over the length (average values are usually suitable). Now we recognize that L_i varies linearly in the length coordinate along the boundary in fact,

$$L_i(s) = \left(1 - \frac{y}{l_{ij}}\right)$$

where s is the length coordinate measured from node i to node j. Hence

$$\{F_T\}^i = \int_0^{l_{ij}} L_i \begin{Bmatrix} X_i^* \\ Y_i^* \end{Bmatrix} ds = \int_0^{l_{ij}} \left(1 - \frac{s}{L}\right) ds \begin{Bmatrix} X_i^* \\ Y_i^* \end{Bmatrix}$$

$$= \left(s - \frac{s^2}{2l_{ij}}\right)_0^{l_{ij}} \begin{Bmatrix} X_i^* \\ Y_i^* \end{Bmatrix} = \frac{l_{ij}}{2} \begin{Bmatrix} X_i^* \\ Y_i^* \end{Bmatrix} \tag{6.31}$$

For the whole system, equations like equation 6.31 are assembled for each element sharing node i.

Comments

In the foregoing we derived the element equations for a simple linear triangle in plane stress and plane strain. The first successful applications of the finite element method in solid mechanics relied on these equations. Since these earliest applications, many other formulations for higher-order triangular elements and quadrilateral elements have been used. Also special formulations for anisotropic materials, as well as incompressible materials,† have

---

† Standard potential energy formulations do not hold for incompressible materials because for these cases Poisson's ratio equals 0.5, and because of the term $(1 - 2v)$ in the denominator the usual matrix of elastic constants is undefined. Materials such as rubber, some soils, and solid propellants for rockets are typically considered to be incompressible.

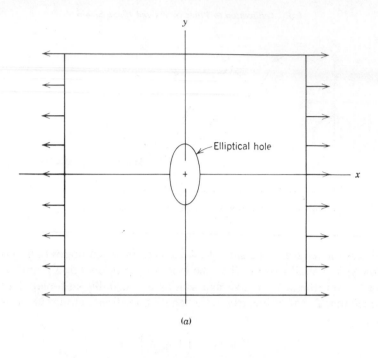

(a)

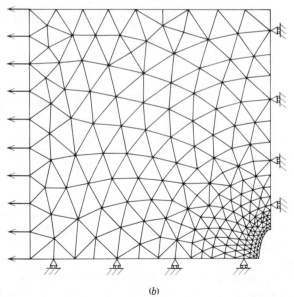

(b)

Figure 6.5. Plane stress problem with an analytical solution [13]. (a) Flat plate experiencing uniform tension. (b) Finite element model of a quarter section.

212

appeared. Readers interested in the explicit equations associated with these special topics should see the standard references [10–12] on the subject.

Example

To demonstrate that plane stress analysis by the finite element method actually works, we present a sample problem and its solution. Consider a flat plate subjected to uniform tension (Figure 6.5a). The presence of the elliptical hole in the plate makes determination of the stress distribution a nontrivial problem. Fortunately, for this problem there is a closed-form analytical solution that we can use to assess the accuracy of the finite element solution. Figure 6.5b shows a finite element model composed of constant-strain triangular elements (linear displacement field with each element). Symmetry has been used to reduce the number of elements needed for the problem. By imposing the roller boundary constraints along the lines of symmetry (interior edges of the model), we need consider only one-fourth of the plate.

Figure 6.6 illustrates qualitatively the close agreement between stresses computed in both ways. Even with a relatively coarse-element mesh the accuracy achievable is remarkable.

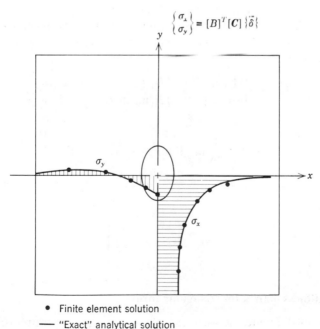

$$\left\{\begin{matrix} \sigma_x \\ \sigma_y \end{matrix}\right\} = [B]^T [C] \{\bar{\delta}\}$$

• Finite element solution

—— "Exact" analytical solution

Figure 6.6. Comparison of centerline stresses computed by the finite element method and the closed-form analytical solution [13].

6.4 APPLICATION TO AXISYMMETRIC STRESS ANALYSIS

The axisymmetric solid with axisymmetric loading offers another example of a three-dimensional elasticity problem which can be described by two independent variables. Recalling our discussion in Section 5.2, we note that axisymmetry prevails whenever derivatives of the dependent variables with respect to the circumferential coordinate are zero. In other words, when axisymmetric problems are described in terms of cylindrical coordinates r, θ, z, all variables are independent of θ.

The simplest element for general axisymmetric stress problems is a toroidal volume swept by a triangle revolved about the z axis (Figure 6.5). The nodes of this element are then viewed as *circles* in space instead of points. We may derive the element equations by careful application of the formulas in equations 6.10–6.19.

Displacement model for triangular toroid

In the axisymmetric problem only the radical and axial displacements are nonzero. For the axial and radial displacements we may use the same linear displacement model we used for the three-node triangle in plane stress and plane strain. The same natural coordinates apply if we recognize that the coordinates r and z correspond to x and y, respectively. Thus we can write the displacement field as

$$\{\tilde{\delta}\}^{(e)} = \begin{Bmatrix} u(r,z) \\ w(r,z) \end{Bmatrix}^{(e)} = \begin{Bmatrix} L_1 u_1 + L_2 u_2 + L_3 u_3 \\ L_1 w_1 + L_2 w_2 + L_3 w_3 \end{Bmatrix}^{(e)} \tag{6.32}$$

where

$$L_i(r,z) = \frac{1}{2\Delta}(a_i + b_i r + c_i z), \quad i = 1, 2, 3$$

$$a_1 = r_2 z_3 - r_3 z_2, \qquad b_1 = z_2 - z_3, \qquad c_1 = r_3 - r_2, \text{ etc.}$$

and

$$2\Delta = \begin{bmatrix} 1 & r_1 & z_1 \\ 1 & r_2 & z_2 \\ 1 & r_3 & z_3 \end{bmatrix}$$

Element stiffness matrix for triangular toroid

An important difference between the plane stress-plane strain problem and the axisymmetric problem is that in the former type we must consider four strains, ϵ_r, ϵ_θ, ϵ_z, and γ_{rz}, instead of just three. Even though the displacement v

in the θ direction is identically zero, the strain ϵ_θ is not zero because it is preceded by radial displacement. The equations relating strain and displacement in cylindrical coordinates are

$$\epsilon_r = \frac{\partial u}{\partial r}, \qquad \epsilon_\theta = \frac{u}{r}, \qquad \epsilon_z = \frac{\partial w}{\partial z}, \qquad \gamma_{rz} = \frac{\partial u}{\partial z} + \frac{\partial w}{\partial r}$$

Hence, with a linear displacement model, we may write the matrix relating strains and displacement as follows:

$$[B]_i^{(e)} = \begin{bmatrix} \dfrac{\partial L_i}{\partial r} & 0 \\[2mm] \dfrac{L_i}{r} & 0 \\[2mm] 0 & \dfrac{\partial L_i}{\partial z} \\[2mm] \dfrac{\partial L_i}{\partial z} & \dfrac{\partial L_i}{\partial r} \end{bmatrix} = \frac{1}{2\Delta} \begin{bmatrix} b_i & 0 \\[2mm] \dfrac{2\Delta L_i}{r} & 0 \\[2mm] 0 & c_i \\[2mm] c_i & b_i \end{bmatrix}, \quad i - 1, 2, 3 \qquad (6.33)$$

Note that the $[B]$ matrix no longer contains only constant terms; it also includes the function $L_i(r, z)/r$.

The matrix of elastic constants for a linear isotropic solid is

$$[C] = \frac{E}{(1 + v)(1 - 2v)} \begin{bmatrix} 1 - v & v & v & 0 \\[2mm] v & 1 - v & v & 0 \\[2mm] v & v & 1 - v & 0 \\[2mm] 0 & 0 & 0 & \dfrac{1 - 2v}{3} \end{bmatrix} \qquad (6.34)$$

Before we can apply equations 6.11–6.19 to determine the element equations we recognize that the area integrals must be changed to volume integrals and the line integral transformed to a circumferential surface integral. The substitutions to accomplish this are

$$t^{(e)} \, dA^{(e)} = 2\pi r \, dr \, dz$$

and

$$ds^{(e)} = 2\pi r \, ds$$

where s is the length coordinate measured along the sides of the triangular cross-section (see Figure 6.7). Substitution of the foregoing results into equation 6.11 then yields the stiffness relations at node i:

$$
\overset{2 \times 2}{[k]^i} = \frac{\pi E}{2\Delta^2(1 + v)(1 - 2v)} \iint_{A^{(e)}} \begin{bmatrix} b_i & \dfrac{2\Delta L_i}{r} & 0 & c_i \\[2mm] 0 & 0 & c_i & b_i \end{bmatrix}
$$

$$
\begin{bmatrix} 1 - v & v & v & 0 \\[2mm] v & 1 - v & v & 0 \\[2mm] v & v & 1 - v & 0 \\[2mm] 0 & 0 & 0 & \dfrac{1 - 2v}{3} \end{bmatrix} \begin{bmatrix} b_i & 0 \\[2mm] \dfrac{2\Delta L_i}{r} & 0 \\[2mm] 0 & c_i \\[2mm] c_i & b_i \end{bmatrix} r \, dr \, dz \qquad (6.35)
$$

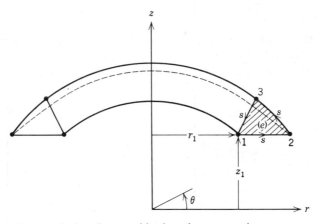

Figure 6.7. Axisymmetric ring element with triangular cross section.

It is evident from equation 6.35 that evaluation of integrals over the element cross section is now complicated by the presence of the radial and axial coordinates in the integrand. This difficulty can be overcome by using numerical integration or by tediously integrating term by term. However, a simpler procedure is to obtain an approximate average value for the matrix

$[B]$ by using centroidal values for r and z; that is, for r and z we substitute

$$\bar{r} = \tfrac{1}{3}(r_1 + r_2 + r_3) \tag{6.36}$$

and

$$\bar{z} = \tfrac{1}{3}(z_1 + z_2 + z_3), \text{ respectively}$$

The accuracy of this approximation deteriorates for elements close to the axis because the relative variation of r is greatest there. However, if we use this simplification, we have

$$[k]^i = \frac{\pi E}{2\Delta(1 + v)(1 - 2v)} \begin{bmatrix} b_i & \dfrac{2\Delta \bar{L}_i}{\bar{r}} & 0 & c_i \\ 0 & 0 & c_i & b_i \end{bmatrix}$$

$$\begin{bmatrix} 1 - v & v & v & 0 \\ v & 1 - v & v & 0 \\ v & v & 1 - v & 0 \\ 0 & 0 & 0 & \dfrac{1 - 2v}{2} \end{bmatrix} \begin{bmatrix} b_i & 0 \\ 2\Delta \bar{L}_i & 0 \\ \bar{r} & \\ 0 & c_i \\ c_i & b_i \end{bmatrix} \tag{6.37}$$

where $\bar{L}_i = a_i + b_i \bar{r} + c_i \bar{z}$.

Equation 6.16 is used to assemble the complete element stiffness matrix from the component parts given by equation 6.37.

Element force matrices for triangular toroid

Evaluation of the nodal forces follows immediately from the application of equations 6.12–6.14 when the substitutions noted previously are made. The body force usually results from gravity or centrifugal action. For the gravity force we may have, for example,

$$\{F^*\}_i^{(e)} = \begin{Bmatrix} \rho g \\ 0 \end{Bmatrix}$$

where ρ is the density of the material, whereas for centrifugal action we have

$$\{F^*\}_i^{(e)} = \begin{Bmatrix} \rho r^2 \omega \\ 0 \end{Bmatrix}$$

where ω is the rotational speed. Nodal forces are easily found by remembering that the components of the surface tractions must be multiplied by $2\pi r$; that is,

$$\{T^*\} = \begin{Bmatrix} 2\pi r R \\ 2\pi r Z \end{Bmatrix}$$

where R and Z are the components of the specified tractions.

Comments

When establishing the element mesh for axisymmetric problems, care should be taken to avoid positioning elements in such a way that two nodes have the same or nearly the same radial coordinates. If two radial coordinates are close, the calculated difference between them may be grossly in error, and if $r_i = r_j$ some of the integrals become infinite. Another problem can arise if nodes lie on the z axis, where $r = 0$, because then infinite terms also result. This can be avoided by introducing a small core hole along the axis and assigning low values to the radial coordinates that would normally be zero. The radial displacements along the core are then set equal to zero to simulate the actual condition of zero radial displacement at $r = 0$. Details of this procedure and other ones for treating nodes with $r = 0$ are described in refs. [14–17].

   If the geometry of a body is axisymmetric, but the loading on the body is not, it is still possible to perform an axisymmetric analysis. When the loads are symmetric about a plane through the axis (see Figure 6.8, for example), the loads as well as the displacement components can be represented by a one-term Fourier series of circular harmonic functions (sine and cosine). Then the corresponding terms of the series are considered individually, and the analysis proceeds in the usual manner. Wilson [14] describes the procedure in detail. For a general nonaxisymmetric loading the displacement components and loads are expressed in terms of a double Fourier series in sine and cosine. The orthogonality property of these trigonometric functions then leaves only $\sin^2 n\theta$ and $\cos^2 n\theta$ terms when integration is performed, and there result $N + 1$ independent *sets* of equations of the form

$$[K]_i\{\delta\}_i = \{F\}_i, \quad i = 0, 1, 2, \ldots, N$$

where N is the number of terms in the assumed series. After these sets of equations are solved, we return to the series expressions for u and v to find the total displacement field.

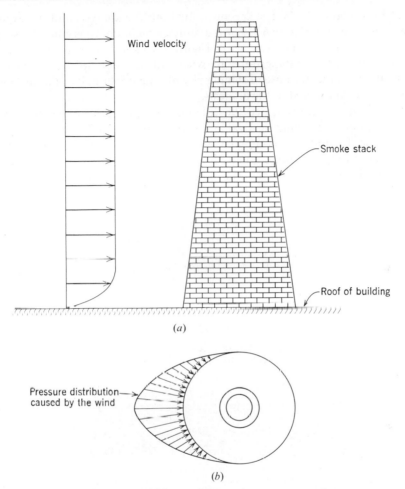

Wind velocity

Smoke stack

Roof of building

(a)

Pressure distribution—
caused by the wind

(b)

Figure 6.8.   Typical situation in which an axisymmetric structure experiences an asymmetric loading. (a) Smoke stack atop a building. (b) Plan view showing the wind loading on the stack.

Example

The stress distribution that occurs around the thread roots of a bolt-nut system is an important factor in the design of these types of mechanical fasteners. To study the fundamental nature of these stress distributions, Maruyama [18] considered a V-grooved rod under tensile loading. He determined the stress distribution surrounding the annular V-groove by first carrying out an axisymmetric finite element analysis and then conducting some special experiments.

The geometries of the test specimen for both the analysis and the experiment are shown in Figure 6.9. For the finite element analysis constant-strain triangular ring elements were used. Figure 6.10 shows the successive mesh refinement used near the groove tip, where stress gradients are expected to be relatively large. Because of symmetry only the upper half of the specimen needs to be considered in the analysis.

Calculations were repeated for different mesh sizes near the stress concentration until the numerical values of stress stabilized. The stress assigned to a particular node was computed as the average of the stresses in all of the elements sharing that node. Figure 6.11 shows the good agreement between the experimental results and the results obtained by the finite element method.

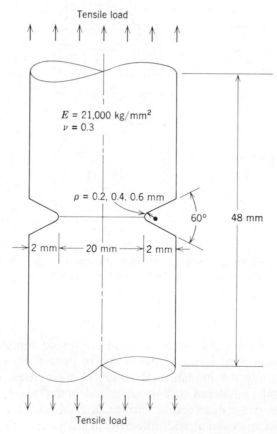

Figure 6.9. Axisymmetric test specimen analyzed by the finite element method [18].

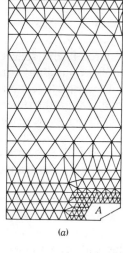

(a)

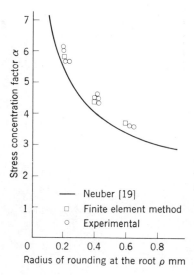

(b)

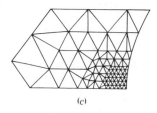

(c)

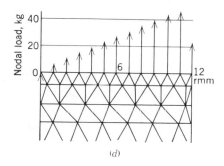

(d)

Figure 6.10. Finite element model and nodal loads for constant-strain ring elements [18]. (a) Overall mesh. (b) Magnified figure of $A(A \times 4)$. (c) Magnified figure of $B(B \times 10)$. (d) Nodal load assignment.

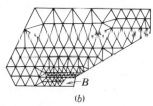

Figure 6.11. Comparison of results for the V-grooved rod in tension [18].

221

6.5   APPLICATION TO PLATE-BENDING PROBLEMS

Many practical structures such as the fuselages of aircraft and spacecraft, the decks, hulls, and bulkheads of ships, and the roofs of buildings contain sections that are thin plates. The shape and loading of these plate sections are often quite complex; but if the thickness is much smaller than the other two dimensions and deflections are small, classical plate theory as described in Appendix C can be applied. Then the problem of determining displacements and stresses in the plates becomes two-dimensional because shear deformation is neglected. The unknown field variable is $w(x, y)$, the transverse deflection of the plate.

In this section we shall show how the finite element method can be used for the analysis of plate-bending problems. Unlike the previous finite element formulations of elasticity problems, the treatment of plate-bending problems involves two levels of approximation. First, the classical plate theory itself is an approximation; second, we have the finite element model, which contains its own built-in approximations. Nevertheless, as we shall see, good prediction of displacement and stress is possible.

As we have seen in Appendix C, it is convenient to express plate-bending problems in terms of plate curvatures and line moments, that is, generalized strains and stresses, rather than the usual strains and stresses. Thus for plate bending we have the following definitions for generalized strains and stresses:

$$\{\epsilon\} \equiv \left\{ \begin{array}{c} -\dfrac{\partial^2 w}{\partial x^2} \\[2mm] -\dfrac{\partial^2 w}{\partial y^2} \\[2mm] 2\dfrac{\partial^2 w}{\partial x\, \partial y} \end{array} \right\} \tag{6.38}$$

$$\{\sigma\} \equiv \left\{ \begin{array}{c} M_{xx} \\ M_{yy} \\ M_{xy} \end{array} \right\} \tag{6.39}$$

Then the special constitutive relation is

$$\{\sigma\} = [D]\{\epsilon\}$$

where for an isotropic material

$$[D] = \frac{Eh^3}{12(1 - v^2)} \begin{bmatrix} 1 & v & 0 \\ v & 1 & 0 \\ 0 & 0 & \dfrac{1 - v}{2} \end{bmatrix} \tag{C.54}$$

We shall use the potential energy formulation and hence develop the displacement method of analysis. From equation C.58 we have, after inserting equations 6.38, 6.39, and C.54, the following potential energy functional for an isotropic plate in bending:

$$\Pi_p = \frac{Eh^3}{24(1 - v^2)} \int_A \left[\left(\frac{\partial^2 w}{\partial x^2} \right)^2 + 2v \frac{\partial^2 w}{\partial x^2} \frac{\partial^2 w}{\partial y^2} + \left(\frac{\partial^2 w}{\partial y^2} \right)^2 + 2(1 - v) \left(\frac{\partial^2 w}{\partial x\, \partial y} \right)^2 \right] dA$$

$$- \int_{S_q} q^* w\, dS - \Sigma F_i^* w_i - \Sigma M_i^* \left\{ \begin{array}{c} -\dfrac{\partial w}{\partial x} \\[2mm] \text{or} \\[2mm] \dfrac{\partial w}{\partial y} \end{array} \right. \tag{6.40}$$

where, as in Appendix C,

$\quad q^* =$ specified surface loading on surface portion S_q,
$\quad F_i^* =$ specified (may also include edge loading) concentrated forces normal to the plate,
$\quad w_i =$ displacements corresponding to forces F_i,
$\quad M_i^* =$ specified concentrated moments,

$$\left\{ \begin{array}{c} -\dfrac{\partial w}{\partial x} \\[2mm] \text{or} \\[2mm] \dfrac{\partial w}{\partial y} \end{array} \right\} = \text{angular displacements corresponding to moments } M_i^*.$$

The loading conditions for an arbitrary plate in bending are illustrated in Figure 6.12. This explicit form of Π_p reveals the order of derivatives that it contains.

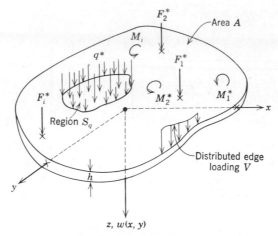

Figure 6.12. Arbitrary flat plate with various loadings defined.

6.5.1 Requirements for the displacement interpolation functions

Following the standard procedure in finite element analysis, we divide the region of the plate into elements and then interpolate the displacement [the transverse deflection $w(x, y)$] in each element. Since the potential energy functional, equation 6.40, contains second-order derivatives of $w(x, y)$, the basic compatibility and completeness requirements to be satisfied by the interpolation functions in the displacement model

$$w^{(e)}(x, y) = [N]\{w\}^{(e)} \tag{6.41}$$

are now more stringent than before. The displacement formulation of plate bending is a C^1 problem; hence we have the following requirements.

Compatibility: The interpolation functions should be such that $w^{(e)}(x, y)$ is C^1 continuous at element interfaces. This means that along element interfaces we must have continuity of w as well as continuity of $\partial w/\partial n$, the derivative of w normal to the interface. In Chapter 5 we saw that a number of different polygonal elements give C^1 continuity. All of these elements require specification of w and some or all of its second partial derivatives at corner nodes (see Section 5.8.2).

Completeness: The interpolations functions should be such that $w^{(e)}(x, y)$ is C^2 continuous within the element. Hence rigid-body, constant-slope, and constant-curvature states of deformation should be represented in the displacement model, equation 6.41. This is possible only if the terms 1, x, y, x^2, xy, and y^2 appear in the interpolation polynomials. Including the term is not essential; but if it is omitted, twisting flexure is not well represented.

If element formulations satisfying these compatibility and completeness requirements are used, we can be sure that successive solutions obtained from element mesh refinement will converge monotonically to the correct $w(x, y)$. Convergence of the first derivatives of w (curvatures), however, occurs only in the mean [19]. Though compatibility and completeness guarantee convergence, it is still possible to obtain convergent solutions using incompatible element formulations. And, in some cases, the solutions with incompatible elements are better because convergence is faster [20].

6.5.2 Stiffness matrix for a rectangular element

To illustrate the general procedure for deriving element equations for plate-bending problems, we shall consider the simplest case of a rectangular element for an isotropic plate. A rectangular element (see Figure 6.13) capable of preserving C^1 continuity must have the following degrees of freedom defined at its four corner nodes:

$$w(x, y), \qquad \frac{\partial w}{\partial x}, \qquad \frac{\partial w}{\partial y}, \qquad \frac{\partial^2 w}{\partial x \, \partial y}$$

As we saw in Section 5.8.2, the assumed displacement model, which satisfies the compatibility and completeness requirements, then takes the form

$$w^{(e)}(x, y) = \sum_{i=1}^{4} \left[{}_1N_i w_i + {}_2N_i \left(\frac{\partial w}{\partial x}\right)_i + {}_3N_i \left(\frac{\partial w}{\partial y}\right)_i + {}_4N_i \left(\frac{\partial^2 w}{\partial x \, \partial y}\right)_i \right] \quad (6.42)$$

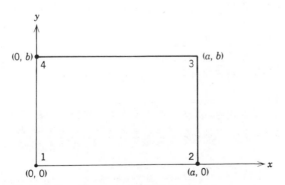

Figure 6.13. Flat rectangular plate element.

Table 6.2 Values of $\tilde{\gamma}_{ij}$, $\tilde{\lambda}_{ij}$, and $\tilde{\mu}_{ij}$ (from Bogner et al. [21])

i	j	$\tilde{\gamma}_{ij}^{(1)}$	$\tilde{\gamma}_{ij}^{(2)}$	$\tilde{\gamma}_{ij}^{(3)}$	$\tilde{\gamma}_{ij}^{(4)}$	$\tilde{\gamma}_{ij}^{(5)}$	$\tilde{\lambda}_{ij}$	$\tilde{\mu}_{ij}$
1	1	156/35	156/35	72/25	−0/1	169/1	0	0
2	1	78/35	22/35	6/25	6/5	143/6	1	0
2	2	52/35	4/35	8/25	−0/1	13/3	2	0
3	1	22/35	78/35	6/25	6/5	143/6	0	1
3	2	11/35	11/35	1/50	6/5	121/36	1	1
3	3	4/35	52/35	8/25	−0/1	13/3	0	2
4	1	11/35	11/35	1/50	1/5	121/36	1	1
4	2	22/105	2/35	1/50	2/15	11/18	2	1
4	3	2/35	22/105	8/75	2/15	11/18	1	2
4	4	4/105	4/105	8/225	−0/1	1/9	0	2
5	1	54/35	−156/35	−72/25	−0/1	117/2	2	0
5	2	27/35	−22/35	−6/25	−6/5	33/4	0	1
5	3	13/35	−78/35	−6/25	−0/1	169/12	1	1
5	4	13/70	−11/35	−1/50	−1/10	143/72	0	2
5	5	156/35	156/35	72/25	−0/1	169/1	2	1
6	1	27/35	−22/35	−6/25	−6/5	33/4	1	0
6	2	18/35	−4/35	−8/25	−0/1	3/2	2	0
6	3	13/70	−11/35	−1/50	−1/10	143/72	1	1
6	4	13/105	−2/35	−2/75	−0/1	13/36	0	0
6	5	78/35	22/35	6/25	6/5	143/6	1	0
6	6	52/35	4/35	8/25	−0/1	13/3	0	1
7	1	−13/70	52/35	6/25	−6/5	−169/12	2	2
7	2	−13/35	11/35	1/50	−6/5	−143/72	1	1
7	3	−3/35	26/35	−6/25	−0/1	−13/24	2	0
7	4	−3/70	11/105	−1/150	1/10	−11/24	2	1
7	5	−22/35	−78/35	−6/25	−0/1	−143/6	0	2
7	6	−11/35	−11/35	−1/50	−1/30	−121/36	1	2
7	7	−13/70	52/35	−8/25	−6/5	13/3	1	1
8	1	−13/105	11/35	−1/50	−6/5	−143/72	2	0
8	2	−3/70	2/35	2/75	−0/1	−13/36	1	2
8	3	−1/35	11/105	−1/150	1/10	−11/24	2	0
8	4	−11/35	2/105	−2/225	−0/1	−1/12	0	2
8	5	−11/35	−11/35	−1/50	−1/5	−121/36	1	1
8	6	−22/105	−2/35	−2/75	−2/15	−11/18	2	1
8	7	2/35	22/105	2/75	2/15	11/18	1	2
8	8	4/105	4/105	8/225	−0/1	1/9	2	2
9	1	−54/35	−54/35	72/25	−0/1	81/4	0	0
9	2	−27/35	−13/35	6/25	−0/1	39/8	1	1
9	3	−13/35	−27/35	6/25	−0/1	39/8	0	1
9	4	−13/70	−13/70	1/50	−0/1	169/144	1	2
9	5	−156/35	54/35	−72/25	−0/1	117/2	0	1
9	6	−78/35	13/35	−6/25	−0/1	169/12	1	0
9	7	22/35	−27/35	−6/25	6/5	−33/4	0	1
9	8	11/35	−13/70	1/50	1/10	−143/72	1	0
9	9	156/35	156/35	72/25	−0/1	169/1	0	1
10	1	27/35	13/35	−6/25	−0/1	−39/8	0	0
10	2	9/35	3/35	2/25	−0/1	−9/8	2	0
10	3	13/70	13/70	−1/50	−0/1	−169/144	1	0
10	4	13/210	3/70	1/150	−1/10	−13/48	2	1
10	5	78/35	−13/35	6/25	1/30	−169/12	1	0
10	6	26/35	−3/35	−2/25	−6/5	−13/4	2	1
10	7	−11/35	13/35	−1/50	−0/1	143/72	2	1
10	8	−11/105	3/70	1/150	−0/1	11/24	1	2
10	9	−78/35	−22/35	6/25	−0/1	−143/6	2	2
10	10	52/35	4/35	1/50	−0/1	13/2	0	1
11	1	13/35	27/35	−8/25	6/5	−39/8	1	1
11	2	13/70	11/35	−2/75	1/10	−169/144	2	0
11	3	3/35	26/35	−6/25	−0/1	−13/48	0	1
11	4	3/70	11/105	1/50	−6/5	−33/4	1	1
11	5	22/35	−78/35	8/25	6/5	−143/72	2	2
11	6	11/35	−11/35	1/50	−0/1	3/2	2	2
11	7	−4/35	52/35	1/50	−0/1	13/36	1	1
11	8	−2/35	11/35	8/25	−6/5	−143/6	1	1
11	9	−22/35	2/35	1/50	6/5	121/36	0	2
11	10	11/35	11/105	8/25	6/5	13/3	1	1
11	11	4/35	2/105	1/50	−0/1	169/144	0	1
12	1	−13/70	−11/35	1/50	−0/1	13/48	1	2
12	2	−13/210	−3/70	−1/150	−0/1	13/48	2	1

Table 6.2 (continued)

i	j	$\tilde{\gamma}_{ij}^{(1)}$	$\tilde{\gamma}_{ij}^{(2)}$	$\tilde{\gamma}_{ij}^{(3)}$	$\tilde{\gamma}_{ij}^{(4)}$	$\tilde{\gamma}_{ij}^{(5)}$	$\tilde{\lambda}_{ij}$	$\tilde{\mu}_{ij}$
12	3	$-3/70$	$-13/210$	$-1/150$	$-0/1$	$13/48$	1	2
12	4	$-1/70$	$-1/70$	$1/450$	$-0/1$	$1/16$	2	2
12	5	$-11/35$	$13/70$	$-1/50$	$-1/10$	$143/72$	1	1
12	6	$-11/105$	$3/70$	$1/150$	$1/30$	$11/24$	2	1
12	7	$2/35$	$-13/105$	$2/75$	$-0/1$	$-13/36$	1	2
12	8	$2/105$	$-1/35$	$-2/225$	$-0/1$	$-1/12$	2	1
12	9	$11/35$	$11/35$	$1/50$	$1/5$	$121/36$	2	1
12	10	$-22/105$	$-2/35$	$-2/75$	$-2/15$	$-11/18$	1	2
12	11	$-2/35$	$-22/105$	$-2/75$	$-2/15$	$-11/18$	2	2
12	12	$4/105$	$4/105$	$8/225$	$-0/1$	$1/9$	2	0
13	1	$-156/35$	$54/35$	$-72/25$	$-0/1$	$117/2$	0	1
13	2	$-78/35$	$13/35$	$-6/25$	$-0/1$	$169/12$	1	1
13	3	$-22/35$	$27/35$	$-6/25$	$-6/5$	$33/4$	2	1
13	4	$-11/35$	$13/70$	$-1/50$	$-1/10$	$143/72$	1	1
13	5	$-54/35$	$-54/35$	$72/25$	$-0/1$	$81/4$	0	1
13	6	$-27/35$	$-13/35$	$-6/25$	$-0/1$	$-39/8$	1	0
13	7	$13/35$	$27/35$	$-1/50$	$-0/1$	$-39/8$	0	0
13	8	$13/70$	$13/70$	$1/50$	$-0/1$	$-169/144$	1	0
13	9	$54/35$	$-156/35$	$-72/25$	$-0/1$	$117/2$	2	1
13	10	$-27/35$	$22/35$	$6/25$	$6/5$	$-33/4$	1	0
13	11	$-13/35$	$78/35$	$-1/50$	$-0/1$	$-169/2$	2	1
13	13	$13/70$	$-11/35$	$72/25$	$-0/1$	$143/72$	1	0
14	1	$156/35$	$156/35$	$72/25$	$-0/1$	$159/1$	0	0
14	2	$78/35$	$-13/35$	$-2/25$	$-0/1$	$-169/12$	1	0
14	3	$26/35$	$-3/35$	$1/50$	$-0/1$	$-13/4$	2	0
14	4	$11/35$	$-13/70$	$-1/150$	$1/10$	$-143/72$	1	1
14	5	$11/105$	$-3/70$	$-6/25$	$-1/30$	$-11/24$	1	0
14	6	$27/35$	$13/35$	$2/25$	$-0/1$	$-39/8$	2	1
14	7	$9/35$	$3/35$	$1/50$	$-0/1$	$-9/8$	2	1
14	8	$-13/210$	$-13/70$	$-1/150$	$-0/1$	$169/144$	2	0
14	9	$-13/210$	$-3/70$	$2/105$	$6/5$	$13/48$	1	1
14	10	$-27/35$	$-4/35$	$-8/25$	$-0/1$	$3/2$	2	2
14	11	$13/70$	$-11/35$	$-1/50$	$-1/10$	$143/72$	1	1

i	$\tilde{\gamma}_{ij}^{(-)}$	$\tilde{\gamma}_{ij}^{(2)}$	$\tilde{\gamma}_{ij}^{(3)}$	$\tilde{\gamma}_{ij}^{(4)}$	$\tilde{\gamma}_{ij}^{(5)}$	$\tilde{\lambda}_{ij}$	$\tilde{\mu}_{ij}$
14	$-13/105$	$2/35$	$2/75$	$-0/1$	$-13/36$	2	1
14	$-78/35$	$-22/35$	$-6/25$	$-6/5$	$-143/6$	1	0
14	$52/35$	$4/35$	$8/25$	$-0/1$	$13/3$	2	0
15	$-22/35$	$27/35$	$-6/25$	$-6/5$	$33/4$	0	1
15	$-11/35$	$13/70$	$-1/50$	$-1/10$	$143/72$	1	2
15	$-4/35$	$18/35$	$-8/25$	$-0/1$	$3/2$	2	2
15	$-2/35$	$13/105$	$-2/75$	$-0/1$	$13/36$	1	1
15	$-13/70$	$-27/35$	$6/25$	$-0/1$	$39/8$	0	2
15	$-13/70$	$-13/70$	$1/50$	$-0/1$	$169/144$	1	1
15	$3/35$	$9/35$	$2/25$	$-0/1$	$-9/8$	2	2
15	$3/70$	$13/210$	$1/150$	$-0/1$	$-13/48$	0	1
15	$13/35$	$-78/35$	$-6/25$	$-0/1$	$169/12$	1	1
15	$-13/70$	$11/35$	$1/50$	$1/10$	$-143/72$	0	2
15	$-3/35$	$26/35$	$-2/25$	$-0/1$	$-13/4$	1	1
15	$13/70$	$-11/105$	$1/150$	$1/30$	$11/24$	0	2
15	$22/35$	$78/35$	$6/25$	$6/5$	$143/6$	1	2
16	$-11/35$	$-11/35$	$-1/50$	$-0/1$	$-121/36$	0	1
16	$4/35$	$52/35$	$8/25$	$-0/1$	$13/3$	1	0
16	$11/35$	$-13/70$	$1/50$	$-0/1$	$13/48$	2	2
16	$11/105$	$-3/70$	$-1/150$	$-0/1$	$1/16$	1	1
16	$2/35$	$-13/105$	$2/75$	$-0/1$	$-1/12$	2	2
16	$2/105$	$13/70$	$1/50$	$-1/5$	$-169/144$	1	1
16	$13/70$	$3/70$	$1/150$	$2/15$	$-13/18$	1	2
16	$13/210$	$-13/210$	$-1/150$	$2/15$	$-11/18$	0	2
16	$-3/70$	$11/35$	$-2/75$	$-2/15$	$-11/18$	2	1
16	$4/105$	$4/105$	$8/225$	$-0/1$	$1/9$	2	2

227

where the $_jN_i$ are the cubic Hermite interpolation functions given by equation 5.51b. When we substitute the displacement model of equation 6.42 into equation 6.40, we obtain the potential energy of the element in terms of its 16 nodal degrees of freedom. Minimization of the resulting potential energy expression $\Pi_p^{(e)}$ then requires that the first partial derivatives of $\Pi_p^{(e)}$ with respect to each of the degrees of freedom be zero, that is,

$$
\begin{Bmatrix}
\dfrac{\partial \Pi_p^{(e)}}{\partial w_i} \\[2ex]
\dfrac{\partial \Pi^{(e)}}{\partial (\partial w/\partial x)_i} \\[2ex]
\dfrac{\partial \Pi^{(e)}}{\partial (\partial w/\partial y)_i} \\[2ex]
\dfrac{\partial \Pi^{(e)}}{\partial (\partial^2 w/\partial x\,\partial y)_i}
\end{Bmatrix} = \{0\}, \quad i = 1, 2, 3, 4 \tag{6.43}
$$

The element equations result from carrying out the manipulations indicated in equations 6.43. We shall omit all details of these tedious manipulations and simply quote the results as given by Bogner et al. [21]. In matrix form equations 6.43 become

$$
\overset{16 \times 16 \;\; 16 \times 1}{[K]^{(e)} \{\delta\}} - \{P\} = \{0\} \tag{6.44}
$$

The coefficients of the stiffness matrix (K) are

$$
k_{ij} = \frac{D}{ab}\left[\tilde{\gamma}_{ij}^{(1)}\left(\frac{b}{a}\right)^2 + \tilde{\gamma}_{ij}^{(2)}\left(\frac{a}{b}\right)^2 + \tilde{\gamma}_{ij}^{(3)} + \tilde{\gamma}^{(4)}v \right] a^{\tilde{\lambda}_{ij}} b^{\tilde{\mu}_{ij}}
$$

where $\tilde{\gamma}_{ij}^{(k)}$ $(k = 1, 2, \ldots, 5)$, $\tilde{\lambda}_{ij}$, *and* $\tilde{\mu}_{ij}$ *are given in Table 6.2.*

The displacement vector $\{\delta\}$ is ordered as follows

$$
\{\delta\}^T = \lfloor \delta \rfloor = \left\lfloor w_1, \left(\frac{\partial w}{\partial x}\right)_1, \left(\frac{\partial w}{\partial y}\right)_1, \left(\frac{\partial^2 w}{\partial x\,\partial y}\right)_1, w_2, \left(\frac{\partial w}{\partial x}\right)_2, \ldots, \left(\frac{\partial^2 w}{\partial x\,\partial y}\right)_4 \right\rfloor
$$

$$
= \lfloor w_1, w_{x_1}, w_{y_1}, w_{xy_1}, w_2, w_{x_2}, \ldots, w_{xy_4} \rfloor
$$

Table 6.3 gives the individual terms of the combined load vector $\{P\}$. Terms of the vector $\{P\}$ are obtained by adding all the components in a given row. For example, the term in $\{P\}$ corresponding to the degree of freedom w_{x_3} is given as

$$
0 + M_x - \frac{qa^2 b}{24} + \frac{bM_x}{2} - a^2 \frac{V}{12}
$$

Table 6.3. Individual terms of combined load vector $\{P\}$ (from Bogner et al. [21])

Degree of freedom	Conc. load P at (a, b)	Conc. moment M_x at (a, b)	Uniformly Dist. load q	Dist. edge moment M_x on $x = a$	Dist. edge load V On $x = 0$	On $x = a$	On $y = 0$	On $y = b$
w_1	0	0	$qab/4$		$bV/2$	0	$aV/2$	0
w_{x_1}	0	0	$qa^2b/24$		0	0	$a^2V/12$	0
w_{y_1}	0	0	$qab^2/24$		$b^2V/12$	0	0	0
w_{xy_1}	0	0	$qa^2b^2/144$		0	0	0	0
w_2	0	0	$qab/4$		$bV/2$	0	0	$aV/2$
w_{x_2}	0	0	$qa^2b/24$		0	0	0	$a^2V/12$
w_{y_2}	0	0	$-qab^2/24$		$-b^2V/12$	0	0	0
w_{xy_2}	0	0	$-qa^2b^2/144$		0	0	0	0
w_3	P	0	$qab/4$	0	0	$bV/2$	0	$aV/2$
w_{x_3}	0	M_x	$-qa^2b/24$	$bM_x/2$	0	0	0	$-a^2V/12$
w_{y_3}	0	0	$-qab^2/24$	0	0	$-b^2V/12$	0	0
w_{xy_3}	0	0	$qa^2b^2/144$	$-b^2M_x/12$	0	0	0	0
w_4	0	0	$qab/4$	0	0	$bV/2$	$aV/2$	0
w_{x_4}	0	0	$-qa^2b/24$	$bM_x/2$	0	0	$-a^2V/12$	0
w_{y_4}	0	0	$qab^2/24$	0	0	$b^2V/12$	0	0
w_{xy_4}	0	0	$-qa^2b^2/144$	$b^2M_x/2$	0	0	0	0

Comments

The foregoing results for the conforming rectangular element in plate bending indicate the complexities involved in formulating the element equations. The more complex conforming elements for C^1 problems discussed in Chapter 5 lead to even greater algebraic complexities. In many cases explicit expressions for the element equations are never fully written out; instead, the terms are obtained by computer as needed. Whether or not the element equations obtained in this way are correct is determined by solving test problems whose exact solutions are known.

Here we have considered the simplest plate-bending element and only one plate flexure formulation, namely, the potential energy formulation. Many alternative specialized formulations such as those detailed in refs. 1, 3, and 6–9 may also be used at times with success, but they are still in the development stage and are not as well established as the potential energy formulations.

Rectangular elements are by no means the only plate-bending elements. We have completely omitted many different elements that have been formulated over the years. The interested reader can find and examine most of these formulations in the references gathered by Singhal [22]. Desai and Abel [23] have collected and presented the results of a number of authors who compared the performances of different plate-bending elements. This comparison is based on the solution of two fundamental problems—a centrally loaded square plate with edges either (1) simply supported or (2) clamped. Such comparisons reveal vividly the diverse behavior of the different element types and suggest which elements are superior performers.

Example

Figure 6.14 shows the performance of the 16-degree-of-freedom element when it is used to analyze a centrally loaded plate with simply supported edges. The percentage of error in the calculation of the center displacement is plotted against the parameter NB^2,[†] where N is the number of equations in the final assemblage and B is the half-bandwidth of the assembled stiffness matrix. Square elements were used, and because of symmetry only one-quarter of the plate was considered.

This displacement element satisfies all the convergence requirements; hence we expect the convergence that is indeed achieved. The solution also gives an upper bound on the potential energy of the system and, for this special example, a lower bound on the displacements.

---

† The computer time required to solve banded equations is conveniently characterized by this parameter [23].

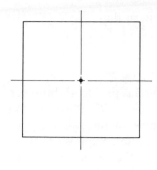

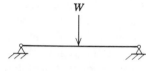

Figure 6.14.   Performance of the 16-degree-of-freedom rectangular element [24]. (*a*) Centrally loaded, simply supported
square plate.

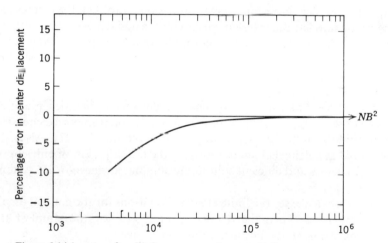

Figure 6.14 (*continued*).   (*b*) Convergence of center displacement.

6.6   THREE-DIMENSIONAL PROBLEMS

6.6.1   Introduction

The finite element analysis of three-dimensional structures follows directly
from the concepts established in Section 6.2. In place of the displacement
vector

$$\{\tilde{\delta}\} = \begin{Bmatrix} u(x, y) \\ v(x, y) \end{Bmatrix}$$

we now consider three components of displacement, so that

$$\{\tilde{\delta}\} = \begin{Bmatrix} u(x, y, z) \\ v(x, y, z) \\ w(x, y, z) \end{Bmatrix}$$

The additional degree of freedom in three-dimensional problems, however, considerably enlarges the magnitude of such problems. To illustrate this point we need only consider the common example of the discretization process for one-, two-, and three-dimensional analyses. Suppose that we pick 10 nodes to define one-dimensional elements for the stress analysis of a beam. Since there is only one degree of freedom per node, we have roughly 10 unknowns in the problem. To perform a stress analysis of a two-dimensional problem to roughly the same degree of accuracy, we would need a 10×10 mesh of nodes to define triangular elements for the region. With two degrees of freedom per node, the number of unknowns for this problem climbs to about 200. Now for a three-dimensional problem the discretization of the region into tetrahedral elements involves a nodal mesh of $10 \times 10 \times 10$, and with three degrees of freedom per node we have almost 3000 unknowns to find. It is obvious from this exercise that the treatment of realistic three-dimensional problems can readily tax the storage capacity of even the largest computers.

Because three-dimensional problems present such difficulties, much effort has been expended to develop three-dimensional isoparametric elements that give accurate results with fewer degrees of freedom [25–27]. Also, three-dimensional problems helped to motivate the development of efficient data-handling schemes and efficient solution techniques for large-order systems of equations.

This section discusses the finite element equations for the general displacement formulation. Then the displacement model for the simplest three-dimensional element, the linear tetrahedron, is given.

6.6.2 Element equations

Following precisely the same procedures for minimizing the potential energy as in Section 6.2, we arrive at the force-displacement relations for *node q* of a general element. These equations are:

$$\underset{3\times 3}{[k]^q}\underset{3\times 1}{\{\tilde{\delta}\}^q} = \underset{3\times 1}{\{F_0\}^q} + \underset{3\times 1}{\{F_B\}^q} + \underset{3\times 1}{\{F_T\}^q} = \underset{3\times 1}{\{F\}^q} \tag{6.45}$$

where

$$\{\tilde{\delta}\}^q = \begin{Bmatrix} u_q \\ v_q \\ w_q \end{Bmatrix}$$

and the remaining matrices are given by equations 6.11–6.15, with $[B]$, $[C]$, $\{\epsilon_0^*\}$, and $\{T^*\}$ given their proper three-dimensional interpretations, and $t^{(e)} \, dA^{(e)}$ replaced by $dV^{(e)}$ (see Appendix C). The element stiffness matrix is then formed as in equation 6.16; that is,

$$
[K]^{(e)} = \underset{3r \times 3r}{}
\begin{bmatrix}
[k]^1 & & & \\
& [k]^2 & & \\
& & \ddots & \\
& & & [k]^r
\end{bmatrix}
$$

where r is, as before, the number of nodes per element. The displacement vector for the element is then

$$
\{\delta\}_{(e)} =
\begin{Bmatrix}
u_1 \\
v_1 \\
w_1 \\
\vdots \\
u_r \\
v_r \\
w_r
\end{Bmatrix}
$$

The remaining equations have the same form as those in Section 6.2 except that they are expanded in dimension.

6.6.3   Formulations for the linear tetrahedral element

The basic solid element is the four-node tetrahedron with the displacement field interpolated linearly between nodes. In terms of the natural coordinates described in Section 5.5, we may write the displacement field within a typical element as

$$
\begin{Bmatrix}
u(x, y, z) \\
v(x, y, z) \\
w(x, y, z)
\end{Bmatrix}
=
\begin{matrix}
L_1 u_1 + L_2 u_2 + L_3 u_3 + L_4 u_4 \\
L_1 v_1 + L_2 v_2 + L_3 v_3 + L_4 v_4 \\
L_1 w_1 + L_2 w_2 + L_3 w_3 + L_4 w_4
\end{matrix}
$$

where the interpolation functions are

$$
L_i = \frac{1}{6V}(a_i + b_i x + c_i y + d_i z), \quad i = 1, 2, 3, 4
$$

V and the coefficients a_i, b_i, c_i, d_i being given by equations 5.26 and 5.27. As mentioned, if the tetrahedron is defined in a right-handed Cartesian co-ordinate system, the nodes must be numbered so that nodes 1, 2, and 3 are ordered counterclockwise when viewed from node 4.

The matrix $[B]_q^{(e)}$ relating strains and displacements at node q takes the form

$$[B]_q^{(e)} = \begin{bmatrix} \dfrac{\partial N_q}{\partial x} & 0 & 0 \\[2mm] 0 & \dfrac{\partial N_q}{\partial y} & 0 \\[2mm] 0 & 0 & \dfrac{\partial N_q}{\partial z} \\[2mm] \dfrac{\partial N_q}{\partial y} & \dfrac{\partial N_q}{\partial x} & 0 \\[2mm] 0 & \dfrac{\partial N_q}{\partial z} & \dfrac{\partial N_q}{\partial y} \\[2mm] \dfrac{\partial N_q}{\partial x} & 0 & \dfrac{\partial N_q}{\partial x} \end{bmatrix} = \frac{1}{6V} \begin{bmatrix} b_q & 0 & 0 \\ 0 & c_q & 0 \\ 0 & 0 & d_q \\ c_q & b_q & 0 \\ 0 & d_q & c_q \\ d_q & 0 & b_q \end{bmatrix}$$

for $q = 1, 2, 3, 4$.

With the matrix of constitutive properties $[C]$ given by equation C.6, we may write the general submatrix of the element stiffness matrix as

$$[k]^q = [B]_q^{T(e)}[C][B]_q^{(e)}V$$

since the integration over the element volume involves only constant terms. Evaluation of the other element matrices, the initial strain matrix, and the load matrices is also straightforward because the strain and stress components are constant within each element.

To visualize a three-dimensional solid constructed from tetrahedral elements is extremely difficult. Thus, in practical applications, the solid is modeled with brick-type or hexahedral shapes and the computer program is then designed to construct automatically the element matrices for the hexahedra by combining the matrices for five tetrahedra. Since hexahedra can be formed from five tetrahedra in two ways, it is desirable to implement both ways and then average the resulting matrices. This leads to a slight improvement in accuracy without much additional effort.

6.6.4 Higher-order elements

As indicated in Chapter 5, any higher-order tetrahedral elements can be easily formed using natural coordinates and the recursion relationship between interpolation functions of consecutive order. Lagrangian interpolation also leads to a family of higher-order hexahedral elements. In actual three-dimensional stress analyses, though, only the lower-order elements (linear, quadratic, and cubic) have been used. The natural question as to which elements give the best accuracy per unit of computation time has been asked and partially answered by three investigators [26,28,29]. Clough [29] studied the performance of several three-dimensional elements and suggested

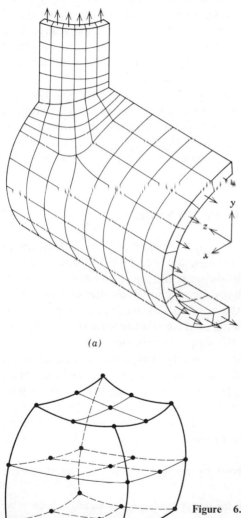

(a)

(b)

Figure 6.15. Three-dimensional finite element analysis of a pipe junction [30]. (a) Finite element model. (b) 27-node isoparametric element used in the analysis.

that the trace† of the element stiffness matrix is a reliable indication of the relative quality of different elements. He found that the best elements are characterized by the lowest trace. When the linear displacement hexahedron was compared on this basis with a hexahedron formed from five tetrahedra, it was found to have only about half the trace of the assembled tetrahedra. In most three-dimensional stress analyses hexahedral elements of various orders are preferred over tetrahedral elements.

As automatic data-handling procedures and efficient equation-solving routines become more widely available, more and more three-dimensional finite element analyses will appear. Many of the published analyses to date treat the problems stemming from the nuclear reactor industry. The design of nuclear reactors for the 1980's is requiring detailed analyses of complex structures such as housings and piping intersections.

Figure 6.15 shows a pipe junction analyzed by Argyris et al. [30]. The part is loaded by uniform internal pressure so that symmetry can be used to reduce the problem size as indicated. Even with this simplification the finite element model required 125 of the elements shown in Figure 6.15*b* and 123 minutes of computation time on an IBM 360-75 computer.

6.7   INTRODUCTION TO STRUCTURAL DYNAMICS

In many practical design situations, it is not sufficient to know just the force-deflection properties of a structure because the loading of the structure may be a function of time. In this case the designer may need to know the natural frequencies of the structure or perhaps the way in which stresses and deflections are propagated through the structure. Just as the finite element method has been found to be a useful tool for the analysis of static structural problems, it is also most helpful in analyzing the dynamic behavior of structures.

When an elastic structure is subjected to a dynamic load, the displacement field within the structure varies with time and two types of distributed body forces must be taken into account [31]. One body force stems from the inertia of the structure, while the other originates because of internal friction or external damping of some sort. Both forces act in the opposite direction to the direction of motion at a given point in the body. In other words, both types of forces oppose the motion.‡

The inertia force is brought about by an acceleration of the structure.

---

† The *trace* of a square matrix is defined as the sum of the diagonal terms and is equal to the sum of the eigenvalues of the matrix.

‡ An exception is negative damping, where energy is actually added to the system, but for simplicity we shall not consider that here.

If $\{\delta\}$ is the displacement field of the elastic body, the distributed inertia force acting throughout the body is given by $\rho\{\ddot{\delta}\}$, where ρ is the mass density of the material and $\{\ddot{\delta}\} \equiv (\partial^2/\partial t^2)\{\delta\}$ is the column vector of distributed accelerations. According to the well-known D'Alembert principle, this inertia force is statically equivalent to the force $\{F_a\} = -\rho\{\ddot{\delta}\}$, that is, $\{F_a\}$ may be treated as a statically imposed body force.

Compared with the inertia force, the damping force is far more difficult to characterize in general. Damping forces may result from friction within a deforming material, from motion through a viscous fluid, or from rubbing with some other bodies in dry contact. Lazan [32] discusses many of the details of these kinds of damping. Usually the damping force is not linearly related to the rate of change of displacement in a body, but rather has a more complicated relationship. However, to simplify the analysis of a dynamically loaded structure, the conventional procedure is to replace the actual damping force, which is usually unknown, by an approximate viscous damping force proportional to velocity. By this means we write the damping force acting on the body as $\{F_d\} = v_d\{\dot{\delta}\}$, where v_d is some known damping coefficient and $\dot{\delta} \equiv (\partial/\partial t)(\delta)$. Determining suitable values for v_d poses a difficult problem, and much research in this area remains to be done. Some suggested approaches are developed by Jacobson [33] and Wilson [34].

Assuming that we may describe the inertia and damping forces as indicated, we can derive the equations of motions governing the dynamic behavior of an elastic structure by referring to the discretized equilibrium equations obtained earlier in this chapter. At any instant of time, the potential energy of the body is given by equation C.22. Within a typical three-dimensional element we assume that the displacement field is expressed as

$$\{\delta\}^{(e)} = \left\{\begin{array}{l} u(x, y, z, t) \\ v(x, y, z, t) \\ w(x, y, z, t) \end{array}\right\}^{(e)} = \left\{\begin{array}{l} \Sigma\, N_i(x, y, z)u_i(t) \\ \Sigma\, N_i(x, y, z)v_i(t) \\ \Sigma\, N_i(x, y, z)w_i(t) \end{array}\right\}^{(e)} \tag{6.46}$$

where now u, v, and w are the time-dependent components of displacement in the three coordinate directions. Minimization of the potential energy with respect to the nodal values of displacement then leads to force-displacement relations for a typical node of the form of equation 6–10 except that all the parameters may now be time dependent. The nodal body force vector, however, now includes the inertia and damping forces. Hence, at node q, we have for the body force term the usual term plus two extra terms:

$$\{F_B\}^q = \int_{\Omega^{(e)}} [N_q\{F^*\}_q - v_d N_q\{\dot{\delta}\}^q - \rho N_q\{\ddot{\delta}\}] \, d\Omega^{(e)} \tag{6.47}$$

The other terms of equation 6.10 remain the same. Assembly of the nodal equations as indicated in Section 6.2.3 then leads to element equations of the form

$$[K]^{(e)}\{\delta\} + [C]^{(e)}\{\dot{\delta}\} + [M]^{(e)}\{\ddot{\delta}\} = \{F(t)\}^{(e)} \qquad (6.48)$$

where, for an element with r nodes, we have

$$[C]^{(e)} = \int_{\Omega^{(e)}} v_d [N]^T [N]\, d\Omega^{(e)} \qquad (6.49)$$

and

$$[M]^{(e)} = \int_{\Omega^{(e)}} \rho [N]^T [N]\, d\Omega^{(e)} \qquad (6.50)$$

in which we define the interpolation function matrix as

$$[N] = [[I]N_1, [I]N_2, ..., [I]N_r]$$

with

$$[I] = \begin{bmatrix} 1 & 0 & 0 \\ 0 & 1 & 0 \\ 0 & 0 & 1 \end{bmatrix}$$

The interpolation functions N_i are the same as those used for the displacement model, and $[C]^{(e)}$ and $[M]^{(e)}$ are called the element damping and mass matrices, respectively. Further assembly gives the system equations of motions:

$$[K]\{\delta\} + [C]\{\dot{\delta}\} + [M]\{\ddot{\delta}\} = \{F(t)\} \qquad (6.51)$$

The matrices $[C]$ and $[M]$ are usually densely populated matrices, which, when formed from equations 6.49 and 6.50, are referred to as *consistent* damping and mass matrices [35]. The term consistent is used to distinguish these matrices from *lumped* mass or damping matrices, which are obtained by assuming that element mass and damping characteristics are concentrated at the nodes. To form lumped matrices, we simply assign to each node of the system an amount of mass or a damping coefficient that can physically be attributed to that location in the body. Usually nonoverlapping yet contiguous regions surrounding the nodes are chosen, and the mass and damping associated with a particular region are assigned to the node in that region. Lumped matrices obtained in this way are diagonal.

Lumped matrices offer computational advances because they are easy to store and invert, and they considerably simplify the resulting equations of motion. However, when we use consistent matrices, we can expect a more accurate calculation of eigenvalues and eigenvectors. In general, the errors incurred by lumping increase as the complexity of the element we use increases. Further details of the relative merits of consistent versus lumped matrices can be found in the thorough discussion by Clough [36].

Once the matrices have been formed, the solution of structural dynamics problems involves the solution of equations of the form of equation 6.51. This is the subject of Section 7.6.

6.8 CLOSURE

In this chapter we examined a general finite element formulation for a variety of elasticity problems, namely, plane stress, plane strain, axisymmetric solids, plate bending, and general three-dimensional solids. Though the element equations derived for these problems were explicitly evaluated only for the simplest types in each case, the equations are general and apply for many element shapes and displacement models.

The treatment in this chapter does not include the particular considerations needed for anisotropic and/or incompressible materials. Such considerations are especially important to civil engineers, who study soil and rock mechanics. The finite element analysis of earth structures can still be based on the potential energy principle of equation 6.1, but the constitutive laws are usually nonlinear. Discussion of these topics can be found in more specialized books such as refs. 23 and 31.

Another advanced topic not considered here is the finite element analysis of shells or thin plates with arbitrary curvature. Such problems are of fundamental importance in structural mechanics, but their treatment poses several difficulties. Zienkiewicz [31] gives a thorough discussion of these difficulties and some approaches to surmount them.

REFERENCES

1. T. H. H. Pian and P. Tong, "Basis of Finite Element Methods for Solid Continua," *Int. J. Numer. Methods Eng.*, Vol. 1, No. 1, 1969, p. 26.

2. R. J. Melosh, "Basis for Derivation of Matrices for the Direct Stiffness Method," *J. Am. Inst. Aeronaut. Astron.*, Vol. 1, No. 7, 1963, pp. 1631–1637.

3. Z. M. Elias, "Duality in Finite Element Methods," *Proc. ASCE, J. Eng. Mech. Div.*, Vol. 94, No. EM 4, 1968, pp. 931–946.

4. L. S. D. Morley, "The Triangular Equilibrium Element in the Solution of Plate Bending Problems," *Aeronaut. Quart.*, 1968, pp. 149–169.

5. B. M. Fraeijs de Veubeke, "Upper and Lower Bounds in Matrix Structural Analysis," *Matrix Methods of Structural Analysis*," AGARD, Vol. 72, pp. 165–201.

6. T. H. H. Pian, "Derivation of Element Stiffness Matrices by Assumed Stress Distribution," *J. Am. Inst. Aeronaut. Astron.*, Vol. 2, No. 7, 1964, pp. 1333–1336.

7. R. Yamamoto, and H. Isshiki, "Variational Principles and Dualistic Scheme for Intersection Problems in Elasticity," *J. Fac. Eng. Univ. Tokyo*, Vol. 30, No. 1, 1969.

8. P. Tong, "New Displacement Hybrid Finite Element Model for Solid Continua," *Int. J. Numer. Methods Eng.*, Vol. 2, No. 1, 1970, pp. 73–84.

9. L. R. Herrmann, "A Bending Analysis for Plates," *Proceedings of 1st Conference on Matrix Methods in Structural Mechanics* (AFFDL-TR-66-80), Wright-Patterson Air Force Base, Dayton, Ohio, 1965.

10. I. Holand, "The Finite Element Method in Plane Stress Analysis," Chapter 2 in *The Finite Element Method in Stress Analysis*, I. Holand and K. Bell (eds.), Tapir Press, Trondheim, Norway, 1969.

11. R. W. Clough, "The Finite Element in Plane Stress Analysis," *Proceedings of 2nd ASCE Conference on Electronic Computation*, Pittsburgh, Pa., September 1960.

12. O. C. Zienkiewicz, *The Finite Element Method in Engineering Science*, McGraw-Hill Book Company, London, 1971, Chapter 4, pp. 48–72.

13. E. L. Wilson, "Finite Element Analysis of Two-Dimensional Structures," *Rept.* 63-2, Dept. of Civil Eng., University of California at Berkeley, 1963.

14. E. L. Wilson, "Structural Analysis of Axisymmetric Solids," *AIAA J.*, Vol. 3, No. 12, December 1965, pp. 2267–2274.

15. R. Clough and Y. Rashid, "Finite Element Analysis of Axisymmetric Solids," *Proc. ASCE, J. Eng. Mech. Div.*, Vol. 91, No. EMI, February 1965, pp. 71–85.

16. S. Vtku, "Explicit Expressions for Triangular Torus Element Stiffness Matrix," *AIAA J.*, Vol. 6, No. 6, June 1968, pp. 1174–75.

17. S. Jordan and E. Helle, "Formulation and Evaluation of a Two Dimensional Core Discrete Element," Bell Aerosystems Company, February 1967.

18. K. Maruyama, "Stress Analysis of a Bolt-Nut Joint by the Finite Element Method and the Copper-Electroplating Method," *Bull. JSME*, Vol. 16, No. 94, April 1973, pp. 671–678.

19. O. C. Zienkiewicz, "The Finite Element Method from Intuition to Generality," *Appl. Mech. Dev.*, Vol. 23, No. 3, March 1970, pp. 249–256.

20. R. H. Gallagher, "Analysis of Plate and Shell Structures," *Proceedings of Symposium on the Application of Finite Element Methods in Civil Engineering*, ASCE-Vanderbilt University, Nashville, Tenn., November 1969.

21. F. K. Bogner, R. L. Fox, and L. A. Schmit, Jr., "The Generation of Interelement-Compatible Stiffness and Mass Matrices by the Use of Interpolation Formulas," *Proceedings of 1st Conference on Matrix Methods in Structural Mechanics* (AFFDL-TR-66-80), Wright-Patterson Air Force Base, Dayton, Ohio, 1965.

22. A. C. Singhal, "775 Selected References on the Finite Element Method and Matrix Methods of Structural Analysis," *Rept.* S-12, Civil Eng. Dept., Laval University, Quebec, January 1969.

23. C. Desai, and J. Abel, *Introduction to the Finite Element Method*, Van Nostrand-Reinhold, New York, 1971, pp. 287–295.

24. R. W. Clough and J. L. Tocher, "Finite Element Stiffness Matrices for Analysis of Plate Bending," *Proceedings of 1st Conference on Matrix Methods in Structural Mechanics*, Wright-Patterson Air Force Base, Dayton, Ohio, November 1965.

25. J. Ergatoudus, B. M. Irons, and O. C. Zienkiewicz, "Three Dimensional Analysis of Arch Dams and Their Foundations," *Proceedings of Symposium on Architecture of Dams*, Institute of Civil Engineering, 1968.

26. S. Fjeld, "Three Dimensional Theory of Elasticity," in *Finite Element Methods in Stress Analysis*, I. Holand and K. Bell (eds.), Tapir Press, Trondheim, Norway, 1969.

27. Y. Rashid, "Three-Dimensional Analysis of Elastic Solids," *Int. J. Solids Struct.*, Part I, Vol. 5, pp. 1311–1332; Part II, Vol. 6, pp. 195–207.

28. R. J. Melosh, "Structural Analysis of Solids," *Proc. ASCE, J. Struct. Div.*, Vol. 89, No. ST-4, August 1963, pp. 205–223.

29. R. W. Clough, "Comparison of Three-Dimensional Finite Elements," *Proceedings of Symposium on Application of Finite Element Methods in Civil Engineering*, ASCE-Vanderbilt University, Nashville, Tenn., November 1969.

30. J. H. Argyris, O. E. Bronlund, and M. Sorensen, "Computer Aided Structural Analysis," The Machine-Independent System ASKA, Nord. Data-70 Conference, Copenhagen, August 1970.

31. O. C. Zienkiewicz, *The Finite Element Method in Engineering Science*, McGraw-Hill Book Company, London, 1971.

32. B. J. Lazan, *Damping of Materials and Members in Structural Mechanics*, Pergamon Press, New York, 1968.

33. L. S. Jacobsen, "Steady Forced Vibration as Influenced by Damping," *Trans. ASME*, APM-52-15, 1930.

34. E. I. Wilson, "A Computer Program for the Dynamic Stress Analysis of Underground Structures," U.S. Army Engineer Waterways Experiment Station, *Contract Rept.* 1-175, January 1968.

35. O. C. Zienkiewicz and Y. K. Cheung, "The Finite Element Method For Analysis of Elastic Isotropic and Orthotropic Slabs," *Proc. Inst. Civ. Eng.*, Vol. 28, 1964, p 471.

36. R. W. Clough, "Analysis of Structural Vibrations and Dynamic Response," in *Recent Advances in Matrix Methods of Structural Analysis and Design*, R. H. Gallagher, Y. Yamada, and J. T. Oden (eds.), University of Alabama Press, Huntsville, Ala., 1971.

7

GENERAL FIELD PROBLEMS

7.1 INTRODUCTION

In this chapter we present finite element formulations for a significant class of physical problems known as field problems. Heat conduction problems, convective-diffusion problems, and irrotational flow problems are just a few of the many important types of field problems that engineers encounter. Field problems share the common characteristic of being governed by similar partial differential equations for the field variable ϕ. Hence we can discuss the solution of these problems by focusing attention on the equations of ϕ without identifying ϕ as a particular physical quantity for a particular problem.

As in Chapter 6, our goal here is to derive element equations in terms of general interpolation functions N_i. We shall do this by applying either a classical variational principle or the method of weighted residuals with Galerkin's criterion. The formulations are for three-dimensional problems, but equations for problems of other dimensions follow immediately. After the element equations are known, we can evaluate them for the element type of our choice. We shall consider some examples of this evaluation process for simple linear elements.

We begin the chapter with a discussion of general steady-state field problems characterized by quasi-harmonic equations and Helmholtz equations. Then we consider the practically important time-dependent field problems whose solution ultimately involves the solution of matrix differential equations. Finally, we make a few remarks concerning the solution of such equations.

7.2 QUASI-HARMONIC EQUATIONS (STEADY-STATE)

Suppose that the field variable ϕ is to be found in a three-dimensional solution domain Ω bounded by surface Γ (Figure 7.1). For steady-state (time-

242

independent) problems the field equation to be solved is the quasi-harmonic
equation, expressed in general terms as

$$\frac{\partial}{\partial x}\left(k_x \frac{\partial \phi}{\partial x}\right) + \frac{\partial}{\partial y}\left(k_y \frac{\partial \phi}{\partial y}\right) + \frac{\partial}{\partial z}\left(k_z \frac{\partial \phi}{\partial z}\right) = f(x, y, z) \qquad (7.1)$$

where k_x, k_y, k_z, and f are given functions independent of ϕ, and the coefficients k_x, k_y, and k_z are bounded away from zero in Ω. The physical
interpretation of the parameters in equation 7.1 depends on the particular
physical problem. Table 7.1 lists a number of typical field problems and
indicates the meaning of ϕ as well as the other parameters for each problem.

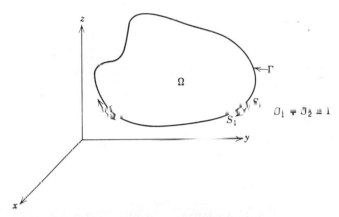

Figure 7.1.   Three-dimensional solution domain.

It is important to note that, for an inhomogeneous and/or nonisotropic
medium, equation 7.1 implies that the coordinates x, y, and z coincide with
the principal coordinates. For a homogeneous medium, k_x, k_y, and k_z are
constant, while for an isotropic medium, $k_x = k_y = k_z = k$.

7.2.1   Boundary conditions

The description of the field problem is not complete until boundary conditions are specified; that is, equation 7.1 must be solved subject to additional
constraints imposed on the bounding surface. Usually, on some part of the
boundary, the value of ϕ is a specified function such as

$$\phi = \Phi(x, y, z) \quad \text{on } S_1 \qquad (7.2)$$

Table 7.1.　Identification of physical parameters

Problem	ϕ	k_x, k_y, k_z	f	g	h
Diffusion flow in porous media	Hydraulic head	Hydraulic conductivity	Internal source flow	Boundary flow	—
Electric conduction	Voltage	Electric conductivity	Internal current source	Externally applied boundary current	—
Electrostatic field	Electric force field intensity	Permitivity	Internal current source	—	—
Fluid-film lubrication	Pressure	$k_z = 0$ k_x, k_y are functions of film thickness and viscosity	Net flow due to various actions	Boundary flow	—
Gravitation	Component of gravitational force vector per unit mass	—	—	—	—
Heat conduction	Temperature	Thermal conductivity	Internal heat generation	Boundary heat generation	Convective heat transfer coefficient
Irrotational flow	Velocity potential or stream function	—	0	Boundary velocity	0
Torsion	Stress function	Reciprocal of shear modulus	Angle of twist per unit length	—	—
Seepage	Pressure	Permeability	Internal flow source	—	—
Magnetostatics	Magnomotive force	Magnetic permeability	Internal magnetic field source	Externally applied magnetic field intensity	—

while on the remaining part of the boundary we have the condition

$$k_x \frac{\partial \phi}{\partial x} n_x + k_y \frac{\partial \phi}{\partial y} n_y + k_z \frac{\partial \phi}{\partial z} n_z + g(x, y, z) + h(x, y, z)\phi = 0 \quad \text{on } S_2 \quad (7.3)$$

where g and h are known a priori and n_x, n_y, and n_z are the direction cosines of the outward normal of the surface. The union of S_1 and S_2 forms the complete boundary Γ, $S_1 \cup S_2 = \Gamma$.

The boundary condition expressed by equation 7.2 is known as the *Dirichlet* condition, and the specified function Φ is sometimes called Dirichlet data. Equation 7.3 is the *Cauchy* boundary condition. If the functions $g = h = 0$, the Cauchy condition reduces to the *Neumann* boundary condition, sometimes known as the "natural" boundary condition. A field problem is said to have mixed boundary conditions when some portions of Γ have Dirichlet boundary conditions while other portions have Cauchy boundary conditions.

Equations 7.1–7.3 comprise a well-posed elliptic boundary value problem whose solution by finite element methods can be based on a classical variational principle. We have formulated this field problem as a general case which can be easily specialized to particular cases. For example, if $k_x = k_y = k_z = k -$ constant, equation 7.1 reduces to

$$\nabla^2 \phi = \frac{f}{k}(x, y, z) \quad (7.4)$$

which is known as *Poisson's* equation. Furthermore, if $f = 0$, we have the well-known *Laplace* equation

$$\nabla^2 \phi = 0 \quad (7.5)$$

We shall encounter this equation again in Chapter 9 when we investigate potential flow problems.

As we pointed out in Chapter 3, the finite element method is only one of many different ways to obtain approximate numerical solutions to these equations. For many years, analysts used finite difference techniques; but, when nonuniform meshes are needed for problems with irregular geometry, the finite element method is better suited. The finite element equations for this general field problem can be easily derived from a variational principle by following the procedures of Chapter 3.†

---

† Identical element equations would be obtained if, instead of the variational approach, we used the method of weighted residuals with Galerkin's criterion (Chapter 4).

7.2.2 Variational principle

It can be shown [1,2] that the function $\phi(x, y, z)$ that satisfies equations 7.1–7.3 also minimizes the functional

$$I(\phi) = \frac{1}{2} \int_\Omega \left[k_x \left(\frac{\partial \phi}{\partial x} \right)^2 + k_y \left(\frac{\partial \phi}{\partial y} \right)^2 + k_z \left(\frac{\partial \phi}{\partial z} \right)^2 + 2f\phi \right] dx\, dy\, dz$$

$$+ \int_{S_2} (g\phi + \tfrac{1}{2}h\phi^2)\, dS_2 \tag{7.6}$$

Following the procedures of Appendix C, it is easy to show that equation 7.1 is the Euler-Lagrange condition for the functional of equation 7.6. The variational principle on which we can base the derivation of the element equation is

$$\delta I(\phi) = 0 \tag{7.7}$$

7.2.3 Element equations

Suppose that the solution domain Ω is divided into M elements of r nodes each. By the usual procedure, we may express the behavior of the unknown function ϕ within each element as

$$\phi^{(e)} = \sum_{i=1}^{r} N_i \phi_i = \lfloor N \rfloor \{\phi\}^{(e)} \tag{7.8}$$

where ϕ_i is the nodal value of ϕ at node i. Equation 7.8 implies that only nodal values of ϕ are taken as nodal degrees of freedom, but derivatives of ϕ may also be used as nodal parameters without changing the procedure to be followed. The quantity ϕ_i may be thought of as a general nodal parameter.

Since the functional $I(\phi)$ contains only first-order derivatives, we have a C^0 problem and the N_i must be chosen to preserve at least continuity of ϕ at element interfaces. If the interpolation functions quarantee C^0 continuity, we may focus attention on one element because the integral $I(\phi)$ can be represented as the sum of integrals over all the elements, that is,

$$I(\phi) = \sum_{e=1}^{M} I(\phi^{(e)}) \tag{7.9}$$

The discretized form of the functional for one element is obtained by substituting equation 7.8 into equation 7.6. Then the minimum condition $\delta I(\phi) = 0$ for one element becomes

$$\frac{\partial I(\phi^{(e)})}{\partial \phi_i} = 0, \quad i = 1, 2, \ldots, r \tag{7.10}$$

For a node i on boundary S_2, from equation 7.6 we have

$$\frac{\partial I(\phi^{(e)})}{\partial \phi_i} = 0$$

$$= \int_{\Omega^{(e)}} \left[k_x \frac{\partial \phi^{(e)}}{\partial x} \frac{\partial}{\partial \phi_i} \left(\frac{\partial \phi^{(e)}}{\partial x} \right) + k_y \frac{\partial \phi^{(e)}}{\partial y} \frac{\partial}{\partial \phi_i} \left(\frac{\partial \phi^{(e)}}{\partial y} \right) \right.$$

$$\left. + k_z \frac{\partial \phi}{\partial z} \frac{\partial}{\partial \phi_i} \left(\frac{\partial \phi^{(e)}}{\partial z} \right) + f \frac{\partial \phi^{(e)}}{\partial \phi_i} \right] d\Omega^{(e)}$$

$$+ \int_{S_2^{(e)}} \left(g \frac{\partial \phi^{(e)}}{\partial \phi_i} + h\phi^{(e)} \frac{\partial \phi^{(e)}}{\partial \phi_i} \right) dS_2 \quad \text{on surface } S_2{}^{(e)} \tag{7.11}$$

If node i does not lie on S_2, the second integral does not appear. Referring to equation 7.8, we may evaluate each of the derivatives in 7.11. These typically become:

$$\frac{\partial \phi^{(e)}}{\partial x} = \sum_{i=1}^{r} \frac{\partial N_i}{\partial x} \phi_i = \left\lfloor \frac{\partial N}{\partial x} \right\rfloor \{\phi\}^{(e)}$$

$$\frac{\partial}{\partial \phi_i} \left(\frac{\partial \phi^{(e)}}{\partial x} \right) = \frac{\partial N_i}{\partial x}$$

$$\frac{\partial \phi^{(e)}}{\partial \phi_i} = N_i$$

Thus we have

$$\frac{\partial I(\phi^{(e)})}{\partial \phi_i} = 0$$

$$= \int_{\Omega^{(e)}} \left[k_x \left\lfloor \frac{\partial N}{\partial x} \right\rfloor \{\phi\} \frac{\partial N_i}{\partial x} + k_y \left\lfloor \frac{\partial N}{\partial y} \right\rfloor \{\phi\} \frac{\partial N_i}{\partial y} \right.$$

$$\left. + k_z \left\lfloor \frac{\partial N}{\partial z} \right\rfloor \{\phi\} \frac{\partial N_i}{\partial z} + f N_i \right] d\Omega^{(e)}$$

$$+ \int_{S_2^{(e)}} [g N_i + h \lfloor N \rfloor N_i \{\phi\}] \, dS_2{}^{(e)} \quad \text{on surface } S_2 \tag{7.12}$$

Combining all of the equations like equation 7.12 for all of the nodes of the element gives the following set of element equations:

$$\left\{\frac{\partial I}{\partial \phi^{(e)}}\right\}^{(e)} = \left\{\begin{array}{c} \dfrac{\partial I(\phi^{(e)})}{\partial \phi_1} \\[2mm] \dfrac{\partial I(\phi^{(e)})}{\partial \phi_2} \\[1mm] \vdots \\[2mm] \dfrac{\partial I(\phi^{(e)})}{\partial \phi_r} \end{array}\right\} = \underset{r \times r}{[K]^{(e)}}\underset{r \times 1}{\{\phi\}^{(e)}} + \underset{r \times 1}{\{R_1\}^{(e)}} + \underset{r \times r}{[K_{S_2}]^{(e)}}\{\phi\}^{(e)} = 0 \quad (7.13)$$

where the coefficients of the matrices $[K]^{(e)}$, $\{R_1\}^{(e)}$, and $[K_s]^{(e)}$ are given by

$$k_{ij} = \int_{\Omega^{(e)}} \left(k_x \frac{\partial N_i}{\partial x}\frac{\partial N_j}{\partial x} + k_y \frac{\partial N_i}{\partial y}\frac{\partial N_j}{\partial y} + k_z \frac{\partial N_i}{\partial z}\frac{\partial N_j}{\partial z} \right) d\Omega^{(e)} \quad (7.14)$$

$$R_{1i} = \int_{\Omega^{(e)}} fN_i \, d\Omega^{(e)} + \int_{S_2^{(e)}} gN_i \, dS_2{}^{(e)} \quad (7.15)$$

$$k_{S_{2ij}} = \int_{S_2^{(e)}} kN_i N_j \, dS_2{}^{(e)} \quad (7.16)$$

We emphasize again that the last term in equation 7.15 and the matrix $[K_s]^{(e)}$ appear only if element (e) contributes to the definition of the boundary portion S_2. We see that $[K_{S_2}]^{(e)}$ is actually a square matrix with the same dimension as $[K]^{(e)}$ that can be added to $[K]^{(e)}$ to form the following element equations:

$$[[K]^{(e)} + [K_{S_2}]^{(e)}]\{\phi\}^{(e)} + \{R_1\} = 0 \quad (7.17)$$

Assembly of these element equations to obtain the system equations then follows the standard procedure.

7.2.4   Examples

Two-dimensional heat conduction

To illustrate the application of the foregoing equations to a practical situation, we shall use linear triangular elements and develop the matrix equations for a two-dimensional heat conduction problem with general boundary conditions. As in Section 4.2.2, the governing equation is

$$\frac{\partial}{\partial x}\left(k_x \frac{\partial T}{\partial x} \right) + \frac{\partial}{\partial y}\left(k_y \frac{\partial T}{\partial y} \right) + \tilde{Q} = 0 \quad (7.18)$$

with the boundary conditions segregated as follows:

$$T = T(x, y) \quad \text{on } C_1 \tag{7.19}$$

$$k_x \frac{\partial T}{\partial x} n_x + k_y \frac{\partial T}{\partial y} n_y = q \quad \text{on } C_2 \tag{7.20}$$

$$k_x \frac{\partial T}{\partial x} n_x + k_y \frac{\partial T}{\partial y} n_y + h(T - T_\infty) = 0 \quad \text{on } C_3 \tag{7.21}$$

where, as before, k_x and k_y are the thermal conductivities in the principal directions; $\tilde{Q}$ is a specified function representing internal heat generation; q is a specified surface heat flow due to conduction; and $h(T - T_\infty)$ is a surface heat flux due to convection. Figure 7.2 shows the two-dimensional solution domain and the various boundary segments.

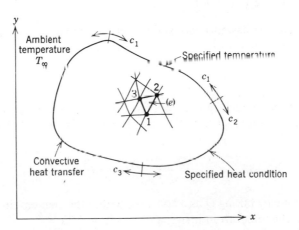

Figure 7.2. Solution domain for general two-dimensional heat conduction problem.

If we divide the region into triangular elements and assume that T varies linearly within each element, we may write immediately

$$T^{(e)}(x, y) = \sum_{i=1}^{3} L_i T_i$$

where the L_i are the natural coordinates for the triangle given by equations 5.17–5.19:

$$L_i = \frac{a_i + b_i x + c_i y}{2\Delta}, \quad i = 1, 2, 3$$

After identifying the interpolation functions N_i with L_i, we may refer to equations 7.14–7.16 or equation 4.7f to write the element matrices. Since

$$\frac{\partial N_i}{\partial x} = \frac{b_i}{2\Delta}, \qquad \frac{\partial N_j}{\partial x} = \frac{b_j}{2\Delta}$$

$$\frac{\partial N_i}{\partial y} = \frac{c_i}{2\Delta}, \qquad \frac{\partial N_j}{\partial y} = \frac{c_j}{2\Delta}$$

we have from equation 7.14

$$k_{ij} = \int_{A^{(e)}} \left(k_x \frac{b_i}{2\Delta} \frac{b_j}{2\Delta} + k_y \frac{c_i}{2\Delta} \frac{c_j}{2\Delta} \right) t \, dx \, dy$$

where t is the thickness of the elements and is usually taken as unity.

Hence

$$k_{ij} = \frac{t}{4\Delta^2} \left(b_i b_j \int_{A^{(e)}} k_x \, dx \, dy + c_i c_j \int_{A^{(e)}} k_y \, dx \, dy \right) \tag{7.22}$$

If k_x and k_y can be assigned constant average values within the element, evaluation of the integrals is trivial and we have

$$k_{ij} = \frac{t}{4\Delta} (k_x b_i b_j + k_y c_i c_j) \tag{7.23}$$

To evaluate the remaining matrices requires some special considerations. For an interior element such as the one shown in Figure 7.2, we need only evaluate

$$R_{1i} = -\int_{A^{(e)}} \tilde{Q} L_i \, dx \, dy$$

This can be done by taking $\tilde{Q}$ as a constant within the element, in which case we have from the integration formula of equation 5.22

$$R_{1i} = -\frac{\Delta \tilde{Q}}{3}$$

or we may linearly interpolate $\tilde{Q}$ in terms of its nodal values as $\tilde{Q} = \tilde{Q}_1 L_1 + \tilde{Q}_2 L_2 + \tilde{Q}_3 L_3$, in which case we have

$$R_{1i} = -\int_{A^{(e)}} (\tilde{Q}_1 L_1 L_i + \tilde{Q}_2 L_2 L_i + \tilde{Q}_3 L_3 L_i) \, dx \, dy$$

Again, employing the formula of equation 5.22, we have

$$\{R_1\}^{(e)} = -\frac{\Delta}{12} \begin{Bmatrix} 2\tilde{Q}_1 + \tilde{Q}_2 + \tilde{Q}_3 \\ \tilde{Q}_1 + 2\tilde{Q}_2 + \tilde{Q}_3 \\ \tilde{Q}_1 + \tilde{Q}_2 + 2\tilde{Q}_3 \end{Bmatrix} \tag{7.24}$$

Now suppose that element (e) forms part of the bounding curve C_2 or C_3. In this case we must evaluate the boundary integrals

$$-\int_{C_2(e)} qL_i \, dC_2, \qquad \int_{C_3(e)} hL_i L_j \, dC_3, \qquad \text{and} \qquad -\int_{C_3(e)} hT_\infty L_i \, dC_3$$

Consider the boundary elements shown in Figure 7.3. Nodes i and j lie on a boundary segment where either convection or conduction is specified. To simplify the computations, we may assume that q, h, and T_∞ are constant over the boundary segment of length l_{ij}. We define the length coordinate s (Figure 7.3) so that integration on the element boundary is in the counterclockwise direction. Then we have at node i

$$-\int_{C_2(e)} qL_I \, dC_2 = -q \int_0^{l_{ij}} \left(1 - \frac{s}{l_{ij}}\right) ds = -q \, \frac{l_{ij}}{2} \tag{7.25}$$

and similarly for node j.

The boundary matrix for convection has the form

$$[K_{C_3}] = h \int_{C_3(e)} \begin{bmatrix} L_i L_i & L_i L_j & L_i \overset{0}{L_k} \\ L_j L_i & L_j L_j & L_i \overset{0}{L_k} \\ \overset{0}{L_k} L_i & \overset{0}{L_k} L_j & \overset{0}{L_k} L_k \end{bmatrix} dC_3 \quad \text{on boundary segment } i \to j$$

Typical terms of this matrix are evaluated as follows:

$$\int_{i \to j} L_i^2 \, ds = \int_0^{l_{ij}} \left(1 - \frac{s}{l_{ij}}\right)^2 ds = \frac{l_{ij}}{3}$$

$$\int_{i \to j} L_i L_j \, ds = \int_0^{l_{ij}} \left(1 - \frac{s}{l_{ij}}\right)\frac{s}{l_{ij}} \, ds = \frac{l_{ij}}{6}$$

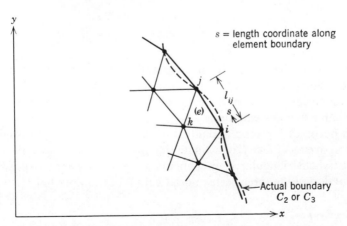

Figure 7.3.   Boundary elements where convection or conduction is specified.

Hence the complete boundary matrix is

$$[K_{C_3}]^{(e)}\{T\} = l_{ij}h \begin{bmatrix} \frac{1}{3} & \frac{1}{6} & 0 \\ \frac{1}{6} & \frac{1}{3} & 0 \\ 0 & 0 & 0 \end{bmatrix} \begin{Bmatrix} T_i \\ T_j \\ T_k \end{Bmatrix} \tag{7.26}$$

For the remaining matrix, we have simply at node i

$$-\int_{C_3^{(e)}} hT_\infty L_i \, dC_3 = -hT_\infty \frac{l_{ij}}{2} \tag{7.27}$$

and similarly for node j.

This completes the evaluation of the element matrices for linear triangular elements. Assembly and solution are the remaining two steps.

Axisymmetric heat conduction

The equation governing axisymmetric heat conduction is also a quasi-harmonic equation, given by

$$\frac{\partial}{\partial r}\left(k_r r \frac{\partial T}{\partial r}\right) + \frac{\partial}{\partial z}\left(k_z r \frac{\partial T}{\partial z}\right) + r\tilde{Q} = 0 \tag{7.28}$$

with boundary conditions

$$T = T(r, z) \quad \text{on } S_1 \tag{7.29}$$

$$k_r r \frac{\partial T}{\partial r} n_r + k_z r \frac{\partial T}{\partial z} n_z + rq = 0 \quad \text{on } S_2 \tag{7.30}$$

$$k_r r \frac{\partial T}{\partial r} n_r + k_z r \frac{\partial T}{\partial z} n_z + rh(T - T_\infty) = 0 \quad \text{on } S_3 \tag{7.31}$$

Here we assume that $\tilde{Q}$, q, and h are rotationally symmetric. Now, if the thermal conductivities are taken as $(k_r r)$ and $(k_z r)$, the previous formulation for two-dimensional heat conduction can be used and the element matrices have the same form as equations 7.14–7.16. The principal difference is that $d\Omega^{(e)}$ now becomes $2\pi r \, dr \, dz$, and integration for the boundary terms requires integration over a surface. To evaluate the integrals appearing in the element equations when triangular ring elements are used, we may express r as a linear function of the three vertex nodal values of r and write [3]

$$r = \sum_{i=1}^{3} L_i r_i \tag{7.32}$$

This useful substitution makes it possible to apply the integration formula of equation 5.22 and avoids the need for numerical integration. Explicit element matrix equations are given by Silvester and Konard [3] for triangular ring elements of arbitrarily high order, and matrices up to sixth order are tabulated.

Readers wishing to solve various forms of the two-dimensional, quasi-harmonic equation can benefit from the experience reported by Emery and Carson [4]. They studied the accuracy and efficiency of the finite element method in comparison to the finite difference method for the computation of temperature. Although they were primarily interested in transient problems, they also investigated methods for computing steady-state temperature distributions. Working with linear, quadratic, and cubic triangular elements, they chose three test problems and concluded that the quadratic element is the most accurate and desirable. When comparing finite element and finite difference models, they demonstrated that for steady-state temperature distributions the finite element models are superior.

7.3 HELMHOLTZ EQUATIONS

Another class of steady-state field equations closely related to the quasi-harmonic equation consists of Helmholtz equations, given generally by†

$$\frac{\partial}{\partial x}\left(k_x \frac{\partial \phi}{\partial x}\right) + \frac{\partial}{\partial y}\left(k_y \frac{\partial \phi}{\partial y}\right) + \frac{\partial}{\partial z}\left(k_z \frac{\partial \phi}{\partial z}\right) + \lambda \phi = 0 \qquad (7.33)$$

with Dirichlet and Neumann type boundary conditions. The constant λ may be known or unknown. Equations of the form of equation 7.33 arise in propagation problems involving wave motion of one type or another. This will become evident when we consider transient field problems in the following sections. Before developing the element equations for the Helmholtz problem, we consider several special forms of equation 7.33.

7.3.1 Special cases

The following are some practical examples of particular forms of Helmholtz equations.

† For inhomogeneous and anisotropic conditions, x, y, and z are taken as the principal coordinates.

Seiche motion

Standing waves on a bounded shallow body of water are governed by the equation [5]

$$\frac{\partial}{\partial x}\left(h\frac{\partial w}{\partial x}\right) + \frac{\partial}{\partial y}\left(h\frac{\partial w}{\partial y}\right) + \frac{4\pi^2}{gT^2}w = 0 \tag{7.34}$$

where h = water depth at the quiescent state,
$\quad\quad w$ = elevation of the free surface above the quiescent level,
$\quad\quad g$ = acceleration of gravity,
$\quad\quad T$ = period of oscillation.
  Equation 7.34 holds under the following assumptions:

 (a)   the flow is frictionless (no damping),
 (b)   the fluid inertia is small,
 (c)   fluid velocities are constant through the depth h.

For equation 7.34, we have the Neumann boundary condition

$$\frac{\partial w}{\partial x}n_x + \frac{\partial w}{\partial y}n_y = 0 \tag{7.35}$$

to be satisfied at solid boundaries.

Electromagnetic waves

The propagation of electromagnetic waves in a waveguide filled with a dielectric material obeys the equation [6]

$$\frac{\partial}{\partial x}\left(\frac{1}{\epsilon_d}\frac{\partial \phi}{\partial x}\right) + \frac{\partial}{\partial y}\left(\frac{1}{\epsilon_d}\frac{\partial \phi}{\partial y}\right) + \frac{\partial}{\partial z}\left(\frac{1}{\epsilon_d}\frac{\partial \phi}{\partial z}\right) + w^2\mu_0\epsilon_0\phi = 0 \tag{7.36}$$

where ϕ = a component of the magnetic field strength vector **H** or a component of the electric field vector **E**,
$\quad\quad \omega$ = wave frequency,
$\quad\quad \mu_0$ = permeability of free space,
$\quad\quad \epsilon_0$ = permittivity of free space,
$\quad\quad \epsilon_d$ = permittivity of the dielectric.
  Equation 7.36 can be derived from Faraday's law and Maxwell's equations, which are explained in the introductory text by Haytt [7]. If ϕ represents an **H** wave component, say $\phi = H_x$, then ϕ must satisfy the Neumann boundary condition at solid boundaries. But, if ϕ represents an **E** wave component, then ϕ satisfies the Dirichlet boundary condition. Equation 7.36 is normally not solved for both the **E** and **H** waves because one vector field can be obtained from the other via the equation

$$\nabla \times \mathbf{E} = -\sqrt{-1}\, w\mu_0\mathbf{H}$$

Acoustic vibrations

A fluid vibrating in a closed volume represents a sonic field of spherical waves governed by the equation [8]

$$\frac{\partial^2 P}{\partial x^2} + \frac{\partial^2 P}{\partial y^2} + \frac{\partial^2 P}{\partial z^2} + \frac{w^2}{c^2} P = 0 \tag{7.37}$$

where P = pressure excess above ambient pressure,
$\quad \omega$ = wave frequency,
$\quad c$ = wave velocity in the medium.

The derivation of equation 7.37 is based on a combination of three basic equations for fluids: (1) the continuity equation; (2) the equation expressing the elastic properties of the fluid; and (3) an elemental force balance equation. Also, the derivation relies on the following assumptions:

(a) the process is adiabatic,
(b) the local density changes are small,
(c) the displacement and velocity of the fluid particles are small.

We note that in each of these special cases the coefficient corresponding to λ in equation 7.33 is nonnegative.

7.3.2 Variational principles

Returning now to the general Helmholtz equation, we can apply the techniques of Appendix C to show that equation 7.33 is the Euler-Lagrange equation for the functional

$$I(\phi) = \int_{\Omega} \left[k_x \left(\frac{\partial \phi}{\partial x} \right)^2 + k_y \left(\frac{\partial \phi}{\partial y} \right)^2 + k_z \left(\frac{\partial \phi}{\partial z} \right)^2 - \lambda \phi^2 \right] d\Omega \tag{7.38}$$

The function $\phi(x, y, z)$ that minimized $I(\phi)$ in equation 7.38 and satisfies the given Dirichlet boundary condition also satisfies equation 7.33. No special consideration is needed for the Neumann boundary condition because this is naturally taken into account in the functional, the additional term being identically zero.

7.3.3 Element equations

The procedure discussed in Section 7.2.3 for deriving element equations from a variational principle applies again in this case. By expressing ϕ in terms of its nodal values within each element (equation 7.8) and then minimizing the corresponding discretized functional, we obtain element equations of the form

$$[K]^{(e)}\{\phi\}^{(e)} - \lambda[H]^{(e)}\{\phi\}^{(e)} = \{0\} \qquad (7.39)$$

where the terms of the matrices $[K]^{(e)}$ and $[H]^{(e)}$ are given by

$$k_{ij} = \int_{\Omega^{(e)}} \left(k_x \frac{\partial N_i}{\partial x} \frac{\partial N_j}{\partial x} + k_y \frac{\partial N_i}{\partial y} \frac{\partial N_j}{\partial y} + k_z \frac{\partial N_i}{\partial z} \frac{\partial N_j}{\partial z} \right) d\Omega^{(e)} \qquad (7.40a)$$

$$h_{ij} = \int_{\Omega^{(e)}} N_i N_j \, d\Omega^{(e)} \qquad (7.40b)$$

For a solution domain of M elements, the systems equations are of the form

$$[[K] - \lambda[H]]\{\phi\} = 0 \qquad (7.41)$$

where

$$[K] = \sum_{e=1}^{M} [K]^{(e)}, \qquad \text{and} \qquad [H] = \sum_{e=1}^{M} [H]^{(e)}$$

from the usual assembly rules. Also, $\{\phi\}$ is the column vector of nodal values of ϕ. Before solving equations 7.41, we must modify them according to the procedures of Section 2.4 to account for the Dirichlet boundary conditions.

Equations 7.41 are a set of, say n,† linear homogeneous algebraic equations in the nodal values of ϕ; but they are different from any that we have encountered thus far because λ is, in general, unknown. The problem we have here is called a general *eigenvalue* or *characteristic value* problem, and the λ values are termed *eigenvalues* or *characteristic values*. For each different value of λ_i, there is a different column vector $\{\phi\}_i$ that satisfies equation 7.41. The vector $\{\phi\}_i$ that corresponds to a particular value λ_i is called an *eigenvector*, characteristic vector, or *modal* vector.

Determining the eigenvalues constitutes part of the problem. From the fundamentals of linear algebra, we know that there will be a nontrivial solution to equation 7.41; in other words, $\{\phi\} \neq \{0\}$ if and only if the determinant (called the *characteristic* determinant) is zero, that is,

$$|[K] - [H]| = 0 \qquad (7.42)$$

In principle, this is the equation used to find the eigenvalues. If we were to expand equation 7.42, we would obtain an nth-order polynomial in λ such as

$$a_n \lambda^n + a_{n-1} \lambda^{n-1} + \cdots + a_1 \lambda + a_0 = 0$$

† Here n is taken as the number of unconstrained degrees of freedom of the problem.

Now the fundamental theorem of algebra assures us that this polynomial has n roots λ_i. When the λ_i are substituted into equation 7.41, we have n *sets* of equations to be solved for the n eigenvectors $\{\phi\}_i$. Essentially the complete solution of an eigenvalue problem requires about n times as much computational effort as is needed to find just one vector $\{\phi\}$.

The finite element discretization of the Helmholtz equation leads to matrices $[K]$ and $[H]$, which are both symmetric and positive definite. In this case, all the eigenvalues λ_i are distinct, real, positive numbers, and the corresponding eigenvectors $\{\phi\}_i$ are all independent. For each λ_i, it is impossible to determine uniquely the n components of $\{\phi\}_i$ because the set of equations is homogeneous. The usual procedure is to assign an arbitrary value to one component of the vector $\{\phi\}_i$ and then solve the remaining $n - 1$ equations for the other components. The consequence of this fact is that the natural modal vectors $\{\phi\}_i$ of the wave motion are known only in relation to one another, not in absolute terms. In other words, we can only determine ratios of modal vectors.

Either direct or iterative numerical methods are available for solving eigenvalue problems such as equation 7.42. When we seek only the first few eigenvalues and eigenvectors, rather than the complete set, iterative methods are best. The interested reader can find discussions of direct and iterative methods in standard textbooks [9,10].

7.3.4 Equations for linear tetrahedral elements

The simplest finite element discretization for the Helmholtz equation in three dimensions is obtained by using four-node linear tetrahedral elements. From Section 5.5 we know that the linear variation of ϕ within a tetrahedron may be expressed in terms of the four nodal values of ϕ as

$$\phi^{(e)}(x, y, z) = \sum_{i=1}^{4} L_i(x, y, z)\phi_i$$

where the interpolation functions L_i are the natural coordinates for the tetrahedron. The element matrices may be found from equations 7.40 after we note that $N_i = L_i$, and from equations 5.29

$$\frac{\partial N_i}{\partial x} = \frac{\partial L_i}{\partial x} = \frac{b_i}{6V}$$

$$\frac{\partial N_i}{\partial y} = \frac{\partial L_i}{\partial y} = \frac{c_i}{6V}$$

$$\frac{\partial N_i}{\partial z} = \frac{\partial L_i}{\partial z} = \frac{d_i}{6V} \tag{7.43}$$

where the b_i, c_i, d_i, and V are defined by equations 5.27. Hence we have, by substituting equations 7.43 into equations 7.40,

$$k_{ij} = \frac{1}{36V^2} \int_{\Omega^{(e)}} (k_x b_i b_j + k_y c_i c_j + k_z d_i d_j) \, d\Omega^{(e)} \tag{7.44}$$

and, if k_x, k_y, and k_z are constant within the element, the integration over the volume becomes trivial and we can write

$$k_{ij} = \frac{1}{36V} (k_x b_i b_j + k_y c_i c_j + k_z d_i d_j) \tag{7.45}$$

Furthermore,

$$h_{ij} = \int_{\Omega^{(e)}} L_i L_j \, d\Omega^{(e)}$$

which can be conveniently evaluated by the formula of equation 5.30 and found to be

$$h_{ij} = \frac{V}{20} \begin{cases} 2, & i = j \\ 1, & i \neq j \end{cases} \tag{7.46}$$

7.3.5 Sample problem [11]

The analysis and control of passenger compartment noise in an automobile constitutes a problem of considerable interest to automotive engineers. The acoustic field in the compartment is governed by the form of Helmholtz equation expressed in equation 7.37, but the solution domain has an exceedingly irregular shape. Shuku and Ishihara [11] recognized that the finite elements method is ideally suited to solve the Helmholtz equation over an irregular domain.

Before directly approaching the problem of noise inside cars, they conducted some accuracy and convergence studies for different calculation techniques. As a test problem, the normal frequencies of a closed rectangular room of dimensions $l_x \times l_y$ were calculated and compared with the exact results given by

$$f_{st} = \frac{c}{2} \sqrt{\left(\frac{s}{l_x}\right)^2 + \left(\frac{t}{l_y}\right)^2}$$

where f_{st} is the natural frequency of the (s, t) mode and c is the speed of sound in air.

Three-node triangular elements with modified cubic interpolation functions were used for the calculations. The pressure within each element was expressed as

$$P(x, y) = a_1 + a_2 x + a_3 y + a_4 x^2 + a_5 xy + a_6 y^2 + a_7 x^3$$
$$+ a_8(x^2 y + xy^2) + a_9 y^3$$

Solution domain (closed room) Mesh configuration (size: 2.0 × 1.1 m)	No. of nodes	No. of elements	Normal frequencies (Hz)		
			(1, 0) mode	(0, 1) mode	(2, 0) mode
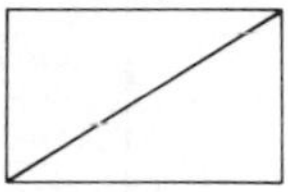	4	2	88.3	165.6	189.9
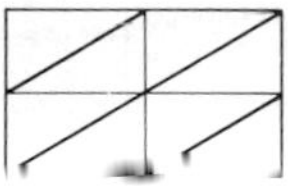	9	8	85.2	150.3	176.2
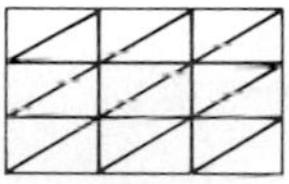	16	18	85.0	154.5	171.8
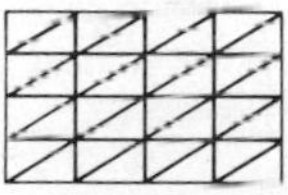	25	32	85.0	154.5	170.7
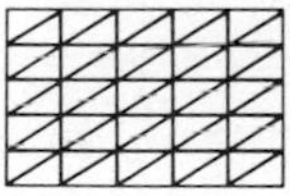	36	50	85.0	154.5	170.3
Exact			85.0	154.5	170.0

Figure 7.4. Effect of mesh size on the computation of normal frequencies of a closed room when modified cubic interpolation is used [11].

with P, $\partial P / \partial x$, and $\partial P / \partial y$ taken as nodal degrees of freedom at the vertices of the triangle. We recall from Chapter 5 that this element model provides C^0 continuity plus continuity of pressure derivatives at the nodes.

Figure 7.4 shows a comparison of exact results with computed results for various mesh sizes, and Figure 7.5 compares a finite difference solution [12]

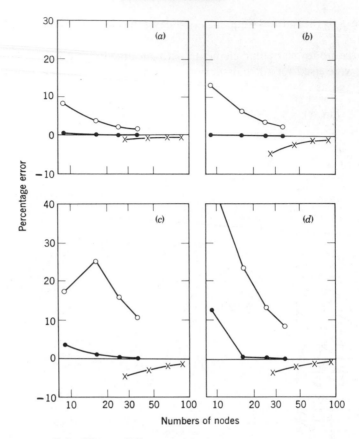

Figure 7.5. Accuracy comparison for different calculation methods [11]. (*a*) (1, 0) mode. (*b*) (0, 1) mode. (*c*) (2, 0) mode. (*d*) (1, 1) mode.

with linear and modified cubic finite element solutions. It can be seen that the cubic element gives decidedly more accurate results.

The cubic finite element model was used to compute the frequencies and modes of an acoustic field in the interior of a car modeled as shown in Figure 7.6. In reality, a car interior has soft boundaries; but in the model the boundaries were assumed to be rigid. Solid and dotted lines indicate the loci of points where the sound-pressure level is a minimum for the first three modes. Good agreement between calculated and experimental results is shown.

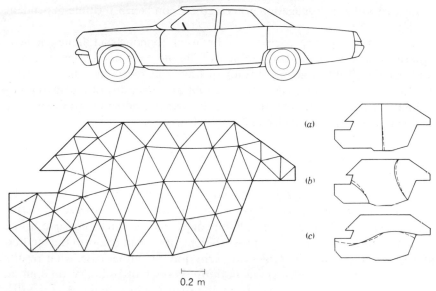

Figure 7.6.   Finite element model of an automobile compartment, with a comparison of calculated and experimental results [11] (*a*) First mode: 86.8 Hz (exp. 87.5 Hz). (*b*) Second mode: 130.0 Hz (exp. 138.5 Hz). (*c*) Third mode: 154.6 Hz (exp. 157 Hz).

7.4   TIME-DEPENDENT WAVE EQUATION

Wave propagation, whether for electromagnetic waves, acoustic waves, surface waves, or any of the many other types, is a time-dependent phenomenon which for a homogeneous, isotropic medium is governed by the equation

$$\frac{\partial^2 \phi}{\partial x^2} + \frac{\partial^2 \phi}{\partial y^2} + \frac{\partial^2 \phi}{\partial z^2} = k_t \, \dot{\phi} + k_{tt} \, \ddot{\phi} \tag{7.47}$$

where

$$\dot{\phi} = \frac{\partial \phi}{\partial t}, \qquad \ddot{\phi} = \frac{\partial^2 \phi}{\partial t^2}$$

and the coefficients k_t and k_{tt} can, in general, be functions of time. Both Dirichlet and Neumann boundary conditions can apply to equation 7.47, depending on the type of boundary considered; and, for the problem statement to be complete, initial conditions on ϕ and $\dot{\phi}$ must be given. The first term on the right-hand side represents the energy dissipation or damping associated with the propagation process; the second term, the time-dependent motion.

Several different approaches can be used to construct finite element equations for time-dependent problems. The most common approach is to consider the problem at one instant of time and to assume that the time derivatives such as $\dot{\phi}$ and $\ddot{\phi}$ at that instant are functions of the spatial co-ordinates only. Then we construct a finite element model by expressing the field variable for a typical element with r degrees of freedom as

$$\phi^{(e)}(x, y, z, t) = \sum_{i=1}^{r} N_i(x, y, z)\phi_i(t) \qquad (7.48)$$

where the N_i are the usual interpolation functions and the $\phi_i(t)$ and the time-dependent nodal parameters. At this point, we can proceed to derive the element equations by using an appropriate variational principle if one exists for our contrived steady-state problem, or we can resort to the method of weighted residuals with Galerkin's criterion.† In either case, what results after assembly of the element equations is a set of ordinary differential equations in $\{\phi\}$ which we can solve by a variety of techniques to be discussed in Section 7.6.

Another approach (suggested by Oden [13]) is to view the general time-dependent problem in a four-dimensional, space-time domain and represent the field variable within a typical element as

$$\phi^{(e)}(x, y, z, t) = \sum_{i=1}^{r} N_i(x, y, z, t)\phi_i \qquad (7.49)$$

where the interpolation functions N_i incorporate both space and time. The idea is simply a natural extension of the interpolation concepts normally used in one-, two-, or three-dimensional problems. Of course, with each added dimension, the cost of computation greatly increases; hence a full four-dimensional problem could easily exceed the realm of the practical. Nevertheless, the idea is intriguing and is receiving some attention.

Once the discretization described by equation 7.49 is assumed, the element equations can be drived by (*a*) the Galerkin procedure applied directly to the governing differential equation, or (*b*) a variational principle—if one can be found. Gurtin [14] has derived a useful variational principle for time-dependent diffusion equations by employing convolution integrals. Usually the Galerkin approach is easier to apply.

For the particular hyperbolic problem described by equation 7.47, we can use another approach which has more limited applicability. For this problem,

---

† This procedure was used without explanation in the example of Section 4.2.3.

it is especially convenient to separate variables and obtain two independent equations which are easily solved [15]. Following the method of separation of variables [16], we assume that the unknown field variable can be written as the product

$$\phi(x, y, z, t) = \psi(x, y, z)T(t) \tag{7.50}$$

Substituting equation 7.50 into equation 7.47 gives

$$T\nabla^2\psi = k_t\psi\dot{T} + k_{tt}\psi\ddot{T}$$

and, by dividing both sides of this equation by ψT, we obtain

$$\frac{\nabla^2\psi}{\psi} = k_t\frac{\dot{T}}{T} + k_{tt}\frac{\ddot{T}}{T} \tag{7.51}$$

Now we recognize that the right-hand side of equation 7.51 is a function only of time, while the left-hand side depends only on the spatial coordinates. This circumstance can prevail only if each side is equal to a constant, say $-\lambda$. Hence

$$\frac{\nabla^2\psi}{\psi} = -\lambda$$

and

$$k_t\frac{\dot{T}}{T} + k_{tt}\frac{\ddot{T}}{T} = -\lambda$$

or

$$\nabla^2\psi + \lambda\psi = 0 \tag{7.52}$$

and

$$k_{tt}\ddot{T} + k_t\dot{T} + \lambda T = 0 \tag{7.53}$$

Equation 7.52 with its boundary conditions is the Helmholtz equation, which we can solve by the finite element techniques described in Section 7.3. Here we see by a separation-of-variables technique how the Helmholtz equation naturally arises from wave propagation problems. Equation 7.53 with initial conditions is simply an ordinary linear differential equation having, in general, variable coefficients. This equation can be solved by standard methods either in closed form or by using a numerical integration scheme such as the Runge-Kutta method or its variants [17]. Once we find the functions ψ and T, the complete solution is obtained from equation 7.50.

As a particular example of the time-dependent wave equation, we consider the Schrödinger equation [18], the heart of quantum mechanics. The wave nature of a single particle is described by the equation

$$\frac{\partial^2 U}{\partial x^2} + \frac{\partial^2 U}{\partial y^2} + \frac{\partial^2 U}{\partial z^2} = \frac{2m(hv - V)}{h^2 v^2}\frac{\partial^2 U}{\partial t^2} \tag{7.54}$$

where U = wave function whose amplitude squared is proportional to the probability of finding the particle at any given point in space,

m = particle mass,

V = potential energy of the particle,

h = Planck's constant,

v = wave frequency.

The central problem in wave mechanics is to assign an appropriate character to the potential energy function V and then solve equation 7.54 for U.

Since, in general, only wave functions U that are harmonic functions of time are physically important [18], we seek solutions in terms of combinations of $\sin 2\pi vt$ and $\cos 2\pi vt$. Noting that

$$e^{i2\pi vt} = \cos 2\pi vt + i \sin 2\pi vt$$

we assume solutions of the form

$$U(x, y, z, t) = u(x, y, z)e^{i2\pi vt} \tag{7.55}$$

where $i = \sqrt{-1}$.

Substituting equation 7.55 into equation 7.54 and noting that

$$\frac{\partial^2 U}{\partial t^2} = -4\pi^2 v^2 e^{i2\pi vt} = -4\pi^2 v^2 U$$

gives

$$\frac{\partial^2 U}{\partial x^2} + \frac{\partial^2 U}{\partial y^2} + \frac{\partial^2 U}{\partial z^2} = -\frac{8m\pi^2(hv - V)}{h^2} U \tag{7.56a}$$

or, for the wave amplitude only, we have

$$\frac{\partial^2 u}{\partial x^2} + \frac{\partial^2 u}{\partial y^2} + \frac{\partial^2 u}{\partial z^2} = -\frac{8m\pi^2(hv - V)}{h^2} u \tag{7.56b}$$

Again, this is the familiar Helmholtz equation, and we see how it arises in wave propagation problems.

7.5 GENERAL TIME-DEPENDENT FIELD PROBLEMS

A large class of transient field problems is governed by equations in domain Ω of the form

$$\frac{\partial}{\partial x}\left(k_x \frac{\partial \phi}{\partial x}\right) + \frac{\partial}{\partial y}\left(k_y \frac{\partial \phi}{\partial y}\right) + \frac{\partial}{\partial z}\left(k_z \frac{\partial \phi}{\partial z}\right) = f(x, y, z, t) + k_t \dot\phi + k_{tt} \ddot\phi \tag{7.57}$$

with boundary conditions

$$\phi = \Phi(x, y, z, t) \quad \text{on } S_1, t > 0 \tag{7.58}$$

$$k_x \frac{\partial \phi}{\partial x} n_x + k_y \frac{\partial \phi}{\partial y} n_y + k_z \frac{\partial \phi}{\partial z} n_z$$

$$+ q(x, y, z, t) + h(x, y, z, t)\phi = 0 \quad \text{on } S_2, t > 0 \tag{7.59}$$

and initial conditions

$$\phi = \phi_0(x, y, z) \quad \text{in } \Omega, t = 0 \tag{7.60}$$

$$\dot\phi = \xi(x, y, z) \quad \text{in } \Omega, t = 0 \tag{7.61}$$

The specified parameters k in equation 7.57 may, in general, be functions of space and time. If $k_{tt} = 0$ and $k_t \neq 0$, equation 7.57 is of parabolic form; while if $k_{tt} \neq 0$, the equation is hyperbolic.

No classical variational principle exists for the problem described by equations 7.57–7.61, so we shall resort to the Galerkin approach to derive finite element equations. Within a typical element we assume

$$\phi^{(e)} = \sum_{i=1}^{r} N_i(x, y, z)\phi_i(t) = \lfloor N \rfloor \{\phi\}^{(e)} \tag{7.48}$$

where the interpolation functions N_i need only preserve C^0 continuity. The Galerkin criterion then requires that at any instant of time

$$\int_{\Omega^{(e)}} N_i \left[\frac{\partial}{\partial x}\left(k_x \frac{\partial \phi^{(e)}}{\partial x} \right) + \frac{\partial}{\partial y}\left(k_y \frac{\partial \phi^{(e)}}{\partial y} \right) \right.$$

$$\left. + \frac{\partial}{\partial z}\left(k_z \frac{\partial \phi^{(e)}}{\partial z} \right) - f - k_t \dot\phi^{(e)} - k_{tt} \ddot\phi^{(e)} \right] d\Omega^{(e)} = 0$$

$$i = 1, 2, \ldots, r \tag{7.62}$$

where $\Omega^{(e)}$ is the domain for element (e).

Now, following the standard procedures fully discussed in Chapter 4, we integrate the terms such as

$$\int_{\Omega^{(e)}} N_i \frac{\partial}{\partial x}\left(k_x \frac{\partial \phi^{(e)}}{\partial x} \right) d\Omega^{(e)}$$

by parts to introduce the boundary conditions on S_2, and finally, after the manipulations, the resulting element equations become

$$[K]^{(e)}\{\phi\}^{(e)} + [K_t]^{(e)}\{\dot\phi\}^{(e)} + [K_{tt}]^{(e)}\{\ddot\phi\}^{(e)} + [K_{s_2}]^{(e)}\{\phi\}^{(e)}$$
$$+ \{R_1(t)\}^{(e)} = \{0\} \quad (7.63)$$

where

$$k_{ij} = \int_{\Omega^{(e)}} \left(k_x \frac{\partial N_i}{\partial x} \frac{\partial N_j}{\partial x} + k_y \frac{\partial N_i}{\partial y} \frac{\partial N_j}{\partial y} + k_z \frac{\partial N_i}{\partial z} \frac{\partial N_j}{\partial z} \right) d\Omega^{(e)} \quad (7.64\text{a})$$

$$k_{t_{ij}} = \int_{\Omega^{(e)}} k_t N_i N_j \, d\Omega^{(e)} \quad (7.64\text{b})$$

$$k_{tt_{ij}} = \int_{\Omega^{(e)}} k_{tt} N_i N_j \, d\Omega^{(e)} \quad (7.64\text{c})$$

$$k_{s_{2ij}} = \int_{S_2^{(e)}} hN_i N_j \, dS_2^{(e)} \quad (7.64\text{d})$$

$$R_{1i} = \int_{\Omega^{(e)}} fN_i \, d\Omega^{(e)} + \int_{S_2^{(e)}} q_i \, dS_2^{(e)} \quad (7.64\text{e})$$

Assembly of the element equations then leads to a system of ordinary differential equations of the same form as equation 7.63. The problem solution is complete when these equations are solved for the nodal parameters $\{\phi\}$ subject to the discretized initial conditions. Solution techniques are discussed in Section 7.7.

Convective diffusion

Another type of time-dependent field problem arises from the combined phenomena of convection and diffusion. If, for example, we wish to study the dispersion of a concentrated pollutant in the atmosphere or in some other moving body of fluid, we must solve an equation of the form

$$\frac{\partial}{\partial x}\left(k_x \frac{\partial \phi}{\partial x} \right) + \frac{\partial}{\partial y}\left(k_y \frac{\partial \phi}{\partial y} \right) + \frac{\partial}{\partial z}\left(k_z \frac{\partial \phi}{\partial z} \right) - \left[\frac{\partial}{\partial x}(u\phi) + \frac{\partial}{\partial y}(v\phi) + \frac{\partial(w\phi)}{\partial z} \right]$$
$$= k_t \dot\phi + f(x, y, z, t) \quad (7.65)$$

with boundary and initial conditions given by equations 7.58–7.60. Here ϕ is a concentration function; k_x, k_y, and k_z are specified functions representing diffusion coefficients; and u, v, and w are the velocity components of the prescribed flow field. The function f describes a possible source-sink field.

Again, no classical variational statement exists for this differential equation. By using an integrating factor and thereby changing the form of the equation to a canonical one, it is possible to construct a minimum principle involving an exponential [19]. However, to derive element equations, a more direct approach is to use Galerkin's method. If we assume the usual space discretization expressed by equation 7.48 and follow through the conventional steps, we obtain element equations of the form

$$[K]^{(e)}\{\phi\}^{(e)} + [C]^{(e)}\{\dot{\phi}\}^{(e)} + [K_t]^{(e)}\{\dot{\phi}\}^{(e)} + [K_{s_2}]^{(e)}\{\phi\}^{(e)} + \{R_1(t)\}^{(e)} = \{0\}$$
(7.66)

where k_{ij}, $k_{t_{ij}}$, $k_{s_{2ij}}$, and R_{1i} are given by equations 7.64a, 7.64b, 7.64d, and 7.64e, respectively. Terms of the matrix $[C]^{(e)}$, however, are of the form

$$c_{ij} = \int_{\Omega^{(e)}} N_i \left[\frac{\partial(uN_j)}{\partial x} + \frac{\partial(vN_j)}{\partial y} + \frac{\partial(wN_j)}{\partial z} \right] d\Omega^{(e)}$$
(7.67)

Note that, for this problem, the matrix $[C]^{(e)}$ is not symmetrical; hence the standard routines for inverting symmetrical matrices cannot be used. One way to avoid this problem will be discussed in Chapter 9.

A simple example in one space dimension

To illustrate the use of the general element equations developed for time-dependent field problems, we shall consider here a problem governed by the following differential equation:

$$\frac{\partial}{\partial x}\left(k_x \frac{\partial \phi}{\partial x} \right) = f(x, t) + k_t \dot{\phi} + k_{tt} \ddot{\phi}$$

with

$$\phi(0, t) = \phi(1, t) = 1$$
$$\phi(x, 0) = 0, \qquad \dot{\phi}(x, 0) = 1$$

The first step toward finding $\phi(x, t)$ by the finite element method is to divide the spatial solution domain into elements. For simplicity, we select M elements with linear interpolation functions (Figure 7.7). For a typical element, the interpolation functions are, from equations 5.11,

$$N_1(x) = L_1(x) = \frac{x - x_1}{x_2 - x_1}$$

$$N_2(x) = L_2(x) = \frac{x_2 - x}{x_2 - x_1}$$

so that the interpolation model is

$$\phi^{(e)}(x, t) = L_1 \phi_1(t) + L_2 \phi_2(t)$$

Equation 7.64 allows us to write the terms of the element equations at once.

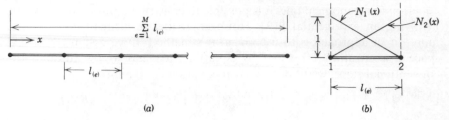

Figure 7.7. Finite element model for a one-dimensional field problem. (*a*) Discretized solution domain containing *M* elements. (*b*) Linear interpolation functions over a typical element.

These are as follows:

$$k_{ij} = \int_{x_1}^{x_2} k_x \frac{\partial L_i}{\partial x} \frac{\partial L_j}{\partial x} \, dx = \frac{(-1)^{i+j}}{(x_2 - x_1)^2} \int_{x_1}^{x_2} k_x \, dx$$

$$k_{t_{ij}} = \int_{x_1}^{x_2} k_t L_i L_j \, dx$$

$$k_{tt_{ij}} = \int_{x_1}^{x_2} k_{tt} L_i L_j \, dx$$

$$k_{s_{2ij}} = 0 \quad \text{(since } h \equiv 0\text{)}$$

$$R_{1i} = \int_{x_1}^{x_2} f(x, t) L_i \, dx$$

If we assume that k_x, k_t, and k_{tt} are constant within an element and $f(x, t)$ varies linearly, that is,

$$f(x, t) = f(x_1, t) L_1 + f(x_2, t) L_2 = f_1 L_1 + f_2 L_2$$

then the above equations reduce to

$$k_{ij} = \frac{k_x^{(e)}}{x_2 - x_1} (-1)^{i+j}$$

$$k_{t_{ij}} = \frac{k_t^{(e)}(x_2 - x_1)}{6} (1 + \delta_{ij})$$

$$k_{tt_{ij}} = \frac{k_{tt}^{(e)}(x_2 - x_1)}{6} (1 + \delta_{ij})$$

$$k_{s_2} = 0$$

$$R_{1_1} = (x_2 - x_1)(\tfrac{1}{3} f_1 + \tfrac{1}{6} f_2)$$

$$R_{1_2} = (x_2 - x_1)(\tfrac{1}{6} f_1 + \tfrac{1}{3} f_2)$$

where δ_{ij} is the Kronecker delta,

$$\delta_{ij} = \begin{cases} 1, & i = j \\ 0, & i \neq j \end{cases}$$

and the superscript (e) designates the value for element (e).

From equation 7.63 and the above equations, we can write the equations for element (e) as

$$\frac{k_x^{(e)}}{x_2 - x_1}\begin{bmatrix} 1 & -1 \\ -1 & 1 \end{bmatrix}\begin{Bmatrix} \phi_1 \\ \phi_2 \end{Bmatrix}^{(e)} = k_t^{(e)}(x_2 - x_1)\begin{bmatrix} \frac{1}{3} & \frac{1}{6} \\ \frac{1}{6} & \frac{1}{3} \end{bmatrix}\begin{Bmatrix} \dot{\phi}_1 \\ \dot{\phi}_2 \end{Bmatrix}^{(e)}$$

$$+ \frac{k_{tt}^{(e)}(x_2 - x_1)}{6}\begin{bmatrix} \frac{1}{3} & \frac{1}{6} \\ \frac{1}{6} & \frac{1}{3} \end{bmatrix}\begin{Bmatrix} \ddot{\phi}_1 \\ \ddot{\phi}_2 \end{Bmatrix}^{(e)} + (x_2 - x_1)\begin{Bmatrix} \frac{1}{3}f_1 + \frac{1}{6}f_2 \\ \frac{1}{6}f_1 + \frac{1}{3}f_2 \end{Bmatrix}^{(e)}$$

The problem solution is complete when the number of elements is selected, the element equations are evaluated and assembled, and the resulting system equations are solved.

7.6 SOLVING THE DISCRETIZED TIME-DEPENDENT EQUATIONS

We saw in the preceding sections that solving time-dependent field problems by the finite element method reduces ultimately to the solution of matrix differential equations of motion of the form:

$$[K_\phi]\{\phi\} + [K_{\dot\phi}]\{\dot\phi\} + [K_{\ddot\phi}]\{\ddot\phi\} = \{R(t)\} \qquad (7.68)$$

The physical interpretation of the various matrices in this equation depends on the particular type of problem, that is, the nature of the field variable ϕ. Generally, $[K_\phi]$ is a type of stiffness matrix; $[K_{\dot\phi}]$ is a capacitance or damping matrix; and $[K_{\ddot\phi}]$ represents system inertia of some type. In some problems either $[K_{\dot\phi}] = [0]$ or $[K_{\ddot\phi}] = [0]$. For example, in heat conduction problems, $[K_{\ddot\phi}] \equiv [0]$. Of course, if both are zero, the dynamic character of the problem is absent even though the right-hand side may remain a function of time. Also, depending on the type of problem, the discretized forcing function $\{R(t)\}$ may be zero, harmonic, periodic, aperiodic, or random [20]. Problems with random or nondeterministic forcing functions will not be discussed here. The interested reader can find an introductory treatment of this subject in Hurty and Rubinstein [20].

We will assume that, after the formulation of equations 7.68, the matrices were modified to account for any boundary conditions not already included as natural boundary conditions. In addition to the boundary conditions, before attempting a solution, we must know both initial conditions $\{\phi(0)\}$ and $\{\dot\phi(0)\}$, when $[K_{\ddot\phi}] \neq [0]$. If $[K_{\ddot\phi}] = [0]$, we need to know only $\{\phi(0)\}$.

Equation 7.68 with its initial conditions is sometimes called an *initial value problem* or a *propagation problem*.

There are many particular methods and special techniques for solving matrix differential equations—so many, in fact, that it is impractical to try and develop most of them here. Instead, we shall consider only three fundamental approaches that have *general applicability*. More specialized techniques and further details can be found in books on structural dynamics, such as refs. [20–22], or on numerical computation, such as refs. [23–26]. As a preliminary to developing one general solution procedure for propagation problems (a procedure known as mode superposition), we shall first review the conventional means for finding natural frequencies and natural modes for a vibrating system. Then, after discussing the mode superposition technique, we shall focus attention on solution methods that rely on recurrence relations which permit time-stepping or time-marching procedures. Clearly, these solution techniques are not related to the finite element method. They apply to equations 7.68 regardless of how these equations were derived.

7.6.1   Finding undamped harmonic motion

If we imagine the ideal case in which the system has no damping and no external forcing functions, the equations of motion reduce to

$$[K]\{\phi\} + [K_{\ddot{\phi}}]\{\ddot{\phi}\} = \{0\} \tag{7.69}$$

This equation expresses the condition of natural vibration (simple harmonic motion), where, at any instant, the restoration influences in the system balance the inertia influences. The states of natural vibration are called *natural modes* or principal modes, and the frequencies of vibration are the natural frequencies. The system will have as many natural modes and natural frequencies as it has unconstrained degrees of freedom.

To find these natural modes and frequencies, we assume that the field variable may be expressed as

$$\{\phi\} = \{\Phi\}e^{i\omega t} = \{\Phi\}(\cos \omega t + i \sin \omega t)$$

where $\{\Phi\}$ is the vector of unknown amplitude at the nodes (modal vector) and ω is one natural frequency. Noting that

$$\{\ddot{\phi}\} = -\omega^2\{\Phi\}e^{i\omega t}$$

we find upon substitution that equation 7.69 reduces to

$$[[K] - \omega^2[K_{\ddot{\phi}}]]\{\Phi\} = \{0\} \tag{7.70}$$

Equation 7.70 we recognize as an eigenvalue problem similar to that which arose from the solution of the Helmholtz equation (Section 7.3.3). As before the equation has a nontrivial solution only when the determinant $|[K] - \omega^2[K_{\ddot{\phi}}]| = 0$. This is equivalent to the polynomial

$$(\omega^2)^n + (\quad)(\omega^2)^{n-1} + \cdots + (\quad)(\omega^2) + (\quad) = 0 \qquad (7.71)$$

For matrices of dimension $n \times n$, there will be n values of ω_i^2 satisfying equation 7.71 and, hence, n vectors $\{\Phi\}$ that satisfy equation 7.70.

When the ideal system is impulsively excited, it may vibrate in any one of its natural modes, depending on just what initial conditions were imposed. Real systems, however, never actually vibrate in a natural mode because they always possess some degree of damping. Nevertheless, the concept of natural modes and frequencies is crucially important in the solution of dynamics problems. As we shall see subsequently, once the natural modes of a system are known, it is possible to use them to uncouple the equations of motion and obtain a solution for the dynamic response to many kinds of forcing functions $\{R(t)\}$.

The weighted orthogonality of the natural modes or eigenvectors is an important property that we can turn to advantage in solving real dynamics problems. It is easy to show [20] that for two different eigenvectors corresponding to two different frequencies ω_i and ω_j we have

$$\lfloor \Phi \rfloor_i [K] \{\Phi\}_j = 0, \quad i \neq j$$

or

$$\lfloor \Phi \rfloor_i [K_{\ddot{\phi}}] \{\Phi\}_j = 0, \quad i \neq j \qquad (7.72)$$

where $[K]$ and $[K_{\ddot{\phi}}]$ are weighting matrices. At a given frequency, say ω_i, we have

$$\lfloor \Phi \rfloor_i [K] \{\Phi\}_i = C_{K_i}$$
$$\lfloor \Phi \rfloor_i [K_{\ddot{\phi}}] \{\Phi\}_i = C_{K_{\ddot{\phi}_i}} \qquad (7.73)$$

where C_{K_i} and $C_{K_{\ddot{\phi}}}$ are constants different from zero. Often it is convenient to normalize the eigenvectors so that $C_{K_i} = 1$ or $C_{K_{\ddot{\phi}_i}} = 1$.

7.6.2 Finding transient motion via mode superposition

Once the eigenvalue problem is solved for the natural modes and frequencies, we can use the method of mode superposition to determine the solution to the complete set of equations with $[K_{\ddot{\phi}}] \neq [0]$ and $\{R(t)\} \neq \{0\}$. The method relies on the basic premise that the solution vector $\{\phi\}$ may be expressed as a linear combination of all n eigenvectors of the system. Hence we set

$$\{\phi(t)\} = [\{\Phi\}_1, \{\Phi\}_2, \{\Phi\}_3, \ldots, \{\Phi\}_n]\{\Lambda(t)\}$$
$$= [A]\{\Lambda(t)\} \tag{7.74}$$

where $[A]$ is a square matrix whose columns are the eigenvectors and $\{\Lambda(t)\}$ is the vector of unknown nodal amplitudes. If we substitute equation 7.74 into equation 7.68 and then premultiply the resulting equation by the transpose of the eigenvector matrix $[A]$, we obtain

$$[A]^T[K][A]\{\Lambda\} + [A]^T[K_{\dot{\phi}}][A]\{\dot{\Lambda}\} + [A]^T[K_{\ddot{\phi}}][A]\{\ddot{\Lambda}\} = [A]^T\{R\} \tag{7.75}$$

or

$$[K^*]\{\Lambda\} + [K_{\dot{\phi}}^*]\{\dot{\Lambda}\} + [K_{\ddot{\phi}}^*]\{\ddot{\Lambda}\} = \{R^*\} \tag{7.76}$$

where

$$[K^*] = [A]^T[K][A]$$
$$[K_{\dot{\phi}}^*] = [A]^T[K_{\dot{\phi}}][A]$$
$$[K_{\ddot{\phi}}^*] = [A]^T[K_{\ddot{\phi}}][A]$$
$$\{R^*\} = [A]^T\{R\}$$

in which the individual terms are given by

$$K_{ij}^* = \lfloor\Phi\rfloor_i[K]\{\Phi\}_j$$
$$K_{\dot{\phi}_{ij}}^* = \lfloor\Phi\rfloor_i[K_{\dot{\phi}}]\{\Phi\}_j$$
$$K_{\ddot{\phi}_{ij}}^* = \lfloor\Phi\rfloor_i[K_{\ddot{\phi}}]\{\Phi\}_j$$
$$R_i^* = \lfloor\Phi\rfloor_i\{R\}$$

Now, by definition,

$$[K]\{\Phi\}_j = \omega_j^2[K_{\ddot{\phi}}]\{\Phi\}_j$$

Hence

$$K_{ij}^* = \omega_j^2 K_{\ddot{\phi}_{ij}}^*$$

In view of the orthogonality relations expressed by equations 7.72, we see that the matrices $[K^*]$ and $[K_{\ddot{\phi}}^*]$ are diagonal because $K_{ij}^* = 0$, $i \neq j$, and so forth. Furthermore, the equations are *uncoupled* if we assume that $[K_{\dot{\phi}}]$ is proportional to either $[K]$ or $[K_{\ddot{\phi}}]$. This assumption is usually justified for structural systems because the actual coupling produced by damping is often negligible. When we assume that $[K_{\dot{\phi}}]$ is proportional to $[K_{\ddot{\phi}}]$, a typical uncoupled equation for the ith mode has the form

$$\omega_i^2 K_{\ddot{\phi}_{ii}}^* \Lambda_i + 2c_i\omega_i K_{\ddot{\phi}_{ii}}^* \dot{\Lambda}_i + K_{\ddot{\phi}_{ii}}^* \ddot{\Lambda}_i = R_i^* \tag{7.77}$$

where c_i is the damping ratio for mode i. A discussion of how the c_i can be determined is given in Clough [27]. Dividing through by $K^*_{\phi_{ii}}$ yields the more classical form of the equation of motion to be solved for $\Lambda_i(t)$:

$$\omega_i^2 \Lambda_i + 2c_i \omega_i \dot{\Lambda}_i + \ddot{\Lambda}_i = \frac{R^*_i}{K^*_{\phi_{ii}}}, \quad i = 1, 2, \ldots, n \qquad (7.78)$$

To achieve the uncoupled equations of motions (such as equation 7.78), we had to assume that $[K_\phi]$ was proportional to $[K_{\ddot{\phi}}]$, but this assumption is not essential. Uncoupling is possible even for the case of nonproportional damping if a special technique introduced by Foss [28] is used.

The solution of the complete dynamics problem now involves solving independently n linear, ordinary, uncoupled differential equations like equation 7.78 for $\Lambda_i(t)$, and then combining the results according to equation 7.74. Equation 7.78 may be solved by several different direct numerical integration schemes [17]; however, the Duhamel integral [20] provides a particularly convenient way to obtain the solution. For any forcing function $(R^*_i/K^*_{\phi_{ii}})$ the time history of Λ_i can be found by evaluating

$$\Lambda_i(t) = \frac{1}{K^*_{\phi_{ii}} \omega_n \sqrt{1 - c_i^2}} \int_0^t \left[\frac{R^*_i(t)}{K^*_{\phi_{ii}}} \right] e^{-c_i \omega_i (t - \tau)} \sin\left[(t - \tau) \omega_n \sqrt{1 - c_i^2} \right] d\tau$$

$$i = 1, 2, 3, \ldots, n \qquad (7.79)$$

where t is the particular time at which the response is sought.† By evaluating equation 7.79 at different time increments—$0 < t_1 < t_2 < t_3 < t_4 \ldots$—we can obtain the complete time history for each modal amplitude $\Lambda_i(t)$.

Solving dynamics problems by the method of mode superposition involves considerable computational effort because the complete eigenvalue problem must first be solved for the natural modes and frequencies. However, once these are found, they may be stored and used later to find the dynamic responses of the system to other forcing functions that may be of interest. It should be noted that mode superposition holds only for linear problems, that is, problems in which $[K]$ and/or $[K_{\ddot{\phi}}]$ do not depend on $\{\phi\}$.

7.6.3   Finding transient motion via recurrence relations

Instead of solving equations 7.68 by first using the eigenvectors of the system to derive decoupled equations, an alternative is to apply direct numerical integration in time and to solve the full set of coupled equations.

† Equation 7.79 results from solving equation 7.78 via Laplace transforms. It was assumed that

$$\Lambda_i(0) = \dot{\Lambda}_i(0) = 0, \quad i = 1, 2, \ldots, n$$

A slightly different expression would be obtained for nonhomogeneous initial conditions.

The procedure relies on deriving recursion formulas that relate the values of $\{\phi\}$, $\{\dot\phi\}$, and $\{\ddot\phi\}$ at one instant of time t to the values of these quantities at a later time, $t + \Delta t$, where Δt is a small time increment. The recursion formulas make it possible for the solution to be "marched out" in time, starting with the initial conditions at time $t = 0$ and continuing step by step until the desired length of time is achieved.

We begin the derivation of appropriate recursion formulas for solving dynamics problems by writing series expansions for the vectors $\{\phi\}$, $\{\dot\phi\}$, and $\{\ddot\phi\}$ at time $t + \Delta t$ in terms of their values at time t. These become

$$\{\phi\}_{t+\Delta t} = \{\phi\}_t + \Delta_t\{\dot\phi\}_t + \frac{(\Delta t)^2}{2}\{\ddot\phi\}_t + \frac{(\Delta t)^3}{6}\{\dddot\phi\}_t + \cdots \qquad (7.80a)$$

$$\{\dot\phi\}_{t+\Delta t} = \{\dot\phi\}_t + \Delta t\{\ddot\phi\}_t + \frac{(\Delta t)^2}{2}\{\dddot\phi\}_t + \cdots \qquad (7.80b)$$

$$\{\ddot\phi\}_{t+\Delta t} = \{\ddot\phi\}_t + \Delta t\{\dddot\phi\}_t \qquad (7.80c)$$

The subscripts in these expressions indicate the times at which the vectors are evaluated.

Before proceeding to derive the recurrence formulas, we must make an assumption regarding the variation of the second derivative, $\{\ddot\phi\}$, within the time increment Δt. Usually, $\{\ddot\phi\}$ is assumed to remain constant within the interval or to vary linearly.

First, we consider the case where $\{\ddot\phi\}$ *remains constant* within Δt, that is, $\{\dddot\phi\} = \{0\}$. Then, from equations 7.80b and 7.80c,

$$\{\dot\phi\}_{t+\Delta t} = \{\dot\phi\}_t + \Delta t\{\ddot\phi\}_t \qquad \text{and} \qquad \{\ddot\phi\}_{t+\Delta t} = \{\ddot\phi\}_t$$

Returning to the original equations of motions, we find that the second derivative is given by

$$\{\ddot\phi\}_t = [K_{\ddot\phi}]^{-1}[\{R\}_t - [K_\phi]\{\phi\}_t - [K_{\dot\phi}]\{\dot\phi\}_t] \qquad (7.81)$$

From equation 7.80a, we have

$$\{\phi\}_{t+\Delta t} = \{\phi\} + \Delta t\{\dot\phi\}_t + \frac{(\Delta t)^2}{2}\{\ddot\phi\}_t \qquad (7.82)$$

and, when the first derivative is averaged at two successive intervals,

$$\{\dot\phi\}_t = \frac{1}{2}\left[\frac{\{\phi\}_{t+\Delta t} - \{\phi\}_t}{\Delta t} + \frac{\{\phi\}_t - \{\phi\}_{t-\Delta t}}{\Delta t}\right] = \frac{\{\phi\}_{t+\Delta t} - \{\phi\}_t}{2\Delta t}$$

equation 7.81 becomes

$$\{\phi\}_{t+\Delta t} = 2\{\phi\}_t - \{\phi\}_{t-\Delta t} + (\Delta t)^2\{\ddot\phi\}_t \qquad (7.83)$$

Equation 7.83 provides the necessary recurrence relation for calculating the vector of nodal values $\{\phi\}$ at the end of the time interval from known values at the beginning of the interval. In summary, the calculation sequence is as follows:

> • Set $t = 0$ (the initial condition).
>
> • Calculate $\{\ddot{\phi}\}_t$ from equation 7.81.
>
> • Calculate $\{\phi\}_{t+\Delta t}$ from equation 7.83.
>
> • Calculate $\{\dot{\phi}\}_{t+\Delta t} = \{\dot{\phi}\}_t + \Delta t\{\ddot{\phi}\}_t$.
>
> • Set $t = t + \Delta t$.
>
> • Stop if desired time is reached.

It is obvious from this scheme that, when the initial conditions $\{\phi(0)\} = \{\phi\}_0$ and $\{\dot{\phi}(0)\} = \{\dot{\phi}\}_0$ are specified, the above procedure is not self-starting because $\{\phi\}_{0-\Delta t}$ is unknown. To avoid this difficulty, we may assume that $\{\ddot{\phi}\}$ varies linearly from $t = 0$ to $t = \Delta t$, in which case, from equation 7.80a, we have

$$\{\phi\}_{\Delta t} = \frac{(\Delta t)^2}{6}\left[2\{\ddot{\phi}\}_0 + \{\ddot{\phi}\}_{\Delta t}\right]$$

This formula permits us to start the calculation sequence one step later with some assumed value for $\{\ddot{\phi}\}_{\Delta t}$. At the first time step, iteration may be used to find the correct value for $\{\ddot{\phi}\}_{\Delta t}$. Solutions obtained by assuming that $\{\ddot{\phi}\}$ is constant within each interval are usually stable and sufficiently accurate when Δt is chosen to be less than one-tenth the shortest period of natural vibration expected.

Another set of recurrence formulas can be derived if we assume that $\{\ddot{\phi}\}$ *varies linearly* within Δt. For this case,

$$\{\dddot{\phi}\} = \frac{1}{\Delta t}\left[\{\ddot{\phi}\}_{t+\Delta t} - \{\ddot{\phi}\}_t\right]$$

and the expressions for the zeroth and first-order derivatives become

$$\{\phi\}_{t+\Delta t} = \{\phi\}_t + \Delta t\{\dot{\phi}\}_t + \frac{(\Delta t)^2}{6}\left[2\{\ddot{\phi}\}_t + \{\ddot{\phi}\}_{t+\Delta t}\right] \tag{7.84}$$

$$\{\dot{\phi}\}_{t+\Delta t} = \{\dot{\phi}\}_t + \frac{\Delta t}{2}\left[\{\ddot{\phi}\}_t + \{\ddot{\phi}\}_{t+\Delta t}\right] \tag{7.85}$$

The next step is to find an expression for $\{\ddot{\phi}\}_{t+\Delta t}$. For this, we turn to the equations of motion and write

$$\{\ddot{\phi}\}_{t+\Delta t} = [K_{\ddot{\phi}}]^{-1}\left[\{R\}_{t+\Delta t} - [K_{\phi}]\{\phi\}_{t+\Delta t} - [K_{\dot{\phi}}]\{\dot{\phi}\}_{t+\Delta t}\right] \tag{7.86}$$

Substituting equations 7.84 and 7.85 into equation 7.86 and solving for $\{\ddot{\phi}\}_{t+\Delta t}$ then gives (after some manipulation)

$$\{\ddot{\phi}\}_{t+\Delta t} = [M_1]^{-1}[\{R\}_{t+\Delta t} - [K_{\dot{\phi}}]\{m_2\} - [K_{\phi}]\{m_3\}] \qquad (7.87)$$

where

$$[M_1] = [K_{\ddot{\phi}}] + \frac{\Delta t}{2}[K_{\dot{\phi}}] + \frac{(\Delta t)^2}{6}[K_{\phi}]$$

$$\{m_2\} = \{\dot{\phi}\}_t + \frac{\Delta t}{2}\{\ddot{\phi}\}_t$$

$$\{m_3\} = \{\phi\}_t + \Delta t\{\dot{\phi}\}_t + \frac{(\Delta t)^2}{3}\{\ddot{\phi}\}_t$$

To start the time-stepping procedure, we need the initial value, $\{\ddot{\phi}\}_0$. This can be obtained by evaluating equation 7.81 at $t = 0$. In summary, the calculation sequence is as follows:

> • Set $t = 0$ (the initial condition).
> • Calculate $\{\ddot{\phi}\}_{t+\Delta t}$ from equation 7.87.
> • Calculate $\{\phi\}_{t+\Delta t}$ from equation 7.84.
> • Calculate $\{\dot{\phi}\}_{t+\Delta t}$ from equation 7.85.
> • Set $t = t + \Delta t$.
> • Stop if desired time is reached.

In contrast to the constant second derivative method, the linear second derivative method allows slightly larger time steps to be taken. Accuracy and stability are usually obtained when Δt is taken to be less than one-sixth the shortest period of natural vibration.

These two time-stepping solution procedures are more general than the method of mode superposition because they can be used regardless of the form of $[K_{\dot{\phi}}]$, and nonlinear problems may also be treated. For nonlinear problems, $[K_{\phi}]$, $[K_{\dot{\phi}}]$, and/or $[K_{\ddot{\phi}}]$ are changing with time and the inversions indicated in equations 7.81 or 7.87 must be recalculated at each time step or at least every few time steps. However, for linear problems, these inversions need only be calculated once.

In addition to these methods, another general method is sometimes used. This involves defining a new solution vector $\{z(t)\}$ to reduce the original second-order differential equations of motion to an expanded set of couple first-order equations. We define

$$\{z(t)\} = \left\{ \begin{matrix} \{\phi(t)\} \\ \hline \{\dot{\phi}(t)\} \end{matrix} \right\}$$

and, by reordering the original system equations, we can write

$$\underset{2m \times 1}{\{\dot{z}\}} = \underset{2m \times 2m}{[B]} \underset{2m \times 1}{\{z\}} + \{F\}$$

which can be solved by a wide variety of standard techniques. See Conte and De Boor [29] for further details.

Zienkiewicz [30] approaches the problem of deriving recurrence relations for equations 7.68 by applying the Galerkin weighting procedure to the vector $\{\phi\}$ and selecting interpolation functions that are functions only of time. The interpolation functions may span one or more time increments, depending on the order of interpolation chosen. This procedure offers the hope of greater solution stability and accuracy because higher-order interpolation can be used if desired.

Further discussion and an overview of the problem of numerically integrating equations 7.68 are presented by Clough [31].

7.7   CLOSURE

The finite element formulation for a wide variety of continuum problems commonly called field problems was explored in this chapter. Rather than considering individual physical problems, we examined classes of problems governed by similar differential equations and their associated boundary conditions. General expressions were derived for the element equations. Clearly, these can be specialized for any particular element type—giving at least C^0 continuity (see Chapter 5).

Our discussion here of steady and unsteady field problems considered only the most commonly encountered situations. Also, we restricted our attention to *linear* second-order partial differential equations. A number of practical situations arise where the governing equations are of the same form as those discussed in this chapter but are inherently nonlinear. For example, problems governed by the quasi-harmonic equation become nonlinear when the coefficients k_x, k_y, and k_z are functions of ϕ or its first derivatives. Heat conduction problems with temperature-dependent thermal conductivities and electromagnetic problems with field-dependent magnetic permeability are just two typical cases. Element equations for problems of this type can be derived by using the method of weighted residuals with Galerkin's criterion. The matrix equations obtained this way have the same forms as those given in this chapter; however, the coefficient matrices $[K_\phi]$, $[K_{\dot{\phi}}]$, and $[K_{\ddot{\phi}}]$ are then functions of ϕ and/or its gradients. Assembly of the element equations to form the system equations is the same, but the resulting systems equations are nonlinear and must be solved iteratively.

In the next chapter, we shall discuss the details of a special type of field problem in fluid mechanics—the fluid-film lubrication problem.

REFERENCES

1. S. G. Mikhlin, *Variational Methods in Mathematical Physics*, Macmillan Company, New York, 1964 (English translation of the 1957 edition).

2. R. S. Schechter, *The Variational Method in Engineering*, McGraw-Hill Book Company, New York, 1967.

3. P. Silvester and A. Konard, "Axisymmetric Triangular Finite Elements for the Scalar Helmholtz Equation," *Int. J. Numer. Methods Eng.*, Vol. 5, No. 3, 1973, pp. 481–497.

4. A. F. Emery and W. W. Carson, "An Evaluation of the Use of the Finite Element Method in the Computation of Temperature," *Sandia Lab. Rept.* SCL-RR-69-83, August 1969 (*ASME Paper* 69-WA/HT-38).

5. C. Taylor, B. S. Patil, and O. C. Zienkiewicz, "Harbor Oscillation: A Numerical Treatment for Undamped Natural Modes," *Proc. Inst. Civil Eng.*, Vol. 43, 1969, pp. 141–155.

6. P. L. Arlett, A. K. Bahrani, and O. C. Zienkiewicz, "Application of Finite Element Method to the Solution of Helmholtz's Equation," *Proc. IEE*, Vol. 115, No. 12, 1968, pp. 1762–1766.

7. W. H. Haytt, Jr., *Engineering Electromagnetics*, McGraw-Hill Book Company, New York, 1958.

8. L. E. Kinsler and A. R. Frey, *Fundamentals of Acoustics*, John Wiley and Sons, New York, 1950.

9. A. Ralston and H. S. Wilf (ed.), *Mathematical Methods for Digital Computers*, Vol. 2, John Wiley and Sons, New York, 1967.

10. J. H. Wilkinson, *The Algebraic Eigenvalue Problem*, Oxford University Press, London, 1965.

11. T. Shuku and K. Ishihara, "The Analysis of the Acoustic Field in Irregularly Shaped Rooms by the Finite Element Method," *J. Sound Vibration*, Vol. 29, 1973, pp. 67–76.

12. T. Shuku, "Finite Difference Analysis of the Acoustic Field in Irregular Rooms," *J. Acoust. Soc. Jap.*, Vol. 28, 1972, pp. 5–12.

13. J. T. Oden, "A General Theory of Finite Elements. II: Applications," *Int. J. Numer. Methods Eng.*, Vol. I, No. 1, 1969, pp. 247–259.

14. M. Gurtin, "Variational Principles for Linear Initial-Value Problems," *Quart. Appl. Math.*, Vol. 22, No. 3, 1964, pp. 252–256.

15. D. H. Norrie and G. de Vries, *The Finite Element Method*, Academic Press, New York, 1973, p. 231.

16. V. Churchill, *Fourier Series and Boundary Value Problems*, McGraw-Hill Book Company, New York, 1963.

17. B. Carnahan, H. Luther, and J. Wilkes, *Applied Numerical Methods*, John Wiley and Sons, New York, 1969.

18. H. Semat, *Introduction to Atomic and Nuclear Physics*, Holt, Rinehart and Winston, New York, 1962, p. 218.

19. M. M. Aral, "Application of Finite Element Analysis in Fluid Mechanics," Ph.D. Thesis, Georgia Institute of Technology, Atlanta, Ga., September 1971.

20. W. C. Hurty and M. F. Rubinstein, *Dynamics of Structures*, Prentice-Hall, Englewood Cliffs, N.J., 1964.

21. J. M. Biggs, *Introduction to Structural Dynamics*, McGraw-Hill Book Company, New York, 1964.

22. L. Meirovitch, *Analytical Methods in Vibrations*, Macmillan Company, New York, 1967.

23. L. Fox and D. F. Mayers, *Computing Methods for Scientists and Engineers*, Oxford University Press, London, 1968.

24. R. Richtmyer and K. Morton, *Difference Methods for Initial Value Problems*, John Wiley-Interscience, New York, 1967.

25. G. Forsythe and W. Wasow, *Finite Difference Methods for Partial Differential Equations*, John Wiley and Sons, New York, 1960.

26. R. L. Ketter and S. Prawel, *Modern Methods of Engineering Computation*, McGraw-Hill Book Company, New York, 1969.

27. R. W. Clough, "Analysis of Structural Vibrations and Dynamic Response," in *Recent Advances in Matrix Methods of Structural Analysis and Design*, R. H. Gallagher, Y. Yamada, and J. R. Oden (eds.), University of Alabama Press, Huntsville, Ala., 1971.

28. K. A. Foss, "Coordinates Which Uncouple the Equations of Motion of Damped Linear Dynamic Systems," *Trans. ASME, Series E: J. Appl. Mech.*, No. 57-8-86; September 1958.

29. S. Conte and C. De Boor, *Elementary Numerical Analysis*, McGraw-Hill Book Company, New York, 1972.

30. O. C. Zienkiewicz, *The Finite Element Method in Engineering Science*, McGraw-Hill Book Company, London, 1971.

31. R. W. Clough, "Numerical Integration of the Equations of Motion," in *Lectures on Finite Element Methods in Continuum Mechanics*, J. T. Oden and E. R. A. Oliveira (eds.), University of Alabama Press, Huntsville, Ala., 1973.

8

THE LUBRICATION PROBLEM

8.1 INTRODUCTION

Chapter 7 reviewed the application of finite element techniques to field problems governed by the quasi-harmonic equation and related equations. Though the governing equation of hydrodynamic lubrication is quasi-harmonic and such equations have already been treated, we devote this chapter to a more detailed discussion of the particular equation for fluid-film lubrication. We begin with a brief historical comment on fluid lubrication and then review the general finite element formulations for incompressible lubricants operating in either the isothermal or the thermohydrodynamic state. Treatments of the compressible lubrication problem are also given. Some sample solutions are drawn from the literature to illustrate the types of problems amenable to finite element analysis.

8.2 HISTORICAL NOTE

Man has been attempting to reduce friction for a period of time reaching far back into antiquity. On the walls of Egyptian tombs, we find paintings showing men pouring lubricant under the runners of heavy sledges. In spite of its long history, lubrication did not emerge from the trial-and-error stage until about 90 years ago, when Beauchamp Tower [1] discovered that a pressure can be generated in the lubricant film between the two surfaces in relative motion. Osborne Reynolds [2] was fascinated by this effect and decided to approach the phenomenon analytically. In 1886, he presented his classical paper on the theory of hydrodynamic lubrication, and the governing partial differential equation for the pressure now bears his name.

Since the early 1900's, tremendous progress has been made and many complex problems in fluid-film lubrication have been studied. This activity in lubrication theory stems from the fact that every machine with moving parts has one or more bearings, and the successful operation of a machine depends crucially on the successful performance of each bearing.

Predicting bearing performance by solving the relevant Reynolds equation usually requires the use of some numerical procedure. The most commonly used procedure is the finite difference method. A number of finite difference schemes for solving Reynolds equations are reviewed by Lloyd and McCallion [3]. In the analysis of practical lubrication problems, however, we often encounter complex geometrical configurations, driving functions, and boundary conditions. We may also have to cope with abrupt changes in field properties such as film thickness. In these cases, the finite difference method becomes inherently difficult to use because irregular meshes should be employed and special auxiliary conditions are required. As we have seen, the finite element method is ideally suited to overcome these difficulties.

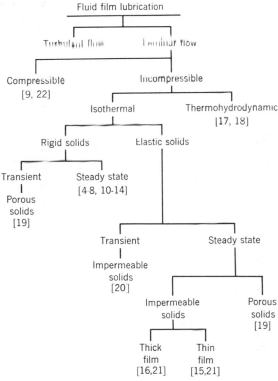

Figure 8.1. Summary of finite element solutions of fluid film lubrication problems. Numbers refer to items in the reference list at the end of the chapter.

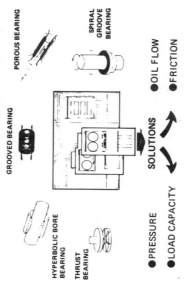

Figure 8.2. The finite element for lubrication problems.

Applications of finite element techniques to lubrication problems began to appear in the mid-1960's [4] after analysts recognized the generality of the method based on a variational principle. Once the appropriate Reynolds equation for a problem is cast into a classical variational form, the procedure for deriving the finite element matrix equations is straightforward. In fact, computer programs for hydrodynamic lubrication problems can follow the closely analogous programs for steady-state heat conduction, mass diffusion, and other phenomena governed by quasi-harmonic equations.

As the literature summary chart of Figure 8.1 indicates, 19 publications on finite element lubrication analysis have appeared in the last 9 years; but the average of about 2 per year is misleading, since most of them emerged rapidly in the period from 1969 to 1973. Here we shall present the unifying principles of these analyses and attempt to tell in greater detail the story outlined in Figure 8.2. The sample solutions are intended to demonstrate the broad range of problems amenable to finite element analysis.

8.3 EQUATIONS GOVERNING INCOMPRESSIBLE HYDRODYNAMIC LUBRICATION

The essence of hydrodynamic lubrication theory is that a pressure and, hence, a load capacity can be developed in the fluid film separating two closely spaced surfaces moving in relation to one another. There are four physical mechanisms that can generate pressure in the fluid film: (1) viscous effects, (2) inertia effects, (3) body force effects, and (4) compressibility effects. In some bearings, only one of these mechanisms is dominant; in others, all four may be equally important. For example, in most oil bearings, the viscous effects caused by shearing and squeezing the lubricant are the major contributors to pressure generation.

The lubricant flow in high-speed bearings operating with low-viscosity lubricants can become "superlaminar" or "turbulent"; however, because most bearings do not operate in these regimes, the discussion here will be restricted to the laminar regime.

8.3.1 Reynolds equation

The differential equation governing the generation of pressure in a bearing is derived from the conservation equations of mass and momentum (see Appendix D). The approximations and assumptions commonly made in the derivation are as follows:

1. The lubricant is a Newtonian fluid.

2. The fluid film is so thin that the derivatives of velocity across the film thickness are far more important than any of the other velocity derivatives (this is the so-called thin-film approximation).

3. Compared to other effects, inertia effects are negligible.

4. Curvature of the bearing components introduces only second-order, negligible effects in journal bearings.

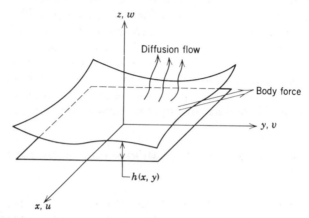

Figure 8.3. Geometry and coordinate system for a lubricated contact.

The general geometry and the coordinate system for a pair of film-lubricated surfaces are shown in Figure 8.3. In this coordinate system and under the foregoing assumptions, the equation expressing mass conservation is

$$\frac{\partial u}{\partial x} + \frac{\partial v}{\partial y} + \frac{\partial w}{\partial z} = 0 \qquad (8.1)$$

while the equations of motion reduce to

$$\frac{\partial P}{\partial x} = \frac{\partial}{\partial z}\left(\mu \frac{\partial u}{\partial z}\right) + B_x \qquad (8.2)$$

$$\frac{\partial P}{\partial y} = \frac{\partial}{\partial z}\left(\mu \frac{\partial v}{\partial z}\right) + B_y \qquad (8.3)$$

$$\frac{\partial P}{\partial z} = 0 \qquad (8.4)$$

Here P is the pressure in the lubricant film and μ is the lubricant viscosity, which is explicitly a function of temperature and implicitly a function of the spatial coordinates x, y, and z. Also, B_x and B_y are components of the body force. Equations 8.1–8.4 are subject to the following boundary conditions:

$$
\begin{aligned}
z = 0, &\quad u = U_1, &\quad v = V_1, &\quad w = W_1 \\
z = h, &\quad u = U_2, &\quad v = V_2, &\quad w = W_2 \\
P = P(x, y) &\ \text{(a specified function on a} \\
&\ \text{nonvanishing segment of the} \\
&\ \text{boundary)}
\end{aligned}
\tag{8.5}
$$

By integrating equations 8.2 and 8.3 twice on z (first from 0 to h and then from 0 to z) and applying the boundary conditions expressed in equation 8.5, we obtain the velocity components

$$
u = \frac{\partial P}{\partial x}\left(\int_0^z \frac{z\,dz}{\mu} - \frac{F_1}{F_0}\int_0^z \frac{dz}{\mu}\right) + U_1 + \frac{U_2 - U_1}{F_0}\int_0^z \frac{dz}{\mu} + \bar{B}_x \tag{8.6}
$$

$$
v = \frac{\partial P}{\partial y}\left(\int_0^z \frac{z\,dz}{\mu} - \frac{F_1}{F_0}\int_0^z \frac{dz}{\mu}\right) + V_1 + \frac{V_2 - V_1}{F_0}\int_0^z \frac{dz}{\mu} + \bar{B}_y \tag{8.7}
$$

where

$$
F_0 = \int_0^h \frac{dz}{\mu}, \qquad F_1 = \int_0^h \frac{z\,dz}{\mu}
$$

$$
\bar{B}_{x,y} = \frac{\int_0^z (dz/\mu)}{F_0}\int_0^h \frac{1}{\mu}\int_0^z B_{x,y}\,dz\,dz - \int_0^z \frac{1}{\mu}\int_0^z B_{x,y}\,dz\,dz
$$

From the continuity equation, the velocity component across the film thickness becomes

$$
w = -\int_0^z \frac{\partial u}{\partial x}\,dz - \int_0^z \frac{\partial v}{\partial y}\,dz + W_1 \tag{8.8}
$$

The average fluid-film velocity in the plane of the film is defined as

$$
\bar{u} = \left(\frac{1}{h}\int_0^h u\,dz\right)\hat{i} + \left(\frac{1}{h}\int_0^h v\,dz\right)\hat{j} \tag{8.9}
$$

or

$$
\bar{u} = \bar{u}\hat{i} + \bar{v}\hat{j}
$$

Hence the volume flow q per unit boundary of the film in the x-y plane is

$$
\mathbf{q} = h\bar{u} = h\bar{u}\hat{i} + h\bar{v}\hat{j} \tag{8.10}
$$

When equations 8.6, 8.7, and 8.9 are combined, we find that

$$\mathbf{q} = G\nabla P + h\mathbf{V}_1 + \frac{\Delta\mathbf{U}}{F_0}\int_0^h\int_0^z \frac{dz}{\mu}\,dz + \tilde{\mathbf{B}} \tag{8.11}$$

where

$$G = \int_0^h\int_0^z \frac{z\,dz}{\mu}\,dz - \frac{F_1}{F_0}\int_0^h\int_0^z \frac{dz}{\mu}\,dz$$

$$\mathbf{V}_1 = U_1\hat{\imath} + V_1\hat{\jmath}$$

$$\Delta\mathbf{U} = (U_2 - U_1)\hat{\imath} + (V_2 - V_1)\hat{\jmath}$$

$$\tilde{\mathbf{B}} = \hat{\imath}\int_0^h \bar{B}_x\,dz + \hat{\jmath}\int_0^h \bar{B}_y\,dz$$

The continuity equation now provides the necessary relation between pressure gradient and volume flow. Integrating equation 8.1 over the film thickness h gives

$$\nabla\cdot\mathbf{q} + \frac{\partial h}{\partial t} + v_d = 0 \tag{8.12}$$

where v_d is the net outward velocity of a diffusion flow through porous boundary surfaces at $z = 0$ and h. The generalized Reynolds equation (which accounts for three-dimensional viscosity variations in the film) results when equation 8.11 is substituted into equation 8.12:

$$-\nabla\cdot G\nabla P = \nabla\cdot(h\mathbf{V}_1) + \nabla\cdot(\Delta\mathbf{U}F_2) + \nabla\cdot\tilde{\mathbf{B}} + \frac{\partial h}{\partial t} + v_d \tag{8.13}$$

where

$$F_2 = \frac{1}{F_0}\int_0^h\int_0^z \frac{dz}{\mu}\,dz$$

The Reynolds equation—equation 8.13—is an elliptic partial differential equation for the pressure in terms of the lubricant properties, density, and viscosity as well as the distributions of film thickness, body forces, and diffusion flows. Determining the pressure distribution in the fluid film is not a well-posed problem until boundary conditions are specified. If Ω is the region over which a solution is desired and S is the boundary of Ω, the pressure and flow boundary conditions take the general form

$$P = \mathbf{P}(x, y) \quad \text{on } S_p$$

and

$$Q = \mathbf{q}\cdot\hat{n} = h\bar{\mathbf{u}}\cdot\hat{n} \quad \text{on } S_q \tag{8.14}$$

where S_p and S_q are boundary segments of S such that $S_p \cup S_q = S$ and n is the outward unit normal to the boundary. Segment S_p is the *nonvanishing* portion of S on which pressure is specified, and S_q is the rest of S, over which volume flow per unit boundary length is specified. Segment S_p must be nonvanishing because, unless some pressure boundary condition is specified, a unique solution is impossible. When sufficient pressure and flow boundary conditions are specified, the Reynolds equation has a unique solution.

8.3.2 Energy equation

If the variation of the fluid properties along and across the film thickness are taken into account, the Reynolds equation must be solved simultaneously with the thermal energy equation and the "equation of state" for the fluid. The thermal energy equation describing the temperature distribution $T(x, y, z)$ in the lubricant takes the form

$$\rho c \mathbf{V} \cdot \nabla T = k \nabla \cdot \nabla T + \Phi \tag{8.15}$$

where ρ, c, and k are the density, specific heat, and thermal conductivity, respectively, of the lubricant, and

$$\mathbf{V} = u\hat{i} + v\hat{j} + w\hat{k} \tag{8.16}$$

When the thin-film approximation is applied along with an order of magnitude analysis, the dissipation function Φ reduces to

$$\Phi = \mu(T)\left[\left(\frac{\partial u}{\partial z}\right)^2 + \left(\frac{\partial v}{\partial z}\right)^2\right] \tag{8.17}$$

Sufficient boundary conditions for equation 8.15 are as follows:

1. $T = T_i$ specified on all surfaces of the control volume through which there is lubricant influx
2. $\dfrac{\partial T}{\partial n} = 0$ on all surfaces through which there is lubricant efflux (n is the outward normal on such surfaces)

8.3.3 Viscosity-temperature characteristic

The formulation of the governing equations for the lubrication problem must include the functional dependence of viscosity on temperature, that is, $\mu = \mu(T)$ must be specified. Often an exponential relation is used, such as

$$\mu(T) = \mu_0 e^{-\beta(T - T_0)} \tag{8.18}$$

where μ_0 is the viscosity at the reference temperature T_0.

8.3.4   Equations for the solids

When heat transfer to the bearing solids is considered, the thermal energy equations for the solids must also be solved simultaneously with the thermal energy equation for the fluid. The lubrication problem becomes further complicated when the bearing solids deform because of thermal and/or mechanical loading. In these cases, the film thickness distribution is not known a priori but rather must be found from a simultaneous solution of the thermal and mechanical energy equations in the solids.

In the following sections, finite element formulations for particular cases of the general lubrication problem are presented.

8.4   LIQUID LUBRICATION IN THE ISOTHERMAL STATE

Most analyses of hydrodynamic lubrication problems assume that the lubricant viscosity remains constant. This is equivalent to an assumption that the bearing operates in the isothermal state at some characteristic operating temperature. Although the isothermal assumption holds for gas-lubricated bearings, for liquid-lubricated bearings it is simply a convenient contrivance that uncouples the Reynolds and energy equations and simplifies the calculations. Isothermal solutions or constant-viscosity solutions to the Reynolds equation can be useful for predicting bearing behavior and developing design data when it is possible to find an "effective" viscosity based on some representative bearing temperature. Usually this representative temperature is found by estimating the work done on the lubricant, determining from this the temperature rise of the lubricant as it passes through the wedge, and then adding one-half the temperature rise to the given inlet temperature. Such approximate procedures (though convenient) rarely lead to accurate predictions of both bearing load capacity and friction torque.

In this section, a special finite element formulation of the Reynolds equation is given for the case in which the lubricant viscosity remains constant across the film thickness. A number of investigators [4–8, 10–14] have applied finite element techniques to this problem, but the treatment here follows that in refs 12 and 13.

8.4.1   Reynolds equation

When the lubricant viscosity is independent of z, the coordinate across the film thickness—the general Reynolds equation 8.13—reduces to

$$\mathbf{\nabla} \cdot \left(\frac{h^3}{12\mu} \mathbf{\nabla}P \right) = \mathbf{\nabla} \cdot \left(h\overline{\mathbf{U}} + \frac{h^3}{12\mu} \mathbf{B} \right) + \frac{\partial h}{\partial t} + v_d \qquad (8.19)$$

where

$$2\overline{U} = (U_1 + U_2)\hat{\imath} + (V_1 + V_2)\hat{\jmath}$$

The boundary conditions in this case become

$$P = P(x, y) \quad \text{on } S_p$$

$$Q = \mathbf{q} \cdot \hat{n} = h\overline{\mathbf{u}} \cdot \hat{n} = h\left(\overline{U} + \frac{h^2}{12\mu}\mathbf{B} - \frac{h^2}{12\mu}\Delta P\right) \cdot \hat{n} \quad \text{on } S_q \qquad (8.20)$$

8.4.2   Equivalent variational principle

The linear self-adjoint elliptic boundary value problem of equations 8.19 and 8.20 has a corresponding variational principle. According to this principle, the solution of equations 8.19 and 8.20 is the function P that satisfies the pressure boundary condition and minimizes the functional

$$I(P) = \int_A \left[\left(\frac{h^3}{12\mu}\nabla P - h\overline{U} - \frac{h^3}{12\mu}\mathbf{B}\right)\cdot\nabla P + \left(\frac{\partial h}{\partial t} + v_d\right)P\right]dA + \int_{S_q} Qp\, ds$$
$$(8.21)$$

8.4.3   Finite element formulation

The isothermal incompressible lubrication problem may be stated as follows:

Given the parameters or forcing functions μ, h, $\partial h/\partial t$, $\overline{v}$, $\mathbf{B}$, V_d, and the bearing geometry, find the pressure distribution $P(x, y)$ in the lubricant film.

The finite element method provides an expedient means of obtaining an approximate solution to this problem regardless of how irregular the bearing geometry may be.

As usual, step 1 of the solution procedure is to divide the solution domain (the region of the bearing over which the pressure distribution is sought) into polygonal subdomains (or elements) which are interconnected at nodes located on the element boundaries. Then the pressure and forcing functions are expressed in terms of assumed interpolation functions within each element. Nodal values and interpolation functions† completely define the behavior of the pressure and the forcing functions within the elements.

---

† The interpolation functions should be chosen so that C^0 continuity is preserved; that is, the pressure field representation should be at least continuous within the elements and across their interfaces.

More explicitly, the finite element discretization process involves sub-dividing the solution domain in the x-y plane into M polygonal elements of r nodes each. Then, within each element, the pressure and the distributions of the various forcing functions are related to their r nodal values by r inter-polating functions $N_i(x, y)$ as follows:

$$P = \lfloor N \rfloor \{P\} = \sum_{i=1}^{r} N_i(x, y) P_i \qquad (8.22)$$

$$\left.\begin{array}{l} \overline{U}^x = \lfloor N \rfloor \{\overline{U}^x\} = \displaystyle\sum_{i=1}^{r} N_i(x, y) \overline{U}_i^x \\[2mm] \overline{U}^y = \lfloor N \rfloor \{\overline{U}^y\} = \displaystyle\sum_{i=1}^{r} N_i(x, y) \overline{U}_i^y \\[2mm] B^x = \lfloor N \rfloor \{B^x\} = \displaystyle\sum_{i=1}^{r} N_i(x, y) B_i^x \\[2mm] B^y = \lfloor N \rfloor \{B^y\} = \displaystyle\sum_{i=1}^{r} N_i(x, y) B_i^y \\[2mm] \dfrac{\partial h}{\partial t} = \dot{h} = \lfloor N \rfloor \{\dot{h}\} = \displaystyle\sum_{i=1}^{r} N_i(x, y) \dot{h}_i \\[2mm] v_d = \lfloor N \rfloor \{v_d\} = \displaystyle\sum_{i=1}^{r} N_i(x, y) v_{d_i} \end{array}\right\} \qquad (8.23)$$

In equation 8.22, the nodal parameters P_i may be nodal pressures and pres-sure derivatives, but for convenience we denote all the nodal degrees of freedom for the pressure simply as P_i.

When equations 8.22 and 8.23 are inserted into the functional of equation 8.21, there results the discretized form of the functional for one element. To minimize the element functional with respect to the nodal pressures of the element, we set

$$\frac{\partial I}{\partial P_i} = 0, \quad i = 1, 2, \ldots, r \qquad (8.24)$$

Equation 8.24 results in a set of element equations of the form

$$\underset{r \times r}{[K_p]} \underset{r \times 1}{\{P\}} = \{q\} - \underset{r \times r}{[K_{\overline{U}}x]} \underset{r \times 1}{\{\overline{U}^x\}} - \underset{r \times r}{[K_{\overline{U}}y]} \underset{r \times 1}{\{\overline{U}^y\}} - \underset{r \times r}{[K_Bx]} \underset{r \times 1}{\{B^x\}}$$

$$- \underset{r \times r}{[K_By]} \underset{r \times 1}{\{B^y\}} - \underset{r \times r}{[K_h]} \underset{r \times 1}{\{\dot{h}\}} - \underset{r \times r}{[K_{vd}x]} \underset{r \times 1}{\{v_d\}} = \underset{r \times 1}{\{Q_R\}} \qquad (8.25)$$

where the coefficients in these fluidity matrices are given by†

$$K_{P_{ij}} = -\int_A \frac{h^3}{12\mu}\left(\frac{\partial N_i}{\partial x}\frac{\partial N_j}{\partial x} + \frac{\partial N_i}{\partial y}\frac{\partial N_j}{\partial y}\right) dA$$

$$K_{\bar{U}_{ij}^x} = \int_A h\frac{\partial N_i}{\partial x} N_j \, dA$$

$$K_{\bar{U}_{ij}^y} = \int_A h\frac{\partial N_i}{\partial y} N_j \, dA$$

$$K_{B_{ij}^x} = \int_A \frac{h^3}{12\mu}\frac{\partial N_i}{\partial x} N_j \, dA$$

$$K_{B_{ij}^y} = \int_A \frac{h^3}{12\mu}\frac{\partial N_i}{\partial y} N_j \, dA \qquad\qquad (8.26)$$

$$K_{h_{ij}} = -\int_A N_i N_j \, dA$$

$$K_{v_{d_{ij}}}^* - K_{h_{ij}}$$

$$q_i = \int_{S_q} Q N_i \, ds$$

†‡

These element equations have a distinct physical meaning. The right-hand side of equation 8.25 may be interpreted as a linear combination of flows caused by shear, body force, squeeze, and diffusion effects. These nodal flows are then balanced by the nodal flows due to pressure and the externally applied flows. Equations 8.25 and 8.26 provide the complete finite element description of the general isothermal incompressible lubrication problem. The fluidity matrices and element equations can be explicitly evaluated after the element geometry and interpolation functions are specified. Particular fluidity matrices for triangular elements with linear interpolation functions are given in ref. 13. More complex elements are also possible. In fact, we may use any of the two-dimensional elements for C^0 problems discussed in Section 5.8.1.

To evaluate the integrals of equations 8.26, it is convenient to interpolate the field properties h and μ between their nodal values. Depending on the

† Here it is understood that the integrals pertain to only one element.
‡ Here $\{q\}$ is a boundary matrix stemming only from elements with portions of their boundaries coinciding with the boundary of the complete domain or elements for which external feeder flows are specified.

shape of the element and the complexity of the integrand, numerical integration may be more feasible than closed-form integration.

The finite element analysis of a system is complete when all of the element equations are assembled to form the system equations, and the system equations are solved for the unknown nodal pressures and flows. Assembly of the system equations is routine and relies on matching element pressures and forcing functions to form system pressures and forcing functions, while element flows are summed to form system flows. The system equations are identical in form to the element equations, except that they are expanded in dimension to include all of the nodes. Hence the system equations become

$$\underset{n \times n}{[K_P]_s} \underset{n \times 1}{\{P\}_s} = \underset{n \times 1}{\{Q_R\}_s} \tag{8.27}$$

where n is the total number of nodes for the system. Solution of equation 8.27 is possible after the system fluidity matrix $[K_P]_s$ is modified to account for the pressure boundary conditions. Flow boundary conditions have already been included in $\{Q_R\}_s$, the resultant or net vector of nodal flows.

Once the nodal pressures and flows have been found for the system, the bearing load capacity W and the friction force $\mathbf{F}$ can be found from the relations

$$W = \int_\Omega P(x, y) \, dx \, dy \tag{8.28}$$

$$\mathbf{F} = \int_\Omega \mu \frac{\partial}{\partial z} (u\hat{i} + v\hat{j}) \bigg|_{z=0,h} dx \, dy \tag{8.29}$$

Examples showing some applications of the foregoing finite element formulation are given in Section 8.6.

8.4.4   Elastic bearing surfaces

When the bearing solids are compliant and deform under the action of hydrodynamic pressure, the film thickness distribution $h(x, y)$ is not known a priori. Hence the solution of isothermal elastohydrodynamic bearing problems involves the simultaneous solution of the Reynolds equation in the fluid film and the linear elasticity equations in the bearing solids. Since realistic bearing configurations often exhibit irregular geometry, the finite element method can be used to advantage in solving the elasticity equations for the solids.

Since the finite element analysis of elasticity problems is detailed in Chapter 6, we shall discuss only the essence of the methodology as it applies to elastohydrodynamic lubrication problems.

Let $J(\Delta)$ be the potential energy of the deformed solid, where Δ is the displacement vector, defined as $\Delta(x, y, z) = \Delta_x(x, y, z)\hat{\imath} + \Delta_y(x, y, z)\hat{\jmath} + \Delta_z(x, y, z)\hat{k}$ with $\Delta \equiv 0$ in the undeformed state. Then, at equilibrium, we have

$$\delta J(\Delta) = 0 \qquad (8.30)$$

Following the usual discretization procedure, we divide the volume of the solid into elements and then approximate the displacement field within each element in terms of known interpolation functions and unknown nodal values of displacement. The minimizing condition of equation 8.30 then leads to matrix equations for the discretized structure of the form

$$[K]\{\Delta\} = \{F\} \qquad (8.31)$$

where $[K]$ is the stiffness matrix and $\{F\}$ is the vector of resultant nodal forces.

Symbolically, the isoviscous elastohydrodynamic lubrication problem reduces to the simultaneous solution of the discretized Reynolds equation:

$$[A_P(\Delta)]\{P\} = \{Q_R(\Delta)\} \qquad (8.32)$$

which contains unknown nodal pressures, and the discretized elasticity equation

$$[K]\{\Delta\} = \{F(P)\} \qquad (8.33)$$

which contains unknown nodal displacements.

Taylor and O'Callaghan [15], as well as Oh and Huebner [16], have investigated the solution of these equations. The solution procedure begins by assuming a film thickness profile for a rigid bearing. Then equation 8.32 is solved for the nodal pressures, which are used to calculate the elastic response of the structure via $\{\Delta\} = [K]^{-1}\{F(P)\}$. These deformations are then applied to obtain a new film thickness profile, and the process is repeated. When two successive displacement fields $\{\Delta^K\}$ and $\{\Delta^{K+1}\}$ are uniformly close, convergence is achieved. This direct iterative scheme converges rapidly for lightly loaded bearings (small eccentricity ratio for journal bearings), but for heavily loaded bearings the scheme diverges unless remedial steps are taken.

The remedial steps involve averaging and limiting nodal displacements for the first few iterations while approaching the heavily loaded cases from the lightly loaded side. Instead of attempting a solution for a heavily loaded case directly, solutions are obtained starting from a lightly loaded case where direct iteration converges. Then, for a slightly heavier load (smaller minimum film thickness), the first guess for the displacement field is taken as the average

$$\{\Delta^{K+1}\} = \frac{1}{n}\left[[K]^{-1}\{F(\Delta^K)\} + \sum_{i=1}^{n-1}\{\Delta^i\} \right]$$

where $n \leq 5$. In addition to this weighting process, the maximum nodal displacement can be limited to avoid excessively large displacement changes if they happen to occur. After a few cycles at a given loading, the constraints are removed and the remaining cycles converge to any desired degree of accuracy. Details of this procedure can be found in Oh and Huebner [16], and an example is given in Section 8.7.

Recently, Day [20] proposed and utilized an alternative procedure for solving the elastohydrodynamic lubrication problem. Instead of iterating between pressures and deflections, he combines the matrix equations for pressure and displacement and considers an initial value problem in time. Starting with an assumed pressure distribution and a corresponding or compatible displacement distribution, the solution marches forward in time until the steady state $\partial P/\partial t = 0$ is approached.

It was found that this procedure works well for highly compliant bearing surfaces, but difficulties are encountered for relatively "stiff" bearing materials because the necessary time steps become small. On the other hand, the procedure of modified iteration between pressure and displacement works well for stiff bearings but encounters difficulties with "soft" materials. The reader is encouraged to see refs. 16 and 20 for further details.

8.5 LIQUID LUBRICATION IN THE THERMOHYDRODYNAMIC STATE

If the bearing solids are assumed to be rigid but the isothermal assumption is dropped, the bearing analyst must take into account viscous heating in the lubricant film and the resulting changes in lubricant viscosity. His analysis must include not only the conservation equations of mass and momentum (Reynolds equation), but also the conservation equation of thermal energy. The Reynolds and energy equations must be solved simultaneously, and an iterative procedure is necessary.

8.5.1 Element equations for pressure

The general Reynolds equation given by equation 8.13 may be cast into variational form to provide a convenient base for deriving the finite element equations. It can be shown that the pressure distribution that satisfies equations 8.13 and 8.14 also minimizes the functional

$$I(P) = \int_{\Omega} \left[-\tfrac{1}{2} G \nabla P \cdot \nabla P + (h \mathbf{V}_1) \cdot \nabla P \right.$$

$$+ \left(\frac{\Delta \mathbf{U}}{F_0} \int_0^h \int_0^z \frac{dz}{\mu} dz \right) \cdot \nabla P + \mathbf{B} \cdot \nabla P + P \frac{\partial h}{\partial t} + P v_d \bigg] d\Omega$$

$$+ \int_{S_q} P \mathbf{q} \cdot \hat{n} \, ds \tag{8.34}$$

After the solution domain is tesselated and the pressure as well as the forcing functions are interpolated in a manner analogous to equations 8.22 and 8.23, the derivation of the finite element matrix equations follows the standard procedure. The element equations become, as before,

$$[K_P]\{P\} = \{q\} - [K_{U_1}]\{U_1\} - [K_{V_1}]\{V_1\} - [K_{\Delta U_x}]\{\Delta U_x\} - [K_{\Delta U_y}]\{\Delta U_y\}$$

$$- [K_{\tilde{B}_x}]\{\tilde{B}_x\} - [K_{\tilde{B}_y}]\{\tilde{B}_y\} - [K_h]\{\dot{h}\} - [K_{v_d}]\{v_d\} \tag{8.35}$$

where

$$K_{P_{ij}} = -\int_A G \left(\frac{\partial N_i}{\partial x} \frac{\partial N_j}{\partial x} + \frac{\partial N_i}{\partial y} \frac{\partial N_j}{\partial y} \right) dA$$

$$K_{U_{1ij}} = \int_A h \frac{\partial N_i}{\partial x} N_j \, dA$$

$$K_{V_{1ij}} = \int_A h \frac{\partial N_i}{\partial y} N_j \, dA$$

$$K_{\Delta U_{x_{ij}}} = \int_A F_2 \frac{\partial N_i}{\partial x} N_j \, dA$$

$$K_{\Delta U_{y_{ij}}} = \int_A F_2 \frac{\partial N_i}{\partial y} N_j \, dA$$

$$K_{\tilde{B}_{x_{ij}}} = \int_A \frac{\partial N_i}{\partial x} N_j \, dA$$

$$K_{\tilde{B}_{y_{ij}}} = \int_A \frac{\partial N_i}{\partial y} N_j \, dA$$

$$K_{h_{ij}} = -\int_A N_i N_j \, dA$$

$$K_{V_{d_{ij}}} = K_{h_{ij}}$$

$$q_i = \int_{S_q} Q N_i \, ds$$

Evaluation of these fluidity matrices requires a known viscosity distribution $\mu(T)$. This is available from either an initially assumed or a previously calculated temperature distribution.

8.5.2   Element equations for temperature

The temperature distribution in the fluid film is obtained from a solution of equation 8.15. Assuming that a set of nodal pressures has been calculated, we can calculate the velocity field in the lubricant film from the expressions for u, v, and w given in equations 8.6, 8.7, and 8.8. This permits evaluation of the coefficients and the dissipation function in the energy equation. The nonlinearity in the energy equation due to the nonlinear viscosity-temperature relation is circumvented by evaluating $\mu(T)$ at the previous temperature distribution.

The energy equation may be expressed in operator form as

$$\mathscr{L}(T) = \rho c \left(u \frac{\partial T}{\partial x} + v \frac{\partial T}{\partial y} + w \frac{\partial T}{\partial z} \right) - k \left(\frac{\partial^2 T}{\partial x^2} + \frac{\partial^2 T}{\partial y^2} + \frac{\partial^2 T}{\partial z^2} \right) - \Phi = 0$$
$$(8.36)$$

Though Tieu [17] has treated two-dimensional problems using the Glansdorff-Prigogine local potential, a particularly convenient way to formulate a finite element model of equation 8.36 is to use the method of weighted residuals with Galerkin's criterion. To apply Galerkin's method, we assume that the unknown temperature distribution within a typical three-dimensional element can be approximated by

$$T^{(e)} = \sum_{i=1}^{s} N_i(x, y, z) T_i = [N]\{T\}$$
$$(8.37)$$

where the N_i are interpolation functions defined over individual elements, s is the number of nodes per element, and the T_i are the unknown nodal temperatures. For the whole solution domain, the complete piecewise representation has the form

$$\tilde{T} = \sum_{e=1}^{M'} T^{(e)}$$
$$(8.38)$$

where M' is the total number of elements. Using the appropriate function at the appropriate node, we may write

$$\tilde{T} = \sum_{i=1}^{n'} N_i T_i$$
$$(8.39)$$

where n' is the total number of nodes in the three-dimensional solution domain.

The trial solution expressed by equation 8.39 is such that $\tilde{T}$ exactly satisfies all of the boundary conditions. But, when $\tilde{T}$ is substituted into equation 8.36, an error or residual results because the trial solution $\tilde{T}$ does not, in general, satisfy the differential equation exactly. The residual becomes

$$R = \mathcal{L}(\tilde{T}) \neq 0 \tag{8.40}$$

Galerkin's error distribution principle states that

$$\int_{\Omega_3} N_i R \, dx \, dy \, dz = 0, \quad i = 1, 2, \ldots, n' \tag{8.41}$$

where the N_i are the same interpolation functions used to represent $\tilde{T}$ and Ω_3 is now the three-dimensional solution domain (the control volume encompassing the lubricant film). Equation 8.41 gives the set of linear simultaneous equations to be solved for the nodal temperature T_i. In matrix form, these equations are

$$[K_t]\{\tilde{T}\} = \{F_T\} \tag{8.42}$$

where

$$K_{T_{ij}} = \int_{\Omega_3} \left[\rho c N_i \left(u \frac{\partial N_j}{\partial x} + v \frac{\partial N_j}{\partial y} + w \frac{\partial N_j}{\partial z} \right) \right.$$
$$\left. + k\left(\frac{\partial N_i}{\partial x} \frac{\partial N_j}{\partial x} + \frac{\partial N_i}{\partial y} \frac{\partial N_j}{\partial y} + \frac{\partial N_i}{\partial z} \frac{\partial N_j}{\partial z} \right) \right] d\Omega_3$$
$$k \int_{\Sigma} N_i \left(\frac{\partial N_j}{\partial x} \hat{\imath} + \frac{\partial N_j}{\partial y} \hat{\jmath} + \frac{\partial N_j}{\partial z} \hat{k} \right) \cdot \hat{n} \, d\Sigma \tag{8.43}$$

on surface

$$F_{T_i} = \int_{\Omega_3} N_i \Phi \, d\Omega \tag{8.44}$$

In the derivation of these element equations, integration by parts was used on the second-order derivatives. This has the effect of reducing the order of the highest-order derivatives appearing in the thermal "stiffness" matrix, and it introduces the boundary conditions for the elements on the bounding surface of the control volume. The surface integral in equation 8.43 appears only for boundary elements.

Since the integrals in equation 8.43 contain only first-order derivatives, the interpolation functions N_i need preserve continuity only of value and not of slope at element interfaces. The same interpolation functions used to represent $\tilde{T}$ can also be employed to express u, v, w, and Φ in terms of their nodal values. Equations 8.42, as they have been derived, hold for the entire solution domain. But, if the N_i are chosen to preserve interelement continuity, the equations may be evaluated for individual elements and then summed in the usual manner to obtain the system equations.

To find compatible pressures and temperatures in the fluid film, equations 8.35 and 8.42 must be solved simultaneously by iteration as follows:

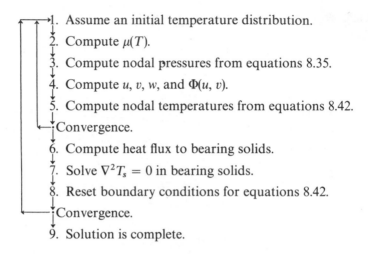

1. Assume an initial temperature distribution.
2. Compute $\mu(T)$.
3. Compute nodal pressures from equations 8.35.
4. Compute u, v, w, and $\Phi(u, v)$.
5. Compute nodal temperatures from equations 8.42.
   Convergence.
6. Compute heat flux to bearing solids.
7. Solve $\nabla^2 T_s = 0$ in bearing solids.
8. Reset boundary conditions for equations 8.42.
   Convergence.
9. Solution is complete.

In some cases, it is permissible to use adiabatic boundary conditions at $z = 0$, h and omit steps 6, 7, and 8.

8.6 GAS LUBRICATION

The derivation of the Reynolds equation for compressible lubrication problems follows the procedure outlined in Section 8.3 except that variable lubricant density must be included in the equations. Variable viscosity need not be considered because, when compared to liquid bearings, gas bearings have far lower shear losses. Also, the viscosity of a gas is relatively unaffected by temperature change in the range of most bearing operations.

For gas lubrication, the relevant Reynolds equation becomes

$$\nabla \cdot \left(\frac{\rho h^3}{12\mu} \nabla P \right) = \nabla \cdot \left(\rho h \overline{U} + \frac{\rho h^3}{12\mu} \mathbf{B} \right) + \frac{\partial}{\partial t}(\rho h) + \rho v_d \tag{8.45}$$

with boundary conditions given by equation 8.20. Though thermal effects are generally not important in gas-lubricated bearings, the Reynolds equation becomes nonlinear because ρ depends on p through the lubricant's equation of state. For an ideal gas, we have $p = \rho RT$, and, since the flow is isothermal, equation 8.45 takes the form

$$\mathbf{V} \cdot \left(\frac{h^3 P}{12\mu} \nabla P \right) = \mathbf{V} \cdot \left(Ph\overline{\mathbf{U}} + \frac{Ph^3}{12\mu} \mathbf{B} \right) + \frac{\partial}{\partial t} (Ph) + Pv_d \qquad (8.46)$$

Equation 8.46 does not have a classical variational statement, and hence it is not amenable to finite element analysis by this means. Reddi and Chu [9], however, have suggested a semivariational formulation for a Reynolds equation similar to equation 8.46—body force, squeeze effects, and diffusion flow were absent. When equation 8.46 is nondimensionalized, an important parameter λ emerges, called the "compressibility number" and defined as $\lambda = 6\mu U L/h_0{}^2 P_a$, where L is some characteristic length, h_0 is a reference dimension across the film thickness, and P_a is the ambient pressure. Using a perturbation technique, Reddi and Chu set up an incremental solution procedure which involves solving a set of linear finite element equations for each incremental step in λ. A starting solution is found by solving the corresponding incompressible lubrication problem. A sample solution obtained by this procedure is given in Section 8.7.

Rohde and Oh [21] devised a different technique based directly of equation 8.46. They used Newton's method on the governing nonlinear equation— equation 8.46—while employing Galerkin's criterion to solve the resulting linear equations for the correction term. This procedure (with bicubic Hermit interpolation functions defined over rectangular elements) led to rapid convergence and high accuracy with relatively few elements. Problems with compressibility numbers as high as $\lambda = 100$ were handled with ease. Experience to date has indicated that the method suggested by Rohde and Oh is the most effective way to treat the general compressible lubrication problem.

8.7   SAMPLE SOLUTIONS

The finite element method has undoubtedly been used in industry to solve many practical, real-life lubrication problems. For proprietary reasons, some of these solutions do not appear in the literature; but enough solutions have appeared to give a good indication of the type of complex lubrication problems being solved. The following sample solutions were selected to demonstrate the versatility, power, and potential application of finite element methods in lubrication. Some of the details of these solutions are presented; however, for a complete description of the problem in each case, the reader should refer to the source material.

8.7.1   Incompressible isothermal solutions

Rigid bearing surfaces

Sometimes it is possible to obtain an accurate prediction of the load capacity and friction of a bearing by neglecting the side leakage of lubricant and performing a one-dimensional analysis. This is possible whenever the

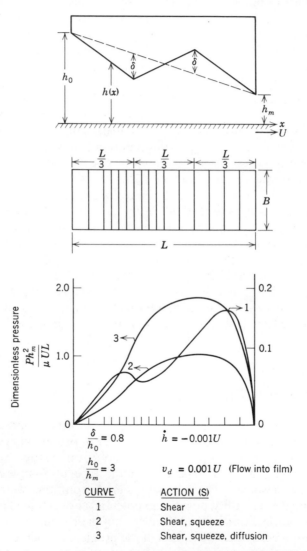

Figure 8.4.   Finite element idealization and solution of a composite slider bearing.

dimensions of the bearing are such that the ratio of its length (the dimension in the direction of sliding) to its width (the dimension perpendicular to the direction of sliding) is greater than one-half. The composite slider bearing [12] depicted in Figure 8.4 satisfies this condition. Linear one-dimensional elements (Figure 8.4b) were used to model the solution domain, and element boundaries were positioned at places where the slope of the film thickness distribution $h(x)$ changes abruptly. Figure 8.4a shows the effect of squeeze action and diffusion flow on the pressure distribution. A squeeze velocity and a diffusion flow velocity of only 0.1 per cent of the sliding velocity drastically increase the generated pressure.

Determining the load capacity of journal bearings with complex oil-feed grooves (Figure 8.5) is a problem of considerable practical interest. Normally, these bearings are designed so that the load vector is out of the region of the feed groove, but in some modes of operation it is possible for the load vector to move into the grooved region. In these cases, the load capacity of the bearing is diminished; but the question is: " By how much?"

With an analysis tool such as the finite element method, this question can be answered without difficulty. The bearing region can be unwrapped and discretized with a regular triangular mesh, as shown in Figure 8.6. Because of symmetry, only half of the bearing need be treated. In the groove region, the film thickness is taken to be larger by three orders of magnitude than the film thickness anywhere outside the groove. Either feed pressure or flow can be specified at nodes along the boundary of the groove. Figure 8.7 shows the results of an analysis using linear interpolation over triangles. Bearing load degradation is seen to be significant even though the grooved area of the bearing represents only 19 per cent of the total area.

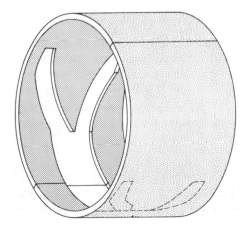

Figure 8.5. Journal bearing with complex oil-feed groove.

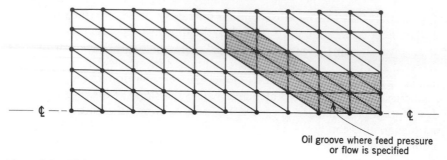

Oil groove where feed pressure
or flow is specified

Figure 8.6. Finite element model of unwrapped grooved bearing. The groove helps promote even oil distribution.

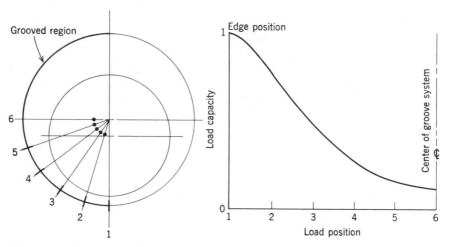

Figure 8.7. Variation of bearing load capacity with groove position.

Wada et al [10, 11] studied the application of finite element techniques to incompressible lubrication problems and reported case studies using linear as well as higher-order elements for one- and two-dimensional problems. They show that, by using bicubic Hermite interpolation for the pressure over rectangular elements, good accuracy can be obtained with only a few elements. To achieve the same accuracy with linear triangular elements requires, by comparison, many elements.

Compliant bearing surfaces

To study the effects of elastic distortion on journal bearing performance, Oh and Huebner [16] used the finite element method to analyze the system shown in Figure 8.8. Linear tetrahedral elements assembled to form hexahedral elements were used to model the bearing housing (Figure 8.9), while linear triangular elements were employed for the fluid film. Without deformation of the bearing, the pressure distribution takes the form shown in Figure 8.10a; but, when the bearing housing is allowed to deform (rigidly fixed only at its base), the pressure distribution changes significantly. Figures 8.10b and

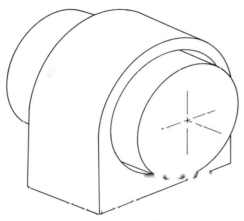

Figure 8.8 Journal bearing with elastic housing.

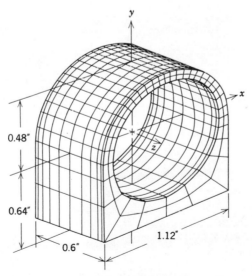

Figure 8.9. Finite element model of the bearing housing.

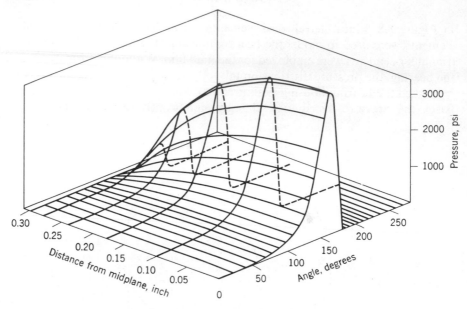

Figure 8.10. Elasticity effects in journal bearings. (*a*) Pressure distribution for a rigid bearing.

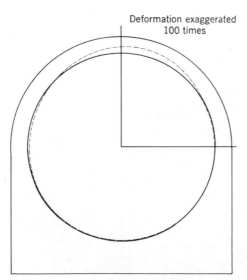

Deformation exaggerated
100 times

Figure 8.10 (*continued*). (*b*) Housing deformation.

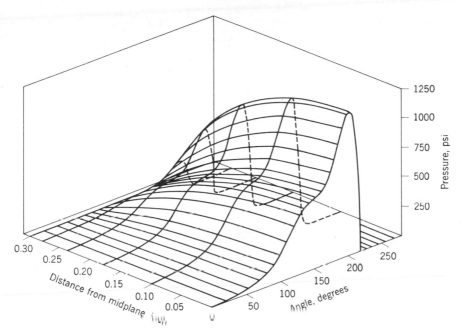

Figure 8.10 (*continued*).   (*c*) Pressure distribution for an elastic bearing.

8.10*c* show the resulting compatible deformations and pressures. The analysis revealed that the elastic housing distributes the load over a wider area and that, for the same minimum film thickness and peak pressure, a bearing with an elastic housing has a higher load capacity than the same bearing with a rigid housing.

Figure 8.11 shows one of the case studies presented by Day [20] and suggests the form of deformation to be expected in slider bearings with one compliant surface.

8.7.2   Incompressible thermohydrodynamic solutions

Examples of finite element techniques applied to thermohydrodynamic problems have been presented by Tieu [17] for the two-dimensional case and by Huebner [18] for the three-dimensional case. The sector thrust bearing studied by Huebner is shown in Figure 8.12. Only the control volume defined by the dotted lines (with the bearing solids immediately

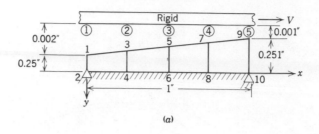

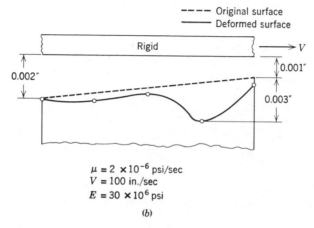

$\mu = 2 \times 10^{-6}$ psi/sec
$V = 100$ in./sec
$E = 30 \times 10^6$ psi

(b)

Figure 8.11. Deformation of a compliant slider bearing [19]. (a) All dimensions are undeformed values. (b) Exaggerated deformation of the bearing surface (very soft case).

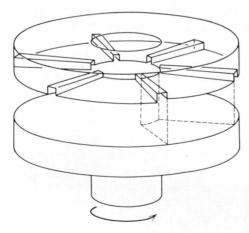

Figure 8.12. Radially grooved sector thrust bearing.

above and below) needs to be considered in the analysis because the complete bearing is actually an assemblage of these subsystems. Figure 8.13 gives some results obtained using simple linear elements. Comparison of the pressure distributions for isothermal and thermohydrodynamic modes of operation indicates that bearing performance can be significantly affected by viscous heating in the lubricant film.

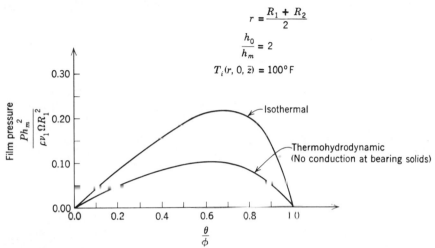

$$r = \frac{R_1 + R_2}{2}$$

$$\frac{h_0}{h_m} = 2$$

$$T_i(r, 0, \bar{z}) = 100°F$$

Isothermal

Thermohydrodynamic
(No conduction at bearing solids)

Figure 8.13. Thermohydrodynamic solution. (*a*) Pad centerline pressure distributions.

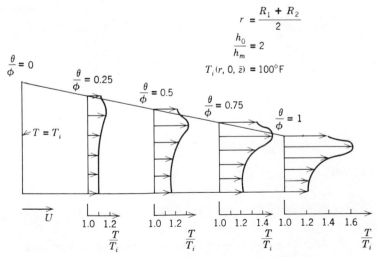

$$r = \frac{R_1 + R_2}{2}$$

$$\frac{h_0}{h_m} = 2$$

$$T_i(r, 0, \bar{z}) = 100°F$$

$$\frac{\theta}{\phi} = 0$$

$$\frac{\theta}{\phi} = 0.25$$

$$\frac{\theta}{\phi} = 0.5$$

$$\frac{\theta}{\phi} = 0.75$$

$$\frac{\theta}{\phi} = 1$$

$T = T_i$

U

Figure 8.13 (*continued*). (*b*) Temperature profiles across the film thickness.

8.7.3 Compressible solutions

Few examples exist for the finite element analysis of compressible lubrication problems. Reddi and Chu [9] studied several configurations, one being the shrouded-step bearing shown in Figure 8.14. The element mesh (Figure 8.14b) consisted of linear, quadrilateral elements refined in areas where pressure variation is expected to be the greatest. Figure 8.15 portrays the computed

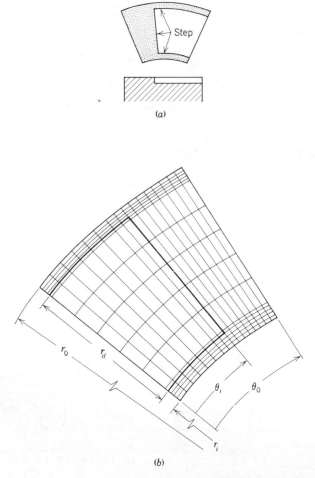

Figure 8.14. Finite element analysis of a shrouded-step bearing [9]. (a) Bearing pad. (b) Finite element mesh.

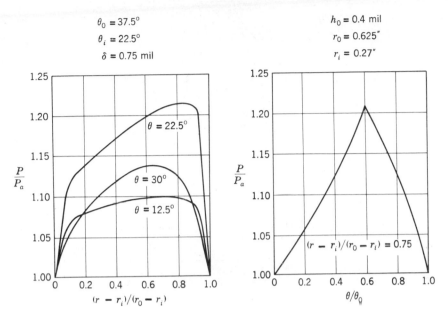

$\theta_0 = 37.5°$ $h_0 = 0.4$ mil

$\theta_i = 22.5°$ $r_0 = 0.625"$

$\delta = 0.75$ mil $r_i = 0.27"$

Figure 8.15. Pressure distributions for the shrouded-step bearing [9].

pressure distributions at various circumferential stations. This example illustrates again that abrupt changes in film thickness at the step cause no special difficulties. Continuity of flow at the step is a natural consequence of the matrix assembly technique.

8.7.4 Lubrication with combined effects

Recently, Eidelberg [19] performed a finite element analysis of lubrication in a joint between two bones. This type of lubrication involves porous as well as elastic bearing surfaces (Figure 8.16). Two different types of elements were used for the poroelastic cartilage material to represent the elastic distortion and the diffusion flow phenomena (see Table 8.1). Figures 8.17 and 8.18 show the compatible pressure and film thickness distributions that result under typical operating conditions. Though these results were obtained neglecting the effects of side leakage, they establish the expected trends and demonstrate the viability of the solution technique. Extension to more realistic three-dimensional problems is straightforward but would involve considerably more computational effort.

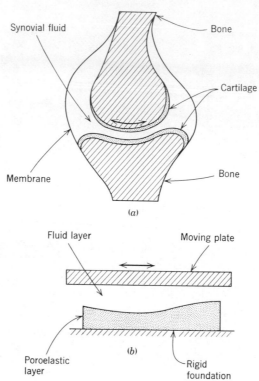

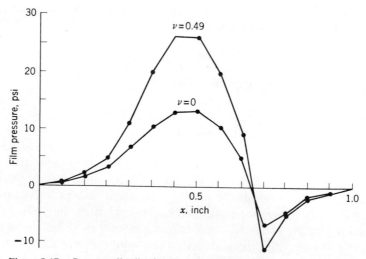

Figure 8.16. The geometry of a natural joint [19]. (*a*) Actual joint. (*b*) Idealized model.

Figure 8.17. Pressure distribution in a natural joint [19].

Table 8.1. Finite element model of a natural joint (After Eidelberg [19])

	No. elements	No. nodes	Type
Elastic	28	87	
Porous	28	30	
Film	14	15	

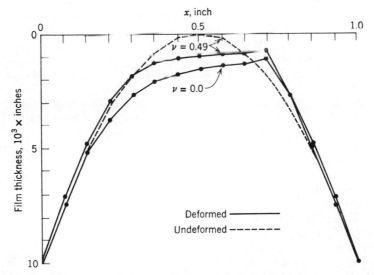

Figure 8.18. Film thickness distribution in a natural joint [19].

8.8 CLOSURE

We have seen in this chapter that the finite element method is now recognized as a potent numerical analysis tool for the solution of complex hydrodynamic lubrication problems. The popularity of the method stems mainly from the ease with which irregular geometries and discontinuous film properties can be handled. Another important advantage is that the method allows development of one general computer program that can solve a variety of problems simply by accepting different input data.

In the following chapter, we consider the application of finite element techniques to far more general fluid mechanics problems.

REFERENCES

1. B. Tower, "First Report on Friction Experiments," *Proc. Inst. Mech. Eng.*, November 1883, p. 632.

2. O. Reynolds, "On the Theory of Lubrication and Its Application to Mr. Beauchamp Tower's Experiments—Including an Experimental Determination of the Viscosity of Olive Oil," *Phil. Trans.*, Vol. 177(i), 1886, p. 157.

3. T. Lloyd and H. McCallion, "Recent Developments in Fluid-Film Lubrication Theory," *Proc. Inst. Mech. Eng.*, Vol. 182, Part 3A, 1967–68, p. 36.

4. D. V. Tansea and I. C. Rao, *Student Project Report on Lubrication*, Royal Naval College, Dartmouth, England, 1966.

5. S. Wada and H. Hayashi, *Conference of Japanese Society of Lubrication Engineers* (in Japanese), Hiroshima, October 1968.

6. M. M. Reddi, "Finite Element Solution of the Incompressible Lubrication Problem," *Trans. ASME*, Series F: *J. Lubr. Technol.*, Vol. 91, No. 3, July 1969, p. 524.

7. T. Fujino, "Analyses of Hydrodynamic and Plate Structure Problems by the Finite Element Method," *Paper* J5-4, Japan-U.S. Seminar on Matrix Methods of Structural Analysis and Design, Tokyo, August 25–30, 1969, p. 725.

8. J. H. Argyris and D. W. Scharpf, "The Incompressible Lubrication Problem," 12th Lanchester Memorial Lecture, Appendix IV, *J. Roy. Aeronaut. Soc.*, Vol. 73, December 1969; p. 1044.

9. M. M. Reddi and T. Y. Chu, "Finite Element Solution of the Steady-State Compressible Lubrication Problem," *Trans. ASME*, Series F: *J. Lubr. Technol.*, Vol. 22, No. 3, July 1970, p. 495.

10. S. Wada, H. Hayashi, and M. Migita, "Application of Finite Element Method to Hydrodynamic Lubrication Problems. Part I: Infinite-Width Bearings," *Bull. Jap. Soc. Mech. Eng.*, Vol. 14, No. 77, November 1971, p. 1222.

11. S. Wada and H. Hayashi, "Application of Finite Element Method to Hydrodynamic Lubrication Problems. Part II: Finite-Width Bearings," *Bull. Jap. Soc. Mech. Eng.*, Vol. 14, No. 77, November 1971, p. 1234.

12. K. H. Huebner, "Finite Element Analysis of Continuum Problems and Its Application to the Incompressible Lubrication Problem," *General Motors Res. Publ.* GMR-1074, February 1971.

13. J. F. Booker and K. H. Huebner, "Application of Finite Element Methods to Lubrication: An Engineering Approach" (*ASME Paper* 72-LUB-N), *Trans. ASME*, Series F: *J. Lubr. Technol.*, Vol. 94, No. 4, October 1972, p. 313.

14. T. Allan, "The Application of Finite Element Analysis to Hydrodynamic and Externally Pressurized Pocket Bearings," *Wear*, Vol. 19, February 1972, p. 169.

15. C. Taylor and J. F. O'Callaghan, "A Numerical Solution of the Elastohydrodynamic Lubrication Problem Using Finite Elements," *J. Mech. Eng. Sci.*, Vol. 14, No. 4, 1972, p. 229.

16. K. P. Oh and K. H. Huebner, "Solution of the Elastohydrodynamic Finite Journal Bearing Problem" (*ASME Paper* 72-LUB-26), *Trans. ASME*, Series F: *J. Lubr. Technol.*, Vol. 95, No. 3, July 1973, p. 342.

17. A. K. Tieu, "Oil-Film Temperature Distribution in an Infinitely Wide Slider Bearing: An Application of the Finite-Element Method," *J. Mech. Eng. Sci.*, Vol. 15, No. 4, 1973, p. 311.

18. K. H. Huebner, "Application of Finite Element Methods to Thermohydrodynamic Lubrication," *Int. J. Numer. Methods Eng.*, Vol. 8, No. 1, 1974, p. 139.

19. B. E. Eidelberg, "Finite Element Analysis of Lubrication in Natural Joints," Ph.D. Thesis, Cornell University, Ithaca, N.Y., January 1974.

20. C. P. Day, "Transient Thermohydrodynamic Lubrication by Finite Element Methods," M.S. Thesis, Cornell University, Ithaca, N.Y., June 1974.

21. S. M. Rohde and K. P. Oh, "A Unified Treatment of Thick- and Thin-Film Elastohydrodynamic Lubrication Problems by Higher-Order Element Methods" (in press).

22. K. P. Oh and S. M. Rohde, "Analysis of Compliant Air Bearing Higher-Order Finite Element Methods" (in press).

9

FLUID MECHANICS PROBLEMS

9.1 INTRODUCTION

During the last 20 years, analysts relied on the traditional finite difference methods to obtain computer-based solutions to difficult flow problems. The progress and success achieved in these pursuits have been, in many cases, noteworthy and remarkable. Slow viscous flows, boundary layer flows, diffusion flows, and variable property (thermohydrodynamic) flows are just a few examples of areas for which analysts have developed refined calculation procedures based on finite difference methods.

Yet there remain a number of problems for which finite difference methods prove inadequate. Problems involving complex geometries, multiply connected domains, and complex boundary conditions always pose perplexing difficulties. Finite element methods can help to alleviate these difficulties, but should not be expected to triumph in every case where finite difference methods have failed. Instead, finite element methods offer easier ways to treat complex geometries requiring irregular meshes, and they provide a more consistent method of using higher-order approximations. In some cases, the finite element method can provide an approximate solution of the same order of accuracy as the finite difference method, but at less expense. Regardless of the method used, the accurate numerical solution of most viscous flow problems requires vast amounts of computer time and data storage. And, of course, problems of stability and convergence can occur with either method.

Only within the last 5 years has the finite element method been recognized as an effective means for solving difficult fluid mechanics problems. Literature on the applications of finite element methods in fluid mechanics is rapidly increasing with contributions being made almost daily. Recently (January 1974) a symposium on the application of finite element methods to flow

314

problems was held at the University College of Swansea† to enable researchers to share their experiences and report their findings. The proceedings of this symposium suggest convincingly that finite element methods offer attractive advantages in a variety of fluid mechanics situations.

This chapter is devoted to the description of finite element methods applied to fluid problems.‡ For incompressible and compressible, inviscid and viscous flows, we lay the theoretical foundations, develop the element equations, and report the most promising findings of recent research efforts. None of the example problems presented really illustrates the full potential of the finite element method in fluid mechanics because all of these problems have been solved using finite difference techniques. Instead, the problems serve as test cases that demonstrate feasibility. Once known solutions are accurately matched and the calculation procedures are proved, the extensions to problems of arbitrary geometry are obvious. Herein lies the real potential of the finite element method in fluid mechanics.

9.2 INVISCID INCOMPRESSIBLE FLOWS

Since all real fluids are viscous, an inviscid fluid is simply a hypothetical concept that simplifies the mathematics of fluid flow problems. Inviscid fluids experience no shearing stresses, and when they come into contact with a solid boundary they slip tangentially along it without resistance. Real fluids, of course, produce shear stresses, and they adhere to flow boundaries so that at the fluid-solid interface no slip occurs.

Despite these differences between viscous and inviscid fluids, there are many practical problems in fluid mechanics which can be analyzed with good accuracy when inviscid flow theory is used. Problems such as flow around streamlined objects, flow through converging or diverging passages, and flow over dams or weirs are just a few significant examples.

9.2.1 Problem statement

The first attempts to extend the finite element method to the solution of fluid mechanics problems dealt with steady, inviscid, incompressible flows [1–12]. As indicated in Appendix D, the problem of predicting flow patterns for this type of flow involves the solution of Laplace's equation. For a two-di-

---

† See footnote on p. 362 of the reference list for this chapter.
‡ A number of fundamental concepts and basic equations from fluid mechanics are reviewed in Appendix D for the reader's convenience.

mensional, irrotational flow (potential flow), the velocity compon... u and v may be expressed in terms of a stream function $\psi(x, y)$ as

$$u = \frac{\partial \psi}{\partial y}, \qquad v = -\frac{\partial \psi}{\partial x} \tag{9.1}$$

or, in terms of a potential function $\phi(x, y)$, as

$$u = -\frac{\partial \phi}{\partial x}, \qquad v = -\frac{\partial \phi}{\partial y} \tag{9.2}$$

Both the stream function and the potential function satisfy Laplace's equation:

$$\nabla^2 \psi = 0 \tag{9.3a}$$

$$\nabla^2 \phi = 0 \tag{9.3b}$$

Dirichlet and Neumann type boundary conditions apply to equation 9.3. At a solid boundary, the velocity component normal to the boundary must be the same as the velocity of the boundary in the normal direction. Hence we have

$$\mathbf{V} \cdot \hat{n} = \mathbf{V}_B \cdot \hat{n} \tag{9.4}$$

$$U n_x + V n_y = V_{B_x} n_x + V_{B_y} n_y \tag{9.5}$$

where $\mathbf{V}_B$ is the velocity of the boundary and n_x, n_y are direction cosines of the outward normal to the boundary. If the boundary is stationary, $\mathbf{V}_B = 0$, equation 9.5 expresses the conditions

$$\frac{\partial \psi}{\partial s} = \frac{\partial \psi}{\partial y} n_x - \frac{\partial \psi}{\partial x} n_y = 0 \tag{9.6a}$$

$$\frac{\partial \phi}{\partial n} = \frac{\partial \phi}{\partial x} n_x + \frac{\partial \phi}{\partial y} n_y = 0 \tag{9.6b}$$

Equations 9.6 state that the *tangential* derivative of the stream function along the wall is zero, while the *normal* derivative of the potential function is zero.

The boundary value problem for potential flow in terms of ϕ may then be stated as follows:

Given a solution domain Ω bounded by curve $C = C_1 + C_2$, find the velocity potential such that

$$\nabla^2 \phi = 0 \quad \text{in } \Omega \tag{9.7}$$

$$\phi = q(x, y) \quad \text{on } C_1 \tag{9.8}$$

$$\frac{\partial \phi}{\partial n} = \mathbf{V}_B \cdot \hat{n} = V^* \quad \text{on } C_2 \tag{9.9}$$

We have seen that this problem is equivalent to the problem of finding the function ϕ that minimizes

$$I(\phi) = \frac{1}{2} \int_{\Omega} \left[\left(\frac{\partial \phi}{\partial x} \right)^2 + \left(\frac{\partial \phi}{\partial y} \right)^2 \right] dx \, dy + \int_{C_2} V^* \phi \, ds \qquad (9.10)$$

subject to the same boundary conditions. Since the details of the finite element formulation of this type of problem are given in Section 7.2, here we shall discuss instead the special considerations applicable to potential flows. The discussion follows that of de Vries and Norrie [3].

Consider the problem of determining the flow pattern about a body of arbitrary shape immersed in a uniflow flow field (Figure 9.1). Since the actual solution domain for this problem is infinite in extent, it is necessary to construct some finite domain. This is done as indicated in Figure 9.1 by taking a boundary sufficiently far from the body so that the flow field at every point on the boundary is known in terms of the given approaching flow. A rectangular boundary is the most convenient choice here.

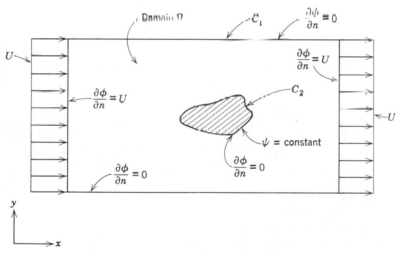

Figure 9.1. Potential flow around a body of arbitrary shape [3].

9.2.2 ϕ and ψ formulations

We shall consider the solution procedure first in terms of the potential function ϕ and then in terms of the stream function ψ. When formulating the problem with reference to ϕ, we note that all the boundary conditions are given in terms of $\partial \phi / \partial n$, and incorporating this Neumann boundary condition requires a special procedure. On solid stationary boundaries, $\partial\phi/\partial n = 0$; but on the outer boundary, $\partial\phi/\partial n \neq 0$, in general. The variational statement—equation 9.10—automatically accounts for these conditions via the line integral. However, the solution of equation 9.6 subject to only Neumann boundary conditions is nonunique. For this reason, when the solution domain is discretized and the element equations are formulated and assembled, the system matrix $[K]$ is nonsingular. To overcome this difficulty, we arbitrarily select one node and specify the value of ϕ at this node—say, $\phi = 0$. Essentially, we impose a Dirichlet boundary condition at one node. When incorporated by one of the means indicated in Chapter 2, this Dirichlet condition removes the singularity and allows us to proceed with the solution as usual.

Formulating the problem in terms of ψ requires a different technique. In this case we know that $\psi = $ constant or $\partial\psi/\partial s = 0$ on all solid boundaries, but the values of the constants are unknown. To proceed, we can use a superposition technique suggested by de Vries and Norrie [3]. On the outer boundary C_1, we recognize that ψ is known in terms of U; hence we can write $\psi = g(x, y)$ on C_1, where $g(x, y)$ is a known function. The next step is to represent the complete solution as a sum of two separate parts, that is,

$$\psi(x, y) = \psi_1(x, y) + b\psi_2(x, y) \tag{9.11}$$

where b is a constant to be determined. The problem then reduces to two separate problems, described as follows:

$$\begin{aligned} \nabla^2\psi_1 &= 0 && \text{in } \Omega \\ \psi_1 &= g(x, y) && \text{on } C_1 \\ \psi_1 &= 0 && \text{on } C_2 \end{aligned} \tag{9.12}$$

$$\begin{aligned} \nabla^2\psi_2 &= 0 && \text{in } \Omega \\ \psi_2 &= 0 && \text{on } C_1 \\ \psi_2 &= 1 && \text{on } C_2 \end{aligned} \tag{9.13}$$

Now equations 9.12 and 9.13 are easily solved using the finite element techniques of Chapter 7. Once $\psi_1(x, y)$ and $\psi_2(x, y)$ are known, we return to equation 9.11 and find the constant b by evaluating $\psi(x, y)$ at one point in Ω close to the boundary C_1 where $\psi(x, y)$ is known. This gives an equation to be solved for b and completes the solution procedure.

Figure 9.2 shows an example of an internal flow studied by Martin [2].

9.2.3   Flow around multiple bodies

The same superposition technique can be used for either ϕ or ψ formulations when we have N arbitrarily shaped bodies immersed in a flow [3] (Figure 9.3).

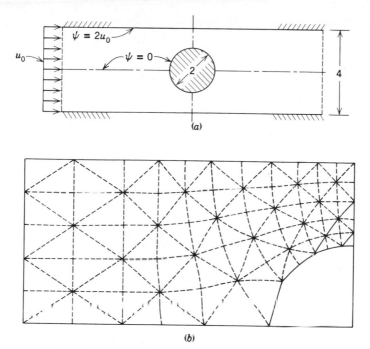

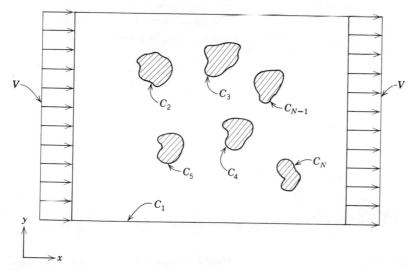

Figure 9.2.   Potential flow around a cylinder between parallel walls [2]. (*a*) Circular cylinder between two parallel walls. (*b*) Finite element model of the cylinder flow problem.

Figure 9.3.   Potential flow around *N* bodies [3].

For the ψ formulation, for example, we would write

$$\psi = \psi_1 + b_2\psi_2 + b_3\psi_3 + \cdots + b_N\psi_N \tag{9.14}$$

and then form N problems as follows.

$$\begin{aligned} \nabla^2\psi_1 &= 0 \quad \text{in } \Omega \\ \psi_1 &= g \quad \text{on } C_1 \\ \psi_1 &= 0 \quad \text{on } C_2, \ldots, C_N \end{aligned} \tag{9.15a}$$

$$\begin{aligned} \nabla^2\psi_2 &= 0 \quad \text{in } \Omega \\ \psi_2 &= 1 \quad \text{on } C_2 \\ \psi_2 &= 0 \quad \text{on } C_1, C_3, C_4, \ldots, C_N \end{aligned} \tag{9.15b}$$

$$\begin{aligned} \nabla^2\psi_3 &= 0 \quad \text{in } \Omega \\ \psi_3 &= 1 \quad \text{on } C_3 \\ \psi_3 &= 0 \quad \text{on } C_1, C_2, C_4, C_5, \ldots, C_N \end{aligned} \tag{9.15c}$$

$$\vdots$$

$$\begin{aligned} \nabla^2\psi_N &= 0 \quad \text{in } \Omega \\ \psi_N &= 1 \quad \text{on } C_N \\ \psi_N &= 0 \quad \text{on } C_1, C_2, \ldots, C_{N-1} \end{aligned} \tag{9.15d}$$

Each of the sets of equations 9.15 is solved by the usual finite element techniques; and, after the ψ_i are found, equation 9.14 is evaluated at $N - 1$ points to form $N - 1$ linear algebraic equations to be solved for the constants $b_2, b_3, \ldots, b_N$. Figure 9.4 shows some sample flow patterns obtained by this method [3].

Once the stream function or potential is known for a given problem, we can find the pressure distribution throughout the flow by using Bernoulli's equation. For a steady, irrotational, inviscid, two-dimensional flow, Bernoulli's equation is

$$\tfrac{1}{2}V^2 + \eta + \frac{P}{\rho} = \text{constant} \tag{9.16}$$

where

$$V^2 = u^2 + v^2 = \left(\frac{\partial \psi}{\partial y}\right)^2 + \left(\frac{\partial \psi}{\partial x}\right)^2 = \left(\frac{\partial \phi}{\partial x}\right)^2 + \left(\frac{\partial \phi}{\partial y}\right)^2$$

and η = body force potential such that $\mathbf{B} = \nabla\eta$, P = pressure, and ρ = fluid density.

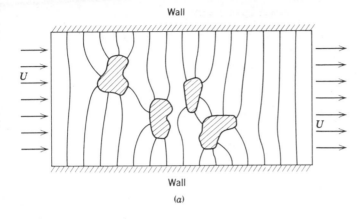

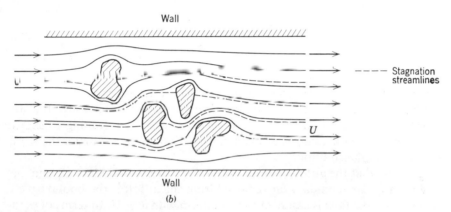

Figure 9.4. Potential flow solution for irregularly shaped bodies in a uniform cross flow [3]. (*a*) Solution for potential function (lines of constant ϕ). (*b*) Solution for stream function (lines of constant ψ).

9.2.4 The Kutta condition

In the foregoing discussion, we showed how to obtain a finite element solution for the flow around an immersed body. But never did we give thought to the physical implications of the results. The results of Figure 9.5, obtained by de Vries and Norrie [3], serve to clarify this comment. The stagnation streamlines, which are a natural result of the solution, attach to the body at points S_1 and S_2. However, the location of these attachment points is not physically realistic because a real flow could not make the

Angle of attack $\alpha = 60°$

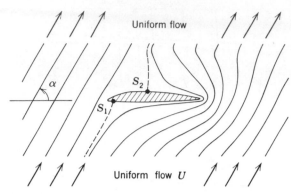

Figure 9.5. Solution for stream function (lines of constant ψ) for a NACA 4412 air foil (Kutta condition not prescribed) [3].

sharp turn around the trailing edge as indicated. The reason for this unrealistic result is that we have not considered the Kutta boundary condition in the solution. It is important to employ the Kutta condition whenever we wish to calculate the lift of a body.† The Kutta condition states that the downstream stagnation point S_2 should be at the sharp trailing edge. The following procedure, suggested in ref. 3, makes it possible to include the Kutta condition in the solution.

Suppose that the airfoil contour is described by the boundary C_1 and that the rectangular boundary far removed from the airfoil is the boundary C_2. Outside C_2 the flow is assumed to be uniform with $u = U$. In terms of ψ, the problem is to solve the following set of equations:

$$\nabla^2 \psi = 0 \qquad\qquad \text{in } \Omega$$

$$\psi = \text{constant} \qquad\qquad \text{on } C_1 \qquad\qquad (9.17)$$

$$\psi = -Uy + \text{constant} \quad \text{on } C_2$$

Kutta condition → trailing edge velocity is zero.

Again, the technique is to use a superposition of different solutions such that

$$\psi = \psi_1 + b_2\psi_2 + b_3\psi_3 \qquad\qquad (9.18)$$

---

† Any body which is not symmetric with respect to the approaching flow direction will have either positive or negative lift.

where b_2 and b_3 are constants to be determined. The individual problems with split Dirichlet boundary conditions are the following:

$$
\begin{aligned}
\nabla^2 \psi_1 &= 0 && \text{in } \Omega \\
\psi_1 &= 0 && \text{on } C_1 \\
\psi_1 &= -Uy + \text{constant} && \text{on } C_2
\end{aligned}
\tag{9.19a}
$$

$$
\begin{aligned}
\nabla^2 \psi_2 &= 0 && \text{in } \Omega \\
\psi_2 &= 0 && \text{on } C_1 \\
\psi_2 &= 1 && \text{on } C_2
\end{aligned}
\tag{9.19b}
$$

$$
\begin{aligned}
\nabla^2 \psi_3 &= 0 && \text{in } \Omega \\
\psi_3 &= 1 && \text{on } C_1 \\
\psi_3 &= 0 && \text{on } C_2
\end{aligned}
\tag{9.19c}
$$

Solving equations 9.19 then gives the streamline distributions ψ_1, ψ_2, and ψ_3, from which the velocity components u and v can be determined by means of equations 9.1. Then, from equation 9.18, we can write

$$
\begin{aligned}
u &= u_1 + b_2 u_2 + b_3 u_3 \\
v &= v_1 + b_2 v_2 + b_3 v_3
\end{aligned}
\tag{9.20}
$$

Evaluating equations 9.20 at the trailing edge, where we insist that $u = v = 0$, then provides two equations to be solved for b_2 and b_3. Figure 9.6 shows the result of the complete solution.

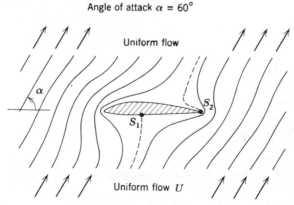

Figure 9.6. Solution for a stream function (lines of constant ψ) for a NACA 4412 airfoil (Kutta condition prescribed) [3].

The work of de Vries and Norrie [3] employed simple triangular elements with linear interpolation functions, but more complicated elements can be used to advantage. Fujino [4] and Argyris and Mariczek [5] used higher-order triangular elements that allowed the specification of ϕ and its derivatives, $\partial\phi/\partial x$ and $\partial\phi/\partial y$, at the nodes. This approach offers the advantage of increased accuracy and the ability to handle Neumann boundary conditions as though they were Dirichlet conditions.

9.2.5   Free-surface flows

In each of the potential flow problems we have discussed thus far, the boundaries of the solution domains were known a priori. The geometry of the flow channel dictated the boundaries for internal flows, and for external flows we chose some boundary sufficiently far removed from the bodies under investigation. However, in a number of important fluid mechanics problems classified as free-surface flows, parts of the flow boundaries (the free surfaces) are unknown a priori. These problems include sluice gate flows, flows over wiers, and ground water flows under dams. The free surface is a streamline, and the problem is to determine the location of this streamline as well as the streamline pattern for the remainder of the flow. Once the streamline pattern is known, flow coefficients and hydraulic forces can be computed.

The procedure for solving free-surface potential flow problems involves solving Laplace's equation as usual plus satisfying the conditions that the velocity normal to the free surface be zero and the pressure along the free surface be constant. Iteration of some kind is used to achieve these conditions. One way is to select a trial free-surface shape, solve $\nabla^2\psi = 0$, and calculate the velocity components along the assumed free-surface profile. If conditions are not met at the free surface, the procedure is repeated until a prescribed error criterion is satisfied. Three to five iterations are usually sufficient for convergence if the initial free surface is judiciously chosen. Examples of these types of calculations can be found in refs. 9–12.

As a particular example, consider the analysis of a sluice gate flow performed by McCorquodal and Li [9]. Figure 9.7 shows the geometry of the flow and the finite element model they used. Assuming that the function $h(x)$ describing the free surface is known, we can formulate the problem as follows:

Find the stream function $\psi(x, y)$ that minimizes

$$I(\psi) = \frac{1}{2g} \iint\limits_{A} \left[\left(\frac{\partial\psi}{\partial x}\right)^2 + \left(\frac{\partial\psi}{\partial y}\right)^2 \right] dx\, dy \qquad (9.21)$$

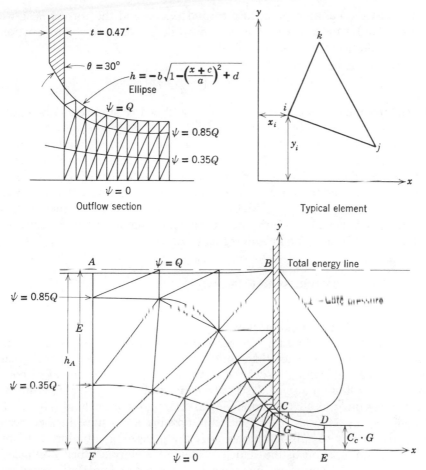

Figure 9.7. Finite element of a sluice gate flow [9].

subject to the imposed boundary conditions:

$$\frac{\partial \psi}{\partial x} = 0 \quad \text{at the inlet and outlet planes} \tag{9.22}$$

$$\frac{1}{2g}\left[\left(\frac{\partial \psi}{\partial x}\right)^2 + \left(\frac{\partial \psi}{\partial y}\right)^2\right]_{\substack{\text{on free} \\ \text{surface}}} + h(x) = E \tag{9.23}$$

$$\psi = Q \quad \text{at } y = h(x) \tag{9.24}$$

$$\psi = 0 \quad \text{at } y = 0 \tag{9.25}$$

To start the iteration procedure to find $\psi(x, y)$ and the correct $h(x)$, we assume that for the upstream flow $h = E$, while for the downstream portion h has an elliptical character described by

$$h(x; a, b) = -b\left[1 - \left(\frac{x + c}{a}\right)^2\right]^{1/2} + d$$

The elliptical curve is chosen so that the outflow free surface can be made tangent to the gate lip by adjusting c and d.

With a trial h, the finite element procedure is used to minimize $I(\psi)$, equation 9.21, subject to the Dirichlet and Neumann conditions of equations 9.22, 9.24, and 9.25. Then the constant-energy requirement of equation 9.23 can be checked and the constants a and b adjusted to minimize the difference between $E_{computed}$ and E_{actual}. McCorquodale and Li [9] found that using a momentum balance to adjust the outflow free surface gave an effective criterion for adjusting h and converging to E.

9.3 INVISCID COMPRESSIBLE FLOWS

Flows with variations in density throughout the flow field are called *compressible flows*. A compressible flow may be classified as subsonic, transonic, or supersonic, depending on whether the Mach number,† M, is less than, about equal to, or greater than unity, respectively. Because of the great mathematical difficulties associated with the general problem of viscous compressible flows, most analyses of compressible two- and three-dimensional flows incorporate the simplifying assumption that the fluid is inviscid. Finite element analyses of inviscid compressible flows have further been restricted mainly to subsonic flows since the governing equation in terms of the stream function or the velocity potential is elliptic, albeit nonlinear. For transonic and supersonic flows, we must deal with a hyperbolic equation and the possibility of shock fronts. Finite element methods to handle these features are as yet not well developed [13], though some work has been done for linearized supersonic flows [14,15].

9.3.1 Governing equations and calculation procedures

The differential equations governing an isentropic, inviscid, steady, irrotational, compressible flow are the following:

$$\frac{1}{2}\left[\left(\frac{\partial\phi}{\partial x}\right)^2 + \left(\frac{\partial\phi}{\partial y}\right)^2 + \left(\frac{\partial\phi}{\partial z}\right)^2\right] + \frac{d}{d\rho}(\rho E) = 0 \qquad (9.26)$$

† Mach number is defined as the ratio of the local flow velocity to the local velocity of sound.

$$P = \rho^2 \frac{dE}{d\rho} \tag{9.27}$$

where ϕ = velocity potential as before, ρ = fluid density, and E = internal energy per unit mass. Equation 9.26 is the equation of motion, while equation 9.27 is the appropriate thermodynamic relation. A relation between pressure and density is also needed. Consider a flow domain bounded by surface $S = S_1 \cup S_2$. Then the potential function ϕ and the pressure P which maximize the functional [16,17]

$$I(\phi, P) = \int_\Omega P d\Omega + \int_{S_2} \phi f \, dS_2 \tag{9.28}$$

satisfy equations 9.26 and 9.27 with the Neumann and Dirichlet conditions

$$\rho \nabla \phi \cdot \hat{n} = f(x, y, z) \quad \text{on } S_2 \tag{9.29}$$
$$\phi = g(x, y, z) \quad \text{on } S_1 \tag{9.30}$$

Equation 9.28 can be used as the basis for deriving finite element equations in the manner described in Chapter 3

If we further restrict our attention to perfect gases, the governing equations have a simpler form. In terms of the velocity potential, the continuity equation is, as before,

$$\nabla \cdot (\rho \nabla \phi) = 0 \tag{9.31}$$

with boundary conditions given by equations 9.29 and 9.30. But the momentum and energy equations can be combined with the equation of state to yield the following relation between the density field and the flow field:

$$\frac{\rho_0}{\rho} = \left(1 + \frac{k-1}{2c_0^2} \nabla \phi \cdot \nabla \phi\right)^{1/(k-1)} \tag{9.32}$$

where ρ_0 = stagnation density, c_0 = velocity of sound at the stagnation state, k = ratio of specific heats.

For a given density field, the velocity potential ϕ which minimizes

$$I(\phi) = \frac{1}{2} \int_\Omega \rho \nabla \phi \cdot \nabla \phi \, d\Omega - \int_{S_2} \phi f \, dS_2 \tag{9.33}$$

and satisfies equation 9.30 also satisfies equations 9.29 and 9.31.

Equations 9.32 and 9.33 provide the basis for a convergent iterative procedure to calculate the density and velocity fields in a compressible flow [18]. To start the calculations, we assume $\rho = \rho_0$ and use the functional of equation 9.33 to derive element equations for the discretized solution domain. After the element equations are assembled and the system equations are solved for the nodal values of ϕ, equation 9.32 is used to calculate a new density field for the next calculation of ϕ. The process of calculating alternately ϕ and ρ continues until convergence is achieved.

For a two-dimensional flow, Gelder [18] shows that a similar formulation is possible in terms of the stream function ψ. The equations are as follows.

Stream function definition:

$$u = \frac{1}{\rho} \frac{\partial \psi}{\partial y}, \qquad v = \frac{1}{\rho} \frac{\partial \psi}{\partial x} \tag{D.9}$$

Continuity:

$$\nabla \cdot \left(\frac{1}{\rho} \nabla \psi \right) = 0 \tag{9.34}$$

Combined momentum, energy, and equation of state:

$$\frac{\rho_0}{\rho} = \left(1 + \frac{k-1}{2c_0^2} \frac{1}{\rho^2} \nabla \psi \cdot \nabla \psi \right)^{1/(k-1)} \tag{9.35}$$

Associated functional:

$$I(\psi) = \frac{1}{2} \int_\Omega \frac{1}{\rho} \nabla \psi \cdot \nabla \psi \, d\Omega - \int_{S_2} \psi f \, dS_2 \tag{9.36}$$

Boundary conditions:

$$\begin{aligned} \psi &= g \quad \text{on } S_1 \\ \nabla \psi \cdot \hat{n} &= f \quad \text{on } S_2 \end{aligned} \tag{9.37}$$

De Vries et al. [19] and Periaux [20] construct a similar iterative finite element method for solving compressible flow problems. For two-dimensional problems, the governing differential equations that they use are as follows.

In terms of (ϕ):

$$\left[\left(\frac{\partial \phi}{\partial x} \right)^2 - c^2 \right] \frac{\partial^2 \phi}{\partial x^2} + 2 \frac{\partial \phi}{\partial x} \frac{\partial \phi}{\partial y} \frac{\partial^2 \phi}{\partial x \partial y} + \left[\left(\frac{\partial \phi}{\partial y} \right)^2 - c^2 \right] \frac{\partial^2 \phi}{\partial y^2} = 0 \tag{9.38}$$

$$c^2 = c_0^2 - \left(\frac{k-1}{2} \right) \left[\left(\frac{\partial \phi}{\partial x} \right)^2 + \left(\frac{\partial \phi}{\partial y} \right)^2 \right] \tag{9.39}$$

In terms of (ψ, ρ):

$$\left[\rho^2 c^2 - \left(\frac{\partial \psi}{\partial y}\right)^2\right]\frac{\partial^2 \psi}{\partial x^2} + \left[\rho^2 c^2 - \left(\frac{\partial \psi}{\partial x}\right)^2\right]\frac{\partial^2 \psi}{\partial y^2} + 2\frac{\partial \psi}{\partial x}\frac{\partial \psi}{\partial y}\frac{\partial^2 \psi}{\partial x \partial y} = 0 \quad (9.40)$$

with ρ given by equation 9.35.

For three-dimensional flows in terms of ϕ:

$$\left[\left(\frac{\partial \phi}{\partial x}\right)^2 - c^2\right]\frac{\partial^2 \phi}{\partial x^2} + \left[\left(\frac{\partial \phi}{\partial y}\right)^2 - c^2\right]\frac{\partial^2 \phi}{\partial y^2} + \left[\left(\frac{\partial \phi}{\partial z}\right)^2 - c^2\right]\frac{\partial^2 \phi}{\partial z^2}$$

$$+ 2\left(\frac{\partial \phi}{\partial x}\frac{\partial \phi}{\partial y}\frac{\partial^2 \phi}{\partial x \partial y} + \frac{\partial \phi}{\partial x}\frac{\partial \phi}{\partial z}\frac{\partial^2 \phi}{\partial x \partial z} + \frac{\partial \phi}{\partial y}\frac{\partial \phi}{\partial z}\frac{\partial^2 \phi}{\partial y \partial z}\right) = 0 \quad (9.41)$$

The usual Dirichlet and Neumann boundary conditions apply to these equations. If the terms of equations 9.38, 9.40, and 9.41 are rearranged so that linear terms are retained on the right-hand side while the nonlinear terms are moved to the left, we obtain equations of the form

$$\nabla^2 \phi = \mathcal{L}_1(\phi) \quad (9.42)$$

and

$$\nabla^2 \psi = \mathcal{L}_2(\psi, \rho) \quad (9.43)$$

where $\mathcal{L}_1$ and $\mathcal{L}_2$ are nonlinear operators. With an iterative calculation scheme, the nonlinear terms in equations 9.42 and 9.43 can be taken as forcing functions in the problem. Thus we can construct recursion formulas by writing

$$\nabla^2 \phi^{(n+1)} = \mathcal{L}_1(\phi^{(n)}) \quad (9.44)$$

$$\nabla^2 \psi^{(n+1)} = \mathcal{L}_2(\psi^{(n)}, \rho^{(n)}) \quad (9.45)$$

where the superscript indicates the iteration number.

In terms of the $(n + 1)$th iteration, equations 9.44 and 9.45 are simply Poisson equations for $\phi^{(n+1)}$ and $\psi^{(n+1)}$. In view of equations 7.1–7.6, we can write the corresponding variational statements immediately. These are as follows:

$$I(\phi^{(n+1)}) = \frac{1}{2}\int_\Omega \left[\nabla\phi^{(n+1)} - 2F_1(\phi^{(n)})\phi^{(n+1)}\right] d\Omega + \int_{C_2} f\phi^{(n+1)} dC_2 \quad (9.46)$$

where

$$F_1(\phi^{(n)}) = \frac{1}{c^2}\left[\left(\frac{\partial \phi^{(n)}}{\partial x}\right)^2\left(\frac{\partial^2 \phi^{(n)}}{\partial x^2}\right) + \left(\frac{\partial \phi^{(n)}}{\partial y}\right)^2\left(\frac{\partial^2 \phi^{(n)}}{\partial y^2}\right) + 2\frac{\partial \phi^{(n)}}{\partial x}\frac{\partial \phi^{(n)}}{\partial y}\frac{\partial^2 \phi^{(n)}}{\partial x \partial y}\right]$$

$$I(\psi^{(n+1)}) = \frac{1}{2}\int_\Omega \left[\nabla\psi^{(n+1)} \cdot \nabla\psi^{(n+1)} + 2F_2(\psi^{(n)})\psi^{(n+1)}\right] d\Omega \quad (9.47)$$

where

$$F_2(\psi^{(n)}) = \left(\frac{1}{c\rho^{(n)}}\right)^2\left[\left(\frac{\partial\psi^{(n)}}{\partial y}\right)^2\left(\frac{\partial^2\psi^{(n)}}{\partial x^2}\right) + \left(\frac{\partial\psi^{(n)}}{\partial x}\right)^2\left(\frac{\partial^2\psi^{(n)}}{\partial y^2}\right)\right.$$

$$\left. - 2\frac{\partial\psi^{(n)}}{\partial x}\frac{\partial\psi^{(n)}}{\partial y}\frac{\partial^2\psi^{(n)}}{\partial x\,\partial y}\right]$$

$$\rho^{(n+1)} = \rho_0\left(1 + \frac{k-1}{2C_0^2\rho(n)^2}\,\nabla\psi^{(n)}\cdot\nabla\psi^{(n)}\right)^{k-1}$$

$$I(\phi^{(n+1)}) = \frac{1}{2}\int_\Omega\left[\nabla\phi^{(n+1)}\cdot\nabla\phi^{(n+1)} + 2F_3(\phi^{(n)})\phi^{(n+1)}\right]d\Omega$$

$$- \int_{S_2} f\phi^{(n+1)}\,dS_2 \tag{9.48}$$

where

$$F_3(\phi^n) = \frac{1}{c^2}\left[\left(\frac{\partial\phi^{(n)}}{\partial x}\right)^2\left(\frac{\partial^2\phi^{(n)}}{\partial x^2}\right) + \left(\frac{\partial\phi^{(n)}}{\partial y}\right)^2\left(\frac{\partial^2\phi^{(n)}}{\partial y^2}\right) + \left(\frac{\partial\phi^{(n)}}{\partial z}\right)^2\left(\frac{\partial^2\phi^{(n)}}{\partial z^2}\right)\right.$$

$$\left. + 2\frac{\partial\phi^{(n)}}{\partial x}\frac{\partial\phi^{(n)}}{\partial y}\frac{\partial^2\phi^{(n)}}{\partial x\,\partial y} + 2\frac{\partial\phi^{(n)}}{\partial x}\frac{\partial\phi^{(n)}}{\partial z}\frac{\partial^2\phi^{(n)}}{\partial x\,\partial z} + 2\frac{\partial\phi^{(n)}}{\partial y}\frac{\partial\phi^{(n)}}{\partial z}\frac{\partial^2\phi^{(n)}}{\partial y\,\partial z}\right]$$

Equations 9.46, 9.47, and 9.48 provide convenient variational statements for deriving the element equations. Following the usual finite element discretization, we require that $\delta I = 0$ with respect to $\phi^{(n+1)}$ or $\psi^{(n+1)}$, and this leads to element equations (see Section 7.2) and system equations of the form

$$[K_\phi]\{\phi^{(n+1)}\} = \{R(\phi^{(n)})\} \tag{9.49}$$

or

$$[K_\psi]\{\psi^{(n+1)}\} = \{R(\psi^{(n)})\} \tag{9.50}$$

depending on the formulation chosen. At each step of the iteration, the nodal forcing functions $\{R\}$ are known in terms of the results of the calculations for the preceding step.

9.3.2   Examples

Periaux [20] used these finite element formulations to solve a number of practical problems involving a steady, inviscid, irrotational, isentropic flow of a perfect gas. Figure 9.8 shows a computer-drawn, two-dimensional streamline solution for a nozzle with an inlet Mach number $M_i = 0.850$ and an outlet Mach number $M_o = 0.234$. With 576 linear triangular elements, convergence was obtained after four iterations and 10 seconds of *CPU* time on an IBM 370-165 computer. A subsonic nozzle of similar shape in three dimensions was analyzed with hexahedral elements assembled from five tetrahedra (Figure 9.9). The velocity distribution in the nozzle is schematically represented as a series of dashed lines of scaled lengths.

As examples of external flows, Periaux [20] studied the flow around different airfoil configurations. For the pair of airfoils shown in Figure 9.10, both linear and second-order isoparametric triangular elements were used for a (ψ, ρ) formulation of the problem. The superposition technique explained in Section 9.2.3 was used to treat the multiply connected solution domain.

The complexity of treating three-dimensional airfoils is evident from Figure 9.11 (p. 338). With 5000 elements and 1500 nodal unknowns, eight iterations and 15 minutes of *CPU* time on an IBM 370-165 computer were required for convergence.

These examples effectively demonstrate the feasibility of using finite element techniques for the solution of complex subsonic compressible flow problems.

9.4   INCOMPRESSIBLE VISCOUS FLOW WITHOUT INERTIA

The preceding sections of this chapter centered on the finite element analysis of an important class of fluid mechanics problems involving inviscid fluids. In this section, we consider the simplest of the viscous flow problems — namely, creeping flow (sometimes also called Stokes flow). If the full Navier-Stokes equations (equations C.4) are made dimensionless, there results a dimensionless group known as the Reynolds number, *Re*, which represents the ratio of inertia forces to viscous forces in a fluid motion. When the Reynolds number is very small (usually $Re < 1$), the inertia forces are insignificant compared to the viscous forces and can be omitted from the governing momentum equations. Small Reynolds numbers characterize slow-moving flows and flows of very viscous fluids. These types of flows occur, for example, in viscometry and polymer processing.

A steady, two-dimensional, isoviscous flow is governed by the following equations.

Continuity:

$$\frac{\partial u}{\partial x} + \frac{\partial v}{\partial y} = 0 \qquad (9.51)$$

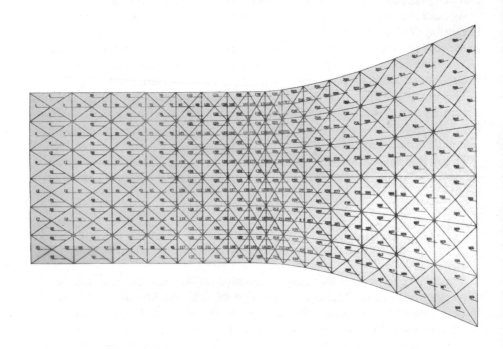

Figure 9.8a. Finite element model and velocities for a subsonic compressible flow in a two-dimensional nozzle [20].

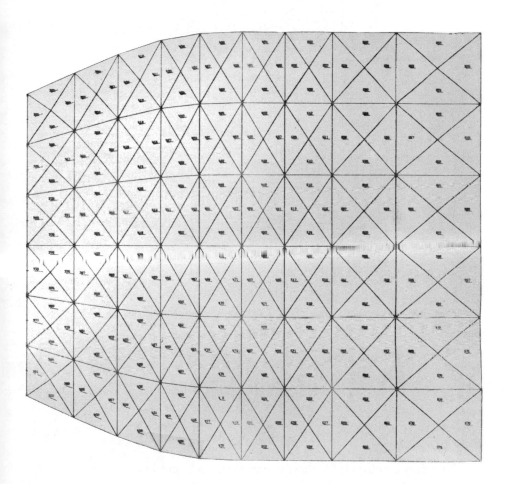

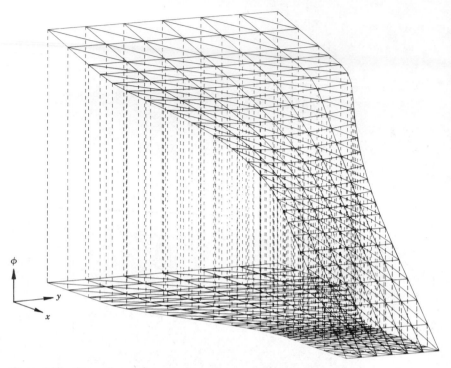

Figure 9.8*b***.**   Streamline surface $\phi(x, y)$ for the two-dimensional nozzle [20].

Momentum:

$$\frac{\partial P}{\partial x} = \mu\left(\frac{\partial^2 u}{\partial x^2} + \frac{\partial^2 u}{\partial y^2}\right) \qquad (9.52a)\dagger$$

$$\frac{\partial P}{\partial y} = \mu\left(\frac{\partial^2 v}{\partial x^2} + \frac{\partial^2 v}{\partial y^2}\right) \qquad (9.52b)$$

Boundary conditions for these equations include specified pressure, velocity, and/or velocity gradient. To solve these equations by the finite element method, we may use one of two different formulations. Either we can introduce a stream function and work with one governing equation of the fourth order, or we can work with velocity and pressure as the field variables. Both approaches will now be outlined because they illustrate important concepts essential for the solution of more complex problems.

---

† Body forces have not been written in these equations because they may be grouped with the pressure terms when the body force can be expressed as the gradient of a potential function, that is, $\mathbf{B} = \nabla \xi$.

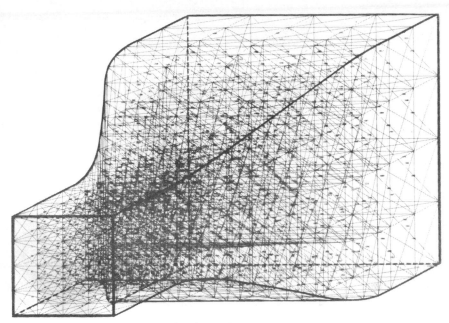

Figure 9.9*a*.   Three-dimensional subsonic nozzle flow modeled with hexahedral elements [20].

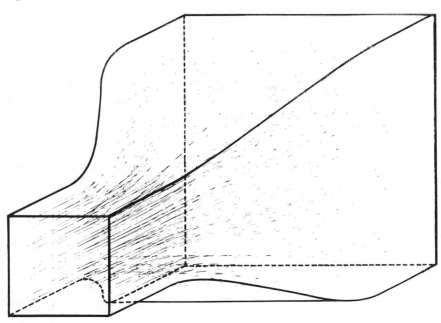

Figure 9.9*b*.   Velocity distribution in a subsonic compressible flow in a three-dimensional nozzle, $M_i = 0.82$ [20].

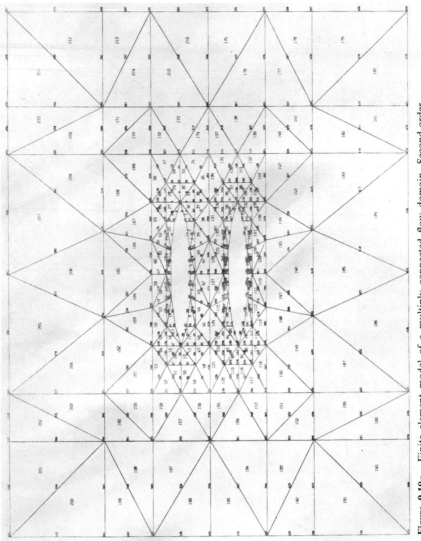

Figure 9.10a. Finite element model of a multiply connected flow domain. Second-order isoparametric triangular elements are used [20].

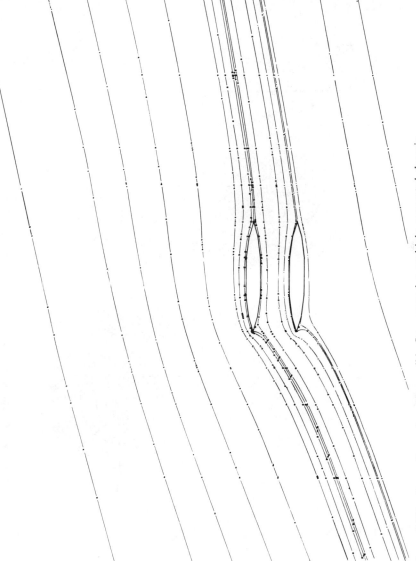

Figure 9.10b. Streamlines of a compressible flow around a multiply connected domain, $M_\infty = 0.3$, $\alpha = 15°$ = angle of attack [20]

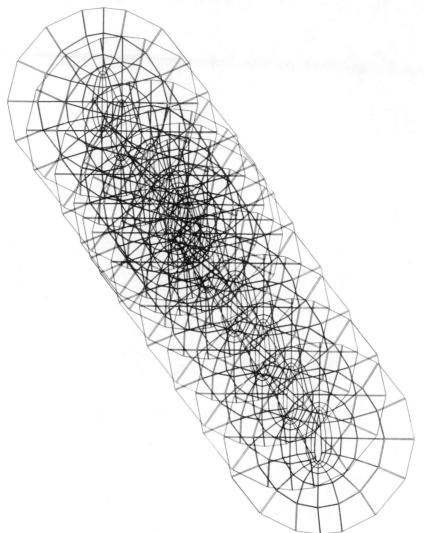

Figure 9.11. Automatically generated hexahedra around a wing [20].

9.4.1 Stream function formulation

As indicated in Appendix D, a stream function defined as

$$u = \frac{\partial \psi}{\partial y}, \qquad v = -\frac{\partial \psi}{\partial x}$$

can be used to combine equations 9.51 and 9.52 into one relation (the biharmonic equation), which is

$$\nabla^4 \psi = \frac{\partial^4 \psi}{\partial x^4} + \frac{\partial^4 \psi}{\partial y^4} + 2\frac{\partial^4 \psi}{\partial x^2\, \partial y^2} = 0 \qquad \text{(C.24)}$$

and which is accompanied by the boundary conditions

$$\psi = g(x, y) \quad \text{on } C_1 \qquad (9.53a)$$

$$\nabla \psi \cdot \hat{n} = f(x, y) \quad \text{on } C_2 \qquad (9.53b)$$

The function ψ which satisfies equations D.24 and 9.53 in domain Ω also minimizes the functional†

$$I(\psi) = \int_\Omega \left[\left(\frac{\partial^2 \psi}{\partial x^2}\right)^2 + \left(\frac{\partial^2 \psi}{\partial y^2}\right)^2 + 4\left(\frac{\partial^2 \psi}{\partial x\, \partial y}\right)^2 - 2\frac{\partial^2 \psi}{\partial x^2}\frac{\partial^2 \psi}{\partial y^2} \right] dx\, dy \quad (9.54)$$

Equation 9.54 provides a variational basis for deriving element equations. After dividing the solution domain Ω into M elements of r nodes each, we approximate $\psi(x, y)$ within each element as

$$\psi^{(e)} = \lfloor N \rfloor \{\tilde{\psi}_i\} \qquad (9.55)$$

where the N_i are the interpolation functions and $\{\tilde{\psi}_i\}$ is a column vector of nodal parameters. Since $I(\psi)$ contains second-order derivatives of ψ, the interpolation functions in equation 9.55 should preserve C^1 continuity. A choice of elements capable of this task is discussed in Section 5.8.2.

An intriguing aspect of formulating two-dimensional, slow, viscous flows in terms of a stream function is that the resulting equation governing for ψ is identical in form to that governing Kirchhoff isotropic plate-bending theory. Furthermore, when plane elasticity problems are formulated in terms of the Airy stress function (see Appendix C), the same biharmonic equation results. This means that the same kinds of elements and, indeed, the same computer programs (with slight changes) can be used to solve all three types of problems.

---

† This functional may be derived from equation D.32 or taken directly from ref 21, where a more general biharmonic equation and its boundary conditions are presented along with the corresponding variational principle.

To indicate the use of this analogy, we refer to Section 6.5, where plate-bending problems were discussed. Comparing the functionals of equations 6.40 and 9.54, we see that the integrands are identical if the parameter v (Poisson's ratio) in equation 6.40 is set equal to -1, and the functionals are identical if, in the plate-bending problem, the coefficient $Eh^3/24(1 - v^2)$ is taken as unity. Hence any element equations derived for plate-bending problems can also be used for the solution of creeping flow problems.

A number of authors [22–24] have used this approach to solve practical problems and to demonstrate the effectiveness of the solution procedure. Atkinson et al. [22,23] considered both axisymmetric and planar two-dimensional flow fields of various geometries. For each of the problems considered, they used three-node triangular elements with the stream function and its first two derivatives specified at each node.† The stream function was interpolated over an element in rectangular coordinates as follows:

$$\psi^{(e)}(x, y) = \sum_{i=1}^{3} \left[{}_1N_i\psi_i + {}_2N_i\left(\frac{\partial\psi}{\partial x}\right)_i + {}_3N_i\left(\frac{\partial\psi}{\partial y}\right)_i \right] \qquad (9.56)$$

where the interpolation functions are given in terms of the natural co-ordinates L_i (see Chapter 5):

$${}_1N_i = L_i + L_i^2(L_j + L_k) - L_i(L_j^2 + L_k^2) \qquad (9.57a)$$

$${}_2N_i = (x_k - x_i)(L_i^2 L_k + \tfrac{1}{2}L_i L_j L_k) - (x_i - x_j)(L_i^2 L_j + \tfrac{1}{2}L_i L_j L_k) \quad (9.57b)$$

$${}_3N_i = (y_k - y_i)(L_i^2 L_k + \tfrac{1}{2}L_i L_j L_k) - (y_i - y_j)(L_i^2 L_j + \tfrac{1}{2}L_i L_j L_k) \quad (9.57c)$$

For axisymmetric problems, the formulation is similar.

The explicit element equations resulting from this type of interpolation are given in refs. 22 and 23. As an example of a solved problem, Figure 9.12 shows the solution domain, the boundary conditions, and the solution for the developing flow between parallel flat plates. Atkinson et al. also studied creeping flow around a sphere, flow through a converging conical section, and developing flow in a circular pipe. Tong and Fung [24] investigated slow viscous flow in a capillary in the presence of moving particles suspended in the flow. This work has direct application to the biomedical problem of determining the influence of red blood cells on the flow in capillary blood vessels.

9.4.2   Velocity and pressure formulation

The stream function formulation we have just discussed offers the advantage that only one governing equation need be considered and the plate-bending

---

† This is one of the so-called incompatible elements referred to in Chapter 5.

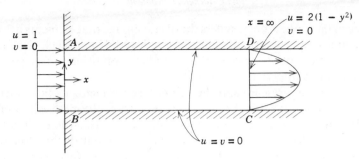

Figure 9.12. Developing laminar flow between parallel plates [23]. (*a*) Solution domain and boundary conditions.

Along AB: $\psi = y,$ $\dfrac{\partial \psi}{\partial x} = 0,$ $\dfrac{\partial \psi}{\partial y} = 1$

Along BC and AD: $\psi = \text{constant},$ $\dfrac{\partial \psi}{\partial x} = \dfrac{\partial \psi}{\partial y} = 0$

Along CD: $\psi = 2y\left(1 - \dfrac{y^2}{3}\right),$ $\dfrac{\partial \psi}{\partial x} = 0,$ $\dfrac{\partial \psi}{\partial y} = 2(1 - y^2)$

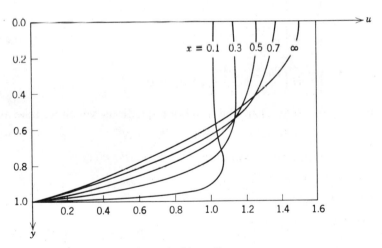

Figure 9.12 (*continued*). (*b*) Velocity profiles.

analogy can be applied, but it suffers the disadvantage that elements achieving or approximating C^1 continuity must be employed. Unless such elements are used, it is impossible to specify the boundary conditions in terms of the normal and tangential derivatives of ψ.

By choosing the pressure and the velocity components as field variables, we can avoid this difficulty. The procedure is to apply the method of weighted residuals with Galerkin's criterion. Consider a two-dimensional flow domain Ω bounded by curve C. For a general element of this domain, we select u, v, and P as nodal variables and interpolate these variables as follows:

$$u^{(e)} = \Sigma N_i^{\,u}(x, y)u_i = \lfloor N^u \rfloor \{u\} \tag{9.58a}$$

$$v^{(e)} = \Sigma N_i^{\,v}(x, y)v_i = \lfloor N^v \rfloor \{v\} \tag{9.58b}$$

$$P^{(e)} = \Sigma N_i^{\,P}(x, y)P_i = \lfloor N^P \rfloor \{P\} \tag{9.58c}$$

where $N_i^{\,u}$, $N_i^{\,v}$, and $N_i^{\,P}$ are the interpolation functions, which need not necessarily be of the same order. To follow the development of Yamada et al. [25], and to simplify the algebraic manipulations, we select $N_i^{\,u} = N_i^{\,v} = N_i$.

The Galerkin procedure (see Chapter 4) applied at node i of an isolated element becomes, in view of equations 9.51 and 9.52,

$$\int_{\Omega^{(e)}} W_i \left[\mu \left(\frac{\partial^2 u^{(e)}}{\partial x^2} + \frac{\partial^2 u^{(e)}}{\partial y^2} \right) - \frac{\partial P^{(e)}}{\partial x} \right] dx\, dy = 0 \tag{9.59a}$$

$$\int_{\Omega^{(e)}} W_i \left[\mu \left(\frac{\partial^2 v^{(e)}}{\partial x^2} + \frac{\partial^2 v^{(e)}}{\partial y^2} \right) - \frac{\partial P^{(e)}}{\partial y} \right] dx\, dy = 0 \tag{9.59b}$$

$$\int_{\Omega^{(e)}} H_i \left(\frac{\partial u^{(e)}}{\partial x} + \frac{\partial v^{(e)}}{\partial y} \right) dx\, dy = 0 \tag{9.59c}$$

where $W_i(x, y)$ and $H_i(x, y)$ are the weighting functions, which we take as

$$W_i = N_i \quad \text{and} \quad H_i = N_i^P \tag{9.60}$$

If we integrate each term of equations 9.59 by parts, we have

$$\int_{\Omega^{(e)}} \mu \left(\frac{\partial N_i}{\partial x} \frac{\partial u^{(e)}}{\partial x} + \frac{\partial N_i}{\partial y} \frac{\partial u^{(e)}}{\partial y} \right) dx\, dy - \int_{\Omega^{(e)}} \frac{\partial N_i}{\partial x} P^{(e)} dx\, dy - \int_C N_i X^* ds = 0 \tag{9.61a}$$

$$\int_{\Omega^{(e)}} \mu \left(\frac{\partial N_i}{\partial x} \frac{\partial v^{(e)}}{\partial x} + \frac{\partial N_i}{\partial y} \frac{\partial v^{(e)}}{\partial y} \right) dx\, dy - \int_{\Omega^{(e)}} \frac{\partial N_i}{\partial y} P^{(e)} dx\, dy - \int_C N_i Y^* ds = 0 \tag{9.61b}$$

where the boundary terms† are given by

$$X^* = \mu \nabla u^{(e)} \cdot \hat{n} - n_x P^{(e)} \tag{9.62a}$$

$$Y^* = \mu \nabla v^{(e)} \cdot \hat{n} - n_y P^{(e)} \tag{9.62b}$$

and $\hat{n} = n_x \hat{i} + n_y \hat{j} \equiv$ outward unit normal to boundary segment ds.

† These terms appear only when the element lies on the boundary.

Zienkiewicz and Taylor [26] and Taylor and Hood [27] perform similar manipulations; however, they neglect to integrate by parts the pressure terms, and consequently they obtain an unsymmetric influence matrix. Integrating the velocity as well as the pressure terms, as suggested by Yamada et al. [25], leads to a symmetric matrix. This will be evident subsequently.

When the approximations of equations 9.58 are substituted into equations 9.60, 9.61, and 9.62, the matrix equations for node i result. These are:

$$\int_{\Omega^{(e)}} \mu \left[\frac{\partial N_i}{\partial x} \frac{\partial \lfloor N \rfloor}{\partial x} + \frac{\partial N_i}{\partial y} \frac{\partial \lfloor N \rfloor}{\partial y} \right] dx\, dy\{u\}$$

$$- \int_{\Omega^{(e)}} \frac{\partial N_i}{\partial x} \lfloor N^P \rfloor\, dx\, dy\{P\} = \int_C N_i X^*\, ds \quad (9.63a)$$

$$\int_{\Omega^{(e)}} \mu \left[\frac{\partial N_i}{\partial x} \frac{\partial \lfloor N \rfloor}{\partial x} + \frac{\partial N_i}{\partial y} \frac{\partial \lfloor N \rfloor}{\partial y} \right] dx\, dy\{v\}$$

$$- \int_{\Omega^{(e)}} \frac{\partial N_i}{\partial y} \lfloor N^P \rfloor\, dx\, dy\{P\} = \int_C N_i Y^*\, ds \quad (9.63b)$$

$$\int_{\Omega^{(e)}} N_i^P \frac{\partial \lfloor N \rfloor}{\partial x} dx\, dy\{u\} + \int_{\Omega^{(e)}} N_i^P \frac{\partial \lfloor N \rfloor}{\partial y} dx\, dy\{v\} = 0 \quad (9.63c)$$

From these equations, we can write the element matrix equations by inspection. Suppose that the velocity components are interpolated at r nodes of the element while the pressure is interpolated at s nodes, where, in general, $r > s$. Then the matrix equations take the form

$$[K]\{\phi\} = \{R\}$$

or

$$
\underbrace{\begin{bmatrix} [K_1] & [0] & [K_2] \\ [0] & [K_1] & [K_3] \\ [K_2] & [K_3] & [0] \end{bmatrix}}_{(2r+s)\ \times\ (2r+s)}
\overbrace{\begin{Bmatrix} u_1 \\ \vdots \\ u_r \\ \hline v_1 \\ \vdots \\ v_r \\ \hline -P_1 \\ \vdots \\ -P_s \end{Bmatrix}}^{(2r+s)\times 1}
=
\overbrace{\begin{Bmatrix} R_1 \\ \vdots \\ R_r \\ \hline R_{r+1} \\ \vdots \\ R_{2r} \\ \hline 0 \\ \vdots \\ 0 \end{Bmatrix}}^{(2r+s)\times 1}
\quad (9.64)
$$

where the blocks are of sizes $r \times r$, $r \times r$, $r \times s$; $r \times r$, $r \times r$, $r \times s$; $s \times r$, $s \times r$, $s \times s$.

where

$$k_{1_{ij}} = \int_{\Omega^{(e)}} \mu \left(\frac{\partial N_i}{\partial x} \frac{\partial N_j}{\partial x} + \frac{\partial N_i}{\partial y} \frac{\partial N_j}{\partial y} \right) dx\, dy \qquad (9.65a)$$

$$k_{2_{ij}} = -\int_{\Omega^{(e)}} \frac{\partial N_i}{\partial x} N_j^P \, dx\, dy \qquad (9.65b)$$

$$k_{3_{ij}} = \int \frac{\partial N_i}{\partial y} N_j^P \, dx\, dy \qquad (9.65c)$$

$$R_i = \int_C N_i X^* \, ds, \quad i = 1, 2, \ldots, r \qquad (9.65d)$$

$$R_i = \int_C N_i Y^* \, ds, \quad i = r + 1, r + 2, \ldots, 2r \qquad (9.65e)$$

We note that the subscript indices in equations 9.65 must be adjusted according to placement position in the complete matrix $[K]$, and that $[K]$ is symmetric.

Serious attention must be given to the choice of interpolation functions for velocity and pressure [28]. This becomes evident when we inspect the variational formulation for this problem. Yamada et al. [25] contend that equations 9.60, 9.61, and 9.62 are directly obtainable from the minimizing condition $\delta I = 0$, where

$$I(u, v, P) = \int_\Omega \left[\frac{1}{2}\mu \left[\left(\frac{\partial u}{\partial x} \right)^2 + \left(\frac{\partial u}{\partial y} \right)^2 + \left(\frac{\partial v}{\partial x} \right)^2 + \left(\frac{\partial v}{\partial y} \right)^2 \right] \right.$$
$$\left. - P\left(\frac{\partial u}{\partial x} + \frac{\partial v}{\partial y} \right) \right] dx\, dy - \int_C (X^* u + Y^* v)\, ds \qquad (9.66)$$

Inspection of equation 9.66 reveals that the functional contains first-order derivatives of velocity and zeroth-order derivatives of pressure. This observation suggests that the interpolation functions for velocity should preserve C^0 continuity and that N_i should be higher by one order than N_i^P to achieve the same order of approximation for u, v, and p. Hood and Taylor [28] arrive at the same conclusion directly from the weighted residual formulation. They argue that two criteria of error consistency must be satisfied:

1. "The maximum order of error associated with the residual of each variable must be equal."

2. "The residuals arising from each equation must be weighted according to the maximum error occurring in each equation."

The first criterion is satisfied when N_i is higher by one order than N_i^P, and the second is satisfied by the Galerkin technique. If the element equations are derived from a variational principle rather than the weighted residual process, the second criterion is automatically satisfied.

Once the element equations have been evaluated, they can be assembled in the usual manner to form the system equations. Two types of boundary conditions must be considered. On one portion of the boundary, the velocity components are prescribed. These boundary conditions are handled in the manner described in Chapter 2 when nodal displacements were prescribed for a structures problem. On the remaining portion of the boundary, Pn_x, Pn_y, $\nabla u \cdot \hat{n}$, and $\nabla v \cdot \hat{n}$ are prescribed. These boundary conditions comprise the forcing functions $\{R\}$ via equations 9.65d and 9.65e, and they are incorporated in the equations for the boundary elements.

To demonstrate the validity of their formulation, Yamada et al. [25] solved several example problems, which included (1) fully developed laminar shear flow with a pressure gradient and (2) developing flow at a channel entrance. Quadratic triangular elements were used for the velocity components, and linear triangular elements with the same vertex nodes for the pressure. Figure 9.13 shows the flow geometry, the corresponding finite

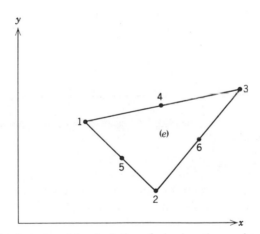

Figure 9.13. Finite element model and solution of a laminar shear and pressure flow between parallel plates [25]. (*a*) Triangular element with linear and quadratic interpolation.

$$u^{(e)} = \sum_{i=1}^{6} N_i u_i, \qquad v^{(e)} = \sum_{i=1}^{6} N_i v_i, \qquad P^{(e)} = \sum_{i=1}^{3} N_i^P P_i$$

$$N_1 = L_1^2 - L_1(L_2 + L_3) \qquad\qquad N_2 = L_2^2 - L_2(L_3 + L_1)$$
$$N_3 = L_3^2 - L_3(L_1 + L_2) \qquad\qquad N_4 = 4L_1 L_3$$
$$N_5 = 4L_1 L_2 \qquad\qquad\qquad\qquad N_6 = 4L_2 L_3$$
$$N_i^P = L_i = \text{natural coordinates}$$

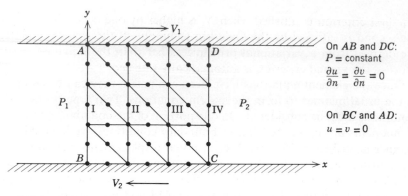

On AB and DC:

$P = \text{constant}$

$\dfrac{\partial u}{\partial n} = \dfrac{\partial v}{\partial n} = 0$

On BC and AD:

$u = v = 0$

Figure 9.13 (*continued*).   (*b*)   Triangular flow domain and boundary conditions.

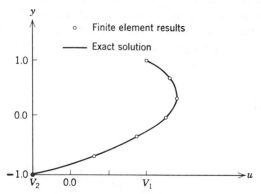

Figure 9.13 (*continued*).   (*c*)   Comparison of calculated and exact solutions for the velocity profiles at sections 1, 2, 3, 4.

element model, and the solution for the velocity profiles in a shear- and pressure-induced flow between parallel plates. The agreement between the calculated and exact solutions is excellent. Yamada et al. [25] also solved the developing flow problem treated by Atkinson et al. [22] (Figure 9.12) and again found excellent agreement.

9.5   INCOMPRESSIBLE VISCOUS FLOW WITH INERTIA

Finite element formulations for fluid mechanics problems classified as inviscid flows ($Re = \infty$) and slow viscous flows ($Re = 0$) were discussed in the preceding sections. In this section, we discuss the more general flow problems that exist when $0 < Re < \infty$. These problems are inherently nonlinear because of the presence of the convective inertia terms.

The full Navier-Stokes equations representing a balance of inertia forces, pressure forces, and viscous forces are capable of describing some of the most interesting phenomena in fluid mechanics; unfortunately they are among the most difficult partial differential equations to solve. In very general, vague (and somewhat reckless) terms, we can say that the mathematical and numerical difficulties involved in solving the governing equations for a particular laminar flow increase as the Reynolds number increases. The nature of the difficulties changes along the way, but we generally encounter stability and convergence problems when the transition Reynolds number range is approached, or even before. This seems to hold regardless of the particular computational scheme employed.

As mentioned earlier, the application of finite element techniques to the solution of nonlinear viscous flow problems is still in its infancy. Hence much research remains to be done. However, what has been accomplished indicates that the finite element method promises to be a very useful tool to solve difficult flow problems.

For steady, incompressible, two-dimensional, isoviscous laminar flows, one faces the problem of solving the following governing equations:

Continuity:

$$\frac{\partial u}{\partial x} + \frac{\partial v}{\partial y} = 0$$

Momentum:

$$u \frac{\partial u}{\partial x} + v \frac{\partial u}{\partial y} = \frac{-1}{\rho} \frac{\partial P}{\partial x} + \frac{\mu}{\rho} \nabla^2 u$$

$$u \frac{\partial v}{\partial x} + v \frac{\partial v}{\partial y} = \frac{-1}{\rho} \frac{\partial P}{\partial y} + \frac{\mu}{\rho} \nabla^2 v$$

Pressure, velocity, and velocity gradient can be specified on flow boundaries. Essentially, there are three different formulations one can use to establish finite element solution procedures for these equations: (1) stream function formulation; (2) stream function and vorticity formulation; and (3) velocity and pressure formulation. Each of these will now be developed by following the work presented in several recent and representative papers. Examples of the use of these formulations will also be presented.

9.5.1 Stream function formulation [29,30]

Introducing the stream function definition—equation D.8, we can combine the continuity and momentum equations as before and obtain one equation for ψ:

$$\frac{\partial \psi}{\partial y} \nabla^2 \left(\frac{\partial \psi}{\partial x}\right) - \frac{\partial \psi}{\partial x} \nabla^2 \left(\frac{\partial \psi}{\partial y}\right) = v \nabla^4 \psi \tag{9.67}$$

where $v = \mu/\rho$. Though no classical variational principle exists for equation 9.67, Olson [29] recognized that, if a pseudo-variational statement could be found, a convenient procedure would be available for deriving element equations. Under certain restrictions such a functional can be found. Consider the triangular solution domain shown in Figure 9.14. If edge 1-2 of this domain is a boundary where either

$$\psi = \text{constant} \quad \text{or} \quad \frac{\partial P}{\partial \xi} = 0 \tag{9.68}$$

and either

$$\frac{\partial \psi}{\partial \eta} = 0 \quad \text{or} \quad \mu\left(\frac{\partial u}{\partial \xi} + \frac{\partial v}{\partial \eta}\right) = 0$$

then the function ψ which satisfies equations 9.67 and 9.68 extremizes the functional

$$I(\psi) = \int_\Omega \left[\frac{v}{2}(\nabla^2\psi)^2 + \left(\frac{\partial \psi}{\partial \eta} \nabla^2\psi\right)\frac{\partial \psi}{\partial \xi} - \left(\frac{\partial \psi}{\partial \xi} \nabla^2\psi\right)\frac{\partial \psi}{\partial \eta}\right] d\xi \, d\eta$$

$$+ \int_{-b}^{a} \left[2v \frac{\partial^2 \psi}{\partial \xi^2}\frac{\partial \psi}{\partial \eta} - \left(\frac{\partial \psi}{\partial \xi}\frac{\partial^2 \psi}{\partial \xi^2} + \frac{\partial \psi}{\partial \eta}\frac{\partial^2 \psi}{\partial \xi \partial \eta}\right)\psi\right]_{\eta=0} d\xi \tag{9.69}$$

when the underscored terms are taken as invariant. The boundary integral appears only when the triangular element lies on a boundary where some of the conditions of equations 9.68 are to be satisfied. The variational statement of equation 9.69 is restricted to triangular elements with particular orientation relative to the specified boundary conditions, but this restriction is not really confining when realistic problems are considered. Since $I(\psi)$ contains second-order derivatives, the interpolation functions for ψ should preserve C^1 continuity. Olson [29] employs the versatile 18-degree-of-freedom triangle with ψ, $\partial\psi/\partial x$, $\partial\psi/\partial y$, $\partial^2\psi/\partial x^2$, $\partial^2\psi/\partial x\,\partial y$, and $\partial^2\psi/\partial y^2$ as nodal variables, and gives explicit expressions for the resulting element equations. The system equations (obtained by the usual assembly procedure)

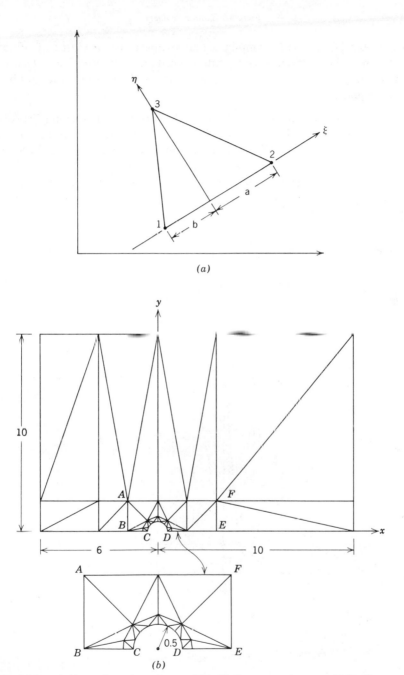

Figure 9.14a. Arbitrary boundary element [29]. (*b*) Forty-two-element grid for flow over a circular cylinder [29].

349

are expressed in terms of a symmetric and an unsymmetric matrix of influence coefficients. The unsymmetric matrix contains the nonlinear terms. A Newton-Raphson iteration technique was used to solve the nonlinear matrix equations.

Sample Solution. To check the validity of the formulation, Olson [29] solved several example problems for which there are known solutions: (1) fully developed parallel flow; (2) circulatory flow in a square cavity; (3) channel entrance flow; and (4) flow over a circular cylinder. Figures 9.14 and 9.15 illustrate the finite element grids he used to analyze the classical problem of a circular cylinder in uniform cross flow. The boundary conditions were as follows:

Along edge AB: $\psi = y_1 \dfrac{\partial \psi}{\partial x} = 0$

Along edges BC and DE: $\psi = 0$

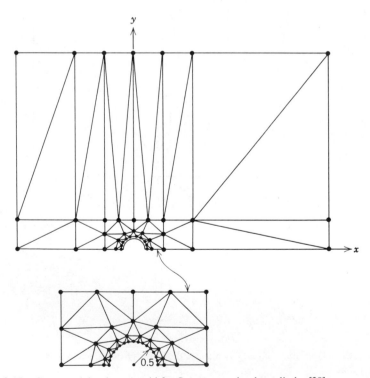

Figure 9.15. Seventy-eight-element grid for flow over a circular cylinder [29].

Along edge $AF: \dfrac{\partial \psi}{\partial y} = 1$

Along edge $EF: \dfrac{\partial \psi}{\partial x} = 0$

On cylinder surface: $\dfrac{\partial \psi}{\partial n} = 0$ (zero normal gradient)

Comparison of the finite element results with an accurate finite difference solution (Figure 9.16) revealed excellent agreement. Even the separation phenomenon behind the cylinder is accurately modeled.

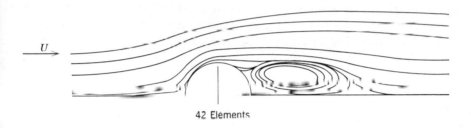

42 Elements

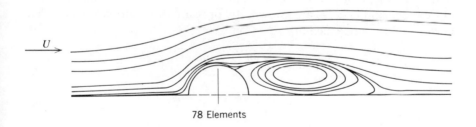

78 Elements

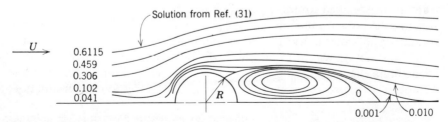

Solution from Ref. (31)

0.6115
0.459
0.306
0.102
0.041

R

0

0.001   0.010

Figure 9.16.   Streamlines over a circular cylinder in cross flow, $Re = \rho UR/\mu = 40$ [29].

Olson [29] found that the Newton-Raphson method of solving the non-linear matrix equations yielded convergent solutions in three to five iterations. He summarizes his experiences in these words [29]: "The computing process consistently exhibits ultrafast convergence for all element grids. Further, the actual finite element predictions of the various flow quantities also converge rapidly to correct results as the element grids are refined. The present method yields accuracies comparable to the finite difference method with an order of magnitude fewer equations."

9.5.2   Stream function and vorticity formulation

Instead of working with one fourth-order partial differential equation in terms of the stream function, several researchers [27,32–41] have chosen to use two coupled equations of second order in terms of the stream function and the vorticity. The governing equations for the general transient case are as follows:

$$\nabla^2 \psi = -\omega \qquad (9.70)$$

and

$$\frac{\partial \omega}{\partial t} + \frac{\partial \psi}{\partial y}\frac{\partial \omega}{\partial x} - \frac{\partial \psi}{\partial x}\frac{\partial \omega}{\partial y} = \frac{\mu}{\rho}\nabla^2 \omega \qquad (9.71)$$

Again, the continuity equation is implicitly satisfied by the definition of ψ. Once the ψ and ω fields are known, the pressure distribution can be calculated from the supplementary equation

$$\nabla^2 P = -2\rho\left(\frac{\partial u}{\partial y}\frac{\partial v}{\partial x} - \frac{\partial u}{\partial x}\frac{\partial v}{\partial y}\right)$$

$$= 2\rho\left[\frac{\partial^2 \psi}{\partial y^2}\frac{\partial^2 \psi}{\partial x^2} - \left(\frac{\partial^2 \psi}{\partial x\,\partial y}\right)^2\right] \qquad (9.72)$$

The boundary conditions for equations 9.70 and 9.71 are specified values of ψ and ω, or specified values of their first derivatives. The principal difficulty in using the stream function and vorticity formulation is that, in general, vorticity is unknown a priori along solid boundaries. A procedure for coping with this situation will be explained in the following development.

The finite element solution of equations 9.70 and 9.71 can be based on the method of weighted residuals [27] or on a quasi-variational† formulation.

---

† By "quasi-variational" we mean that variational techniques are used but no classical variational principle is available.

If the method of weighted residuals is used, there results a set of simultaneous ordinary differential equations to be solved for the nodal values of ψ and ω. The procedure for developing the element equations follows the usual procedure of integrating residuals over the element and then integrating by parts to introduce boundary information. The explicit steps are illustrated, for example, in Ikenouchi and Kimura [41].

Here we present in detail the versatile and widely applicable quasi-variational formulation suggested by Cheng [36]. To circumvent the non-linearity in the governing equations, Cheng considers an unsteady flow problem which is linear in the stream function and vorticity at each time step. Steady-state solutions are achieved by allowing the time-dependent solutions to converge.

To integrate equations 9.70 and 9.71 with respect to time, we consider two solutions $-(\psi_n, \omega_n)$ at the nth time step and $(\psi_{n+1}, \omega_{n+1})$ at a time increment Δt later. Then the governing equations become

$$\nabla^2 \psi_{n+1} = -\omega_n \tag{9.73}$$

and

$$\frac{\partial \omega_n}{\partial t} + \frac{\partial \psi_{n+1}}{\partial y}\frac{\partial \omega_n}{\partial x} - \frac{\partial \psi_{n+1}}{\partial x}\frac{\partial \omega_n}{\partial y} = \frac{\mu}{\rho}\nabla^2 \omega_{n+1} \tag{9.74}$$

Variational statements can now be directly written for these *linear* equations. Corresponding to equation 9.73, we have

$$I_1(\psi_{n+1}) = \frac{1}{2}\int_\Omega \left[\left(\frac{\partial \psi_{n+1}}{\partial x}\right)^2 + \left(\frac{\partial \psi_{n+1}}{\partial y}\right)^2 - 2\omega_n \psi_{n+1}\right] dx\, dy \tag{9.75}$$

and, corresponding to equation (9.74), we have

$$I_2(\omega_{n+1}) = \frac{1}{2}\int_\Omega \left[\frac{\mu}{\rho}\left(\frac{\partial \omega_{n+1}}{\partial x}\right)^2 + \frac{\mu}{\rho}\left(\frac{\partial \omega_{n+1}}{\partial y}\right)^2\right.$$
$$\left. + 2\left(\frac{\partial \psi_{n+1}}{\partial y}\frac{\partial \omega_n}{\partial x} - \frac{\partial \psi_{n+1}}{\partial x}\frac{\partial \omega_n}{\partial y} + \frac{\partial \omega_n}{\partial t}\right)\omega_{n+1}\right] dx\, dy \tag{9.76}$$

The Euler-Lagrange equation for the functional of equation 9.76 is equation 9.74 only when $\partial \omega_n/\partial t$ is taken as invariant. The function ψ_{n+1} which satisfies equation 9.73 and its boundary conditions minimizes $I_1(\psi_{n+1})$; similarly, the function ω_{n+1} satisfying equation 9.74 and its boundary conditions minimizes $I_2(\omega_{n+1})$. Segregating ψ and ω solutions according to different instants of time as indicated reduces the problem to one of consecutively minimizing I_1 and I_2. This can be conveniently accomplished by the finite element method, and the element interpolation functions need only preserve C^0 continuity.

The iterative solution procedure starts by assuming an initial value for ω_n. Then the ψ_{n+1} that minimizes I_1 is found and used as the source function to determine the ω_{n+1} that minimizes I_2. This process is repeated until the steady state is reached.

The general element equations resulting from the minimizing conditions $\delta I_1 = \delta I_2 = 0$ can be written by inspection if we refer to Sections 7.2 and 7.5, where functionals of the same form have been treated. If ψ_{n+1} and ω_{n+1} are interpolated over an element in terms of their nodal values as

$$\psi_{n+1}^{(e)}(x, y, t) = \lfloor N(x, y) \rfloor \{\psi(t)\}_{n+1}^{(e)}$$

and

$$\omega_{n+1}^{(e)}(x, y, t) = \lfloor N(x, y) \rfloor \{\omega(t)\}_{n+1}^{(e)} \tag{9.77}$$

the element equations corresponding to $\delta I_1 = 0$ are

$$[K_\psi]^{(e)}\{\psi\}_{n+1}^{(e)} + \{R_\psi\}_n^{(e)} = \{0\} \tag{9.78}$$

where

$$k_{\psi_{ij}} = \int_{\Omega^{(e)}} \left(\frac{\partial N_i}{\partial x} \frac{\partial N_j}{\partial x} + \frac{\partial N_i}{\partial y} \frac{\partial N_j}{\partial y} \right) d\Omega^{(e)}$$

$$R_{\psi_i} = -\int_{\Omega^{(e)}} \omega_n N_i \, d\Omega^{(e)}$$

From the condition $\delta I_2 = 0$, we have

$$[K_\omega]^{(e)}\{\omega\}_{n+1}^{(e)} + [K_{\dot\omega}]^{(e)}\{\dot\omega\}_{n+1}^{(e)} + \{R_\omega\}_n^{(e)} = \{0\} \tag{9.79}$$

where

$$k_{\omega_{ij}} = \frac{\mu}{\rho} k_{\psi_{ij}}$$

$$k_{\dot\omega_{ij}} = \int_{\Omega^{(e)}} N_i N_j \, d\Omega^{(e)}$$

$$R_{\omega_i} = \int_{\Omega^{(e)}} N_i \left(\frac{\partial \psi_{n+1}^{(e)}}{\partial y} \frac{\partial \omega_n^{(e)}}{\partial x} - \frac{\partial \psi_{n+1}^{(e)}}{\partial x} \frac{\partial \omega_n^{(e)}}{\partial y} \right) d\Omega^{(e)}$$

Then the assembled system equations become

$$[K_\psi]\{\psi\}_{n+1} + \{R_\psi\}_n = \{0\} \tag{9.80}$$

$$[K_\omega]\{\omega\}_{n+1} + [K_{\dot\omega}]\{\dot\omega\}_{n+1} + \{R_\omega\}_n = \{0\} \tag{9.81}$$

Cheng [36] solved equations 9.81 by writing

$$\{\dot\omega\}_{n+1} = \left\{ \frac{d\omega}{dt} \right\}_{n+1} = \left\{ \frac{\omega_{n+1} - \omega_n}{\Delta t} \right\}$$

so that we have the recurrence relation

$$\left[[K_\omega] + \frac{1}{\Delta t} [K_{\dot{\omega}}] \right]\{\omega\}_{n+1} = \frac{1}{\Delta t} [K_{\dot{\omega}}]\{\omega\}_n - \{R_\omega\}_n \qquad (9.82)$$

which can be solved at successive time steps for the column vector of nodal values of vorticity, $\{\omega\}_{n+1}$.

Determining the boundary values for the vorticity requires special attention because application of the no-slip boundary condition alone is insufficient. Equation 9.73 evaluated at the boundary provides the necessary relation. At a point (x_0, y_0) on the wall, the vorticity may be calculated from the relation

$$\omega_n(x_0, y_0) = -\frac{\partial^2 \psi_{n+1}}{\partial \xi^2}(x_0, y_0) \qquad (9.83)$$

where ξ is the coordinate normal to the wall. To evaluate $\partial^2 \psi_{n+1}/\partial \xi^2$ we can use a Taylor series expansion. At a point (x_1, y_1) along the ξ direction, a small distance from the wall, we have

$$\psi_{n+1}(x_1, y_1) = \psi_{n+1}(x_0, y_0) + \xi \frac{\partial \psi_{n+1}}{\partial \xi}(x_0, y_0) + \frac{\xi^2}{2} \frac{\partial^2 \psi_{n+1}}{\partial \xi^2}(x_0, y_0)$$

Since the no-slip condition dictates that

$$\frac{\partial \psi}{\partial \xi}(x_0, y_0) = 0$$

we have

$$\omega_n(x_0, y_0) = -\frac{\partial^2 \psi_{n+1}}{\partial \xi^2}(x_0, y_0) = \frac{2}{\xi^2}[\psi_{n+1}(x_0, y_0) - \psi_{n+1}(x_1, y_1)] \qquad (9.84)$$

Wall vorticity is then given in terms of the stream function evaluated at the wall and a small distance away from the wall.

Sample Solution. Cheng [36] demonstrated the effectiveness of this solution procedure by solving the problem of internal flow in a channel of arbitrary cross section. The geometry of a particular flow passage is shown in Figure 9.17. This type of geometry provides a good test for the calculation scheme because the constriction causes rapid changes in ψ and ω near the constriction region and separation can occur on the downstream side of the constriction. The numerical solution should capture these details. The calculated results using linear triangular elements are shown in Figure 9.18, and it is seen that, indeed, the flow detail is reproduced. To obtain a starting solution,

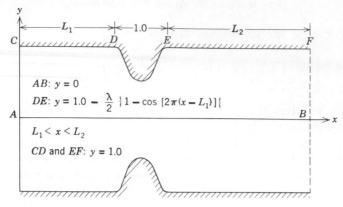

Figure 9.17. Coordinates, geometry, and boundary conditions of a constricting internal passage [36].

Boundary conditions

On AB: $\psi = 0, \omega = 0$

On $CDEF$: $\psi = 1.0, \dfrac{\partial \psi}{\partial \xi} = 0$ (ξ = normal to wall)

On AB and BF: $\left.\begin{array}{l} \psi = 1.5(y - \tfrac{1}{3}y^3) \\ \omega = 3y \end{array}\right\}$ Poiseuille type flow

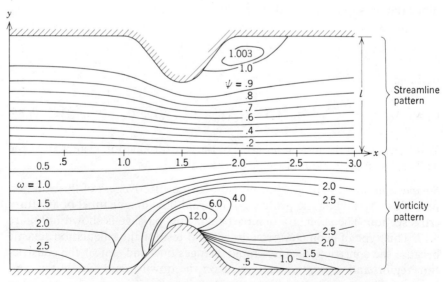

Figure 9.18. Streamlines and equivorticity contours in a 40 % contraction with $Re = \rho U_0 l / \mu = 25$ [36].

356

creeping flow was assumed and ψ and ω fields were calculated by omitting the inertia terms in equation 9.74. After the creeping flow solution was obtained, the iteration process was used to calculate the flow at successively larger Reynolds numbers—the solution at a lower Reynolds number being used as the initial condition for the solution at the next higher Reynolds number. Cheng [36] reports that the iteration procedure was always stable for sufficiently small Δt. As a guideline, Δt should be chosen so that

$$\Delta t < 0.1(\Delta l)^2 \frac{\rho U l}{\mu}$$

where Δl is the characteristic length of an element.

9.5.3 Velocity and pressure formulation

Taylor and Hood [27] and Yamada et al. [25] favor the velocity and pressure formulation for the solution of the nonlinear Navier-Stokes equations for five reasons:

1. The formulation is readily extended to three dimensions.
2. Only C^0 continuity is required of the element interpolation functions.
3. Pressure, velocity, velocity gradient, and stress boundary conditions can be directly incorporated into the matrix equations.
4. Free-surface problems are tractable.
5. The formulation appears to require less computation time than the stream function and vorticity formulation [27].

The approach to deriving the element equations relies on the Galerkin method. Let the velocity and pressure fields be interpolated over an element as indicated in equation 9.58. A procedure similar to that used for creeping flow can again be employed if we linearize the governing equations by approximating the nonlinear convective terms. Suppose that (u_n, v_n, P_n) is some approximate solution to the flow problem. Then the governing equations can be written as

$$\frac{\partial u}{\partial x} + \frac{\partial v}{\partial y} = 0 \qquad (9.85a)$$

$$u_n \frac{\partial u}{\partial x} + v_n \frac{\partial u}{\partial y} + \frac{1}{\rho} \frac{\partial P}{\partial x} - \frac{\mu}{\rho} \nabla^2 u = 0 \qquad (9.85b)$$

$$u_n \frac{\partial v}{\partial x} + v_n \frac{\partial v}{\partial y} + \frac{1}{\rho} \frac{\partial P}{\partial y} - \frac{\mu}{\rho} \nabla^2 v = 0 \qquad (9.85c)$$

The solution (u_n, v_n, P_n) can be the nth iteration or a starting solution based on creeping flow. The Galerkin criterion applied to equations 9.85 then leads to

$$-\int_{\Omega^{(e)}} N_i \left(u_n^{(e)} \frac{\partial u^{(e)}}{\partial x} + v_n^{(e)} \frac{\partial u^{(e)}}{\partial y} + \frac{1}{\rho} \frac{\partial P^{(e)}}{\partial x} - \frac{\mu}{\rho} \nabla^2 u^{(e)} \right) d\Omega^{(e)} = 0 \qquad (9.86a)$$

$$-\int_{\Omega^{(e)}} N_i \left(u_n^{(e)} \frac{\partial v^{(e)}}{\partial x} + v_n^{(e)} \frac{\partial v^{(e)}}{\partial y} + \frac{1}{\rho} \frac{\partial P^{(e)}}{\partial y} - \frac{\mu}{\rho} \nabla^2 v^{(e)} \right) d\Omega^{(e)} = 0 \qquad (9.86b)$$

$$\int_{\Omega^{(e)}} N_i^P \left(\frac{\partial u^{(e)}}{\partial x} + \frac{\partial v^{(e)}}{\partial y} \right) d\Omega^{(e)} = 0 \qquad (9.87)$$

Following the procedure of Section 9.4.2, we integrate by parts all terms in equations 9.86 except those involving u_n and v_n [25]. By this means, the natural boundary conditions applicable to creeping flow remain the same in this case. In fact, the element matrix equations have the same form as equation 9.64, except that the submatrices are slightly different. Instead of the matrix $[K_1]$, we have a substitute matrix $[K_1^*]$, the coefficients of which are given by

$$k_{1_{ij}}^* = \int \left[\frac{\mu}{\rho} \left(\frac{\partial N_i}{\partial x} \frac{\partial N_j}{\partial x} + \frac{\partial N_i}{\partial y} \frac{\partial N_j}{\partial y} \right) - u_n^{(e)} N_i \frac{\partial N_j}{\partial x} - v_n^{(e)} N_i \frac{\partial N_j}{\partial y} \right] dx \, dy$$

$$(9.88)$$

Now the coefficient matrix is unsymmetric because of the presence of the convective inertia terms. The other submatrices in equation 9.64 remain the same. Taylor and Hood [27] and Yamada et al. [25] report that including the inertia terms in the coefficient matrix leads to a more stable solution scheme than results when these terms are evaluated at the previous time step and used as forcing functions in $\{R\}$. Hence the complete system equations have the form

$$[K^*(u_n, v_n)]\{\phi\}_{n+1} = \{R\}_{n+1} \qquad (9.89)$$

where, for a system with r nodes at which velocity is sought and s nodes where pressure is sought, we have

$$\{\phi_{n+1}\}^T = \lfloor u_1, u_2, \ldots, u_r, v_1, v_2, \ldots, v_r, P_1, P_2, \ldots, P_s \rfloor_{n+1}$$

Equations 9.89 are solved successively for $\{\phi\}_{n+1}$ using the last nodal values of u and v to update $[K^*]$. This process is continued until $\{\phi\}_n$ and $\{\phi\}_{n+1}$ are sufficiently close. Usually, convergence is achieved after ten iterations [25,27]. Overrelaxation between iterations helps to speed the convergence.

One of the trial problems considered by Taylor and Hood [27] is the problem of shear-induced flow past a square cavity as shown in Figure 9.19.

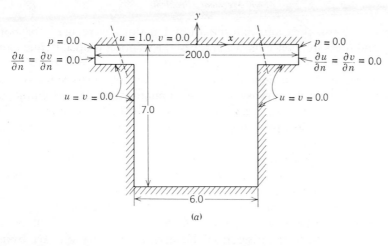

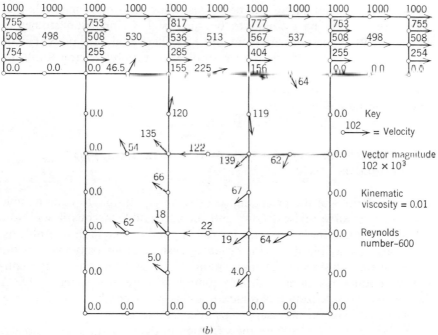

Figure 9.19. Finite element analysis of flow past a square cavity z [27]. (*a*) Geometry, coordinate system, and boundary conditions. (*b*) Finite element mesh and flow velocity vectors.

For this problem, the same interpolation functions were used for both velocity and pressure; hence the accuracy of the solution is open to question. Nevertheless, flow trends are well represented.

In this section, we considered only the solution of steady-state viscous flow problems, though a transient time-stepping technique was employed to obtain solutions for the stream function and vorticity formulation. The analysis of transient flow problems is possible by carrying out a step-by-step analysis between steady-state flows. The process assumes that the values calculated at a particular time step vary in some fashion (constant, linear, quadratic, etc.) during the next small time step. The complete transient solution can be "marched out" in this way. Taylor and Hood [27] discuss the details of a particular procedure they found to be effective.

The methods we have discussed for solving the nonlinear, simultaneous equations arising from the discretized Navier-Stokes equations are by no means the only methods or the best ones. They are simply methods that several researchers have found to be successful. Comparison studies of different solution methods are scant, but Kawahara et al. [42] present some helpful guidelines concerning the use of direct iteration, Newton-Raphson, and perturbation techniques.

For examples of other velocity and pressure formulations, the reader may wish to consult refs. 43–46.

9.6 COMPRESSIBLE VISCOUS FLOWS AND GENERAL FLOW PROBLEMS

The formulations of the preceding sections are applicable to compressible and incompressible inviscid flows and incompressible viscous flows. Now we briefly outline the procedures for treating more general flow problems. The analysis of a fluid flow with variable density and viscosity involves the simultaneous solution of the mass, momentum, and thermal energy conservation equations along with an equation of state (see Appendix D). A finite element discretization of these equations is too lengthy to present here, but we can discuss the necessary steps in general terms.

The first step is to represent the variation of the field variables, velocity, pressure or density, and temperature, over a typical element. The order of the interpolation functions for these variables should be chosen to ensure the same order of approximation as discussed in Section 9.4.2. Then the method of weighted residuals can be used to extend the point approximations to region approximations. Integration by parts is necessary to introduce the boundary conditions in these analog equations. What results from this procedure is a set of element equations which are either nonlinear ordinary differential equations or nonlinear algebraic equations, depending on whether transient or steady-state flow is being considered. The system equations are assembled from these element equations in the usual manner, and solution of the resulting system equations is necessarily iterative because of the nonlinearities. Examples of these general formulations can be found in refs. 47–49. Oden [48] shows how general finite element models can be derived from a global energy principle such as the first law of thermodynamics. This general formulation holds for any type of flow and any type of fluid, Newtonian or non-Newtonian. Such formulations have yet to be fully utilized.

9.7 CLOSURE

This chapter considered fundamental finite element approaches to the solution of incompressible and compressible, inviscid and viscous, Newtonian fluid flows. For the analysis of inviscid flows, we discussed formulations in terms of the stream function for two-dimensional problems and in terms of the velocity potential for more general three-dimensional problems. For viscous fluids, in addition to these formulations, a formulation using velocity and pressure as field variables was presented. Several researchers suggested that this appears to be the best approach.

The exposition here is by no means exhaustive. The brief examples that have been presented are intended only to illustrate the fact that a number of reliable and accurate formulations for some fundamental flow problems are presently available. Developments in the analysis of general fluid flows by the finite element method are still in an early stage, and much research remains to be done before all avenues are fully explored. Readers desiring further study of specialized flow problems and their finite element solutions are encouraged to see the additional references at the end of the numbered reference list for this chapter. References 50 and 51 contain thorough expositions on the solution of Navier-Stokes equations by the finite element method.

REFERENCES

1. I. Fried, "Finite Element Method in Fluid Dynamics and Heat Transfer," *Rept.* 38, Institut für Statik und Dynamik der Luft- and Reumfahrkonstruktionen, University of Stuttgart, West Germany, April 1967.
2. H. C. Martin, "Finite Element Analysis of Fluid Flows," *Proceedings of Second Conference on Matrix Methods in Structural Mechanics* (AFFDL-TR-68-150), Wright-Patterson Air Force Base, Dayton, Ohio, October 1968.

3. G. de Vries and D. H. Norrie, "The Application of the Finite-Element Technique to Potential Flow Problems," *Trans. ASME*, Series E: *J. Appl. Mech.*, Vol. 38, 1971.

4. T. Fujino, "Analyses of Hydrodynamic and Plate Structures Problems by the Finite Element Methods," in *Recent Advances in Matrix Methods of Structural Analysis and Design*, R. H. Gallagher, Y. Yamada, and J. T. Oden (eds.), University of Alabama Press, Huntsville, Ala., 1971.

5. J. H. Argyris and G. Mareczek, "Potential Flow Analysis by Finite Elements," *Ing.-Arch.*, Vol. 42, No. 1, December 12, 1972.

6. V. Meissner, "A Mixed Finite Element Model for Use in Potential Flow Problems," *Int. J. Numer. Methods Eng.*, Vol. 6, No. 4, 1973.

7. D. S. Thompson, "Finite Element Analysis of the Flow Through a Cascade of Aerofoils," Turbo/TR 45, Eng. Dept., Cambridge University, 1973.

8. J. L. Doctors, "An Application of the Finite Element Technique to Boundary Value Problems of Potential Flow," *Int. J. Numer. Methods Eng.*, Vol. 2, No. 2, 1970.

9. J. A. McCorquodale and C. Y. Li, "Finite Element Analysis of Sluice Gate Flow," *Trans. Eng. Inst. Can.*, Vol. 14, No. C-2, March 1971.

10. K. J. Bai, "A Variational Method in Potential Flows with a Free Surface," *Rept.* NA 72-2, College of Engineering, University of California at Berkeley, 1972.

11. M. Ikegawa and K. Washizu, "Finite Element Method Applied to Analysis Flow over a Spillway Crest," *Int. J. Numer. Methods Eng.*, Vol. 6, No. 2, 1973.

12. G. Hiriart and T. Sarpkaya, "Jet Impingement on Axisymmetric Curved Deflectors," Flow Symposium, 1974.†

13. S. F. Shen, "An Aerodynamicist Looks at the Finite Element Method," Flow Symposium, 1974.†

14. J. W. Leonard, "Linearized Compressible Flow by the Finite Element Method," TNTCTN-9500-920156, Bell Aerosystems Company, December 1969.

15. J. W. Leonard, "Finite Element Analysis of Perturbed Compressible Flow," *Int. J. Numer. Methods Eng.*, Vol. 4, No. 1, 1972.

16. H. Bateman, "Notes on a Differential Equation Which Occurs in the Two-Dimensional Motion of a Compressible Fluid and the Associated Variational Problems," *Proc. Roy. Soc.*, Vol. A125, 1929.

17. P. E. Lush and T. M. Cherry, "The Variational Method in Hydrodynamics," *Quart. J. Mech. Appl. Math.*, Vol. 9, 1956.

18. D. Gelder, "Solution of the Compressible Flow Equations," *Int. J. Numer. Methods Eng.*, Vol. 3, No. 1, 1971.

19. G. de Vries, G. P. Berard, and D. H. Norrie, "Application of the Finite Element Technique to Compressible Flow Problems," *Mech. Eng. Rept.* 18, Dept. of Mech. Eng., University of Calgary, Alberta, Canada, 1974.

20. J. Periaux, "Three-Dimensional Analysis of Compressible Potential Flows with the Finite Element Methods," Flow Symposium, 1974.†

21. L. V. Kantorovich and V. I. Krylov, *Approximate Methods of Higher Analysis* (translated by C. D. Benster), John Wiley-Interscience, New York, 1958.

† The notation "Flow Symposium, 1974" refers to the International Symposium on Finite Element Methods in Flow Problems, held January 7–11, 1974, at the University of Wales, Swansea, U.K. A volume containing papers and extended abstracts of papers presented at this meeting is available from University of Alabama Press, Huntsville, Ala.

22. B. Atkinson, M. P. Brocklebank, C. C. H. Card, and J. M. Smith, "Low Reynolds Number Developing Flows," *Am. Inst. Chem. Eng. J.*, Vol. 15, July 1969.

23. B. Atkinson, C. C. H. Card, and B. M. Irons, "Application of the Finite Element Method to Creeping Flow Problems," *Trans. Inst. Chem. Eng.*, Vol. 48, 1970.

24. P. Tong and Y. C. Fung, "Slow Particulate Viscous Flow in Channels and Tubes—Applications to Biomechanics," *Trans. ASME*, Series E: *J. Appl. Mech.*, Vol. 38, December 1971.

25. Y. Yamada, K. Ito, Y. Yokouchi, T. Lamano, and T. Ohtsubo, "Finite Element Analysis of Steady Fluid and Metal Flow," Flow Symposium, 1974.†

26. O. C. Zienkiewicz and C. Taylor, "Weighted Residual Processes in Finite Element Method with Particular Reference to Some Transient and Coupled Problems," *Lectures on Finite Element Methods in Continuum Mechanics*, University of Alabama Press, Huntsville, Ala., 1973.

27. C. Taylor and P. Hood, "A Numerical Solution of the Navier-Stokes Equations Using the Finite Element Technique," *Computers Fluids*, Vol. 1, No. 1, 1973.

28. P. Hood and C. Taylor, "Navier-Stokes Equations Using Mixed Interpolation," Flow Symposium, 1974.†

29. M. D. Olson, "A Variational Finite Element Method for Two-Dimensional Steady Viscous Flows," Joint McGill University-Engineering Institute of Canada Conference, 1972.

30. M. D. Olson, "Variational-Finite Element Methods for Two-Dimensional and Axisymmetric Navier-Stokes Equations," Flow Symposium, 1974.†

31. S. C. R. Dennis and G. Z. Chang, "Numerical Solutions for Steady Flow Past a Circular Cylinder at Reynolds Numbers Up to 100," *J. Fluid Mech.*, Vol. 42, Part 3, 1970.

32. A. J. Baker, "Finite Element Theory for Viscous Fluid Dynamics," *Bell Aerospace Co. Res. Rept.* 9500-920189, 1970.

33. E. Skiba, "A Finite Element Solution of General Fluid Dynamics Problems—Natural Convection in Rectangular Cavities," M.S. Thesis, Dept. of Civil Eng., University of Waterloo, Ontario, Canda, April 1970.

34. A. J. Baker, "A Numerical Solution Technique for Two-Dimensional Problems in Fluid Dynamics Formulated with the Use of Discrete Elements," TN-TCTN1005, Bell Aerosystems Company, 1970.

35. E. Skiba, T. E. Unny, and D. S. Weaver, "A Finite Element Solution for a Class of Two-Dimensional Viscous Fluid Flow Dynamics Problems," symposium held at the University of Waterloo, 1971.

36. R. T. Cheng, "Numerical Solution of the Navier-Stokes Equations by the Finite Element Method," *Phys. Fluids*, Vol. 15, No. 12, 1972.

37. A. J. Baker, "Finite Element Solution Algorithm for Viscous Incompressible Fluid Dynamics," *Int. J. Numer. Methods Eng.*, Vol. 6, No. 1, 1973.

38. T. Bratanow, A. Ecer, and M. Kobiske, "Finite-Element Analysis of Unsteady Incompressible Flow Around an Oscillating Obstacle of Arbitrary Shape," *AIAA J.*, Vol. 11, No. 11, November 1973.

† The notation "Flow Symposium, 1974" refers to the International Symposium on Finite Element Methods in Flow Problems, held January 7–11, 1974, at the University of Wales, Swansea, U.K. A volume containing papers and extended abstracts of papers presented at this meeting is available from University of Alabama Press, Huntsville, Ala.

39. P. Tong, "On the Solution of the Navier-Stokes Equations in Two-Dimensional and Axial Symmetric Problems," Flow Symposium, 1974.†

40. A. B. Huang and V. Y. C. Young, "A Non-variational Finite Element Analysis for the Navier-Stokes Equations," Flow Symposium, 1974.†

41. M. Ikenouchi and N. Kimura, "An Approximate Numerical Solution of the Navier-Stokes Equations by Galerkin Method," Flow Symposium, 1974.†

42. M. Kawahara, N. Yoshimura, and K. Nakagawa, "Analysis of Steady Incompressible Viscous Flow," Flow Symposium, 1974.†

43. P. Tong, "The Finite Element Method for Fluid Flow," in *Recent Advances in Matrix Methods of Structural Analysis and Design*, R. H. Gallagher, Y. Yamada, and J. T. Oden (eds.), University of Alabama Press, Huntsville, Ala., 1971.

44. P. Tong, *Finite Element Methods for Fluid Problems* (a series of ten lectures given at the International Center for Mechanical Science, 1972) (to be published by Springer-Verlag).

45. J. T. Oden and D. Somogyi, "Finite Element Applications in Fluid Dynamics," *Proc. ASCE*, Vol. 95, No. EM3, June 1969.

46. M. Kawahara, N. Yoshimura, K. Nakagawa, et al., "Steady Flow Analysis of Incompressible Viscous Fluid by the Finite Element Method," in *Theory and Practice in Finite Element Structural Analysis*, Y. Yamada and R. H. Gallagher (eds.), University of Tokyo Press, 1973.

47. W. Chu, "Generalized Finite Element Method for Compressible Viscous Flow," Technical Note, *AIAA J.*, Vol. 9, No. 11, November 1971.

48. J. T. Oden and L. C. Wellford, Jr., "Analysis of Flow of Viscous Fluids by the Finite Element Method," *AIAA J.*, Vol. 10, No. 12, December 1972.

49. J. T. Oden, "The Finite Element Method in Fluid Mechanics," in *Lectures on Finite Element Methods in Continuum Mechanics*, J. T. Oden and E. R. A. Oliveira (eds.), University of Alabama Press, Huntsville, Ala., 1973.

50. P. Hood, "A Finite Element Solution of the Navier-Stokes Equations for Incompressible Contained Flow," M.S. Thesis, University of Wales, Swansea, U.K., 1970.

51. P. Hood, Ph.D. Thesis, University of Wales, Swansea, U.K. (in print; January 1974).

ADDITIONAL REFERENCES

M. M. Aral, "Application of Finite Element Analysis in Fluid Mechanics," Ph.D. Thesis, Georgia Institute of Technology, Atlanta, Ga., September 1971.

J. H. Argyris and G. Mareczek, "Finite Element Analysis of Slow Incompressible Viscous Fluid Motion," *Ingenieur-Archiv.*, Vol. 43, Springer-Verlag, 1974.

A. J. Baker, "Finite Element Computational Theory for Three-Dimensional Boundary Layer Flow," *AIAA Paper* 72-108, 1972.

R. Balasubramanian, "An Iterative Finite Element Method for Viscous Flow," M.S. Thesis, University of Calgary, Alberta, Canada, 1973.

† The notation "Flow Symposium, 1974" refers to the International Symposium on Finite Element Methods in Flow Problems, held January 7–11, 1974, at the University of Wales, Swansea, U.K. A volume containing papers and extended abstracts of papers presented at this meeting is available from University of Alabama Press, Huntsville, Ala.

W. W. Bowley and J. F. Prince, "Finite Element Analysis of General Fluid Flow Problems," AIAA 4th Fluid and Plasma Dynamics Conference, June 1971.

T. T. Bramlette, "Plane Poiseuille Flow of a Rarefied Gas Based on the Finite Element Method," *Phys. Fluids*, Vol. 14, No. 2, 1971.

G. Bug and P. Blair, "Finite Element Solution of the Flow Around a Two-Dimensional Descoid," paper presented to ASCE National Meeting, New Orleans, La., February 1969.

R. T. Cheng, "Numerical Investigation of Lake Circulation Around Islands by the Finite Element Method," *Int. J. Numer. Methods Eng.*, Vol. 5, No. 1, 1972.

R. T. Cheng, "Transient Free-Surface Flow in Porous Media by Finite Element Method," paper presented at American Geophysical Union 53rd Annual Meeting; Washington, D.C., April 1972.

W. Chu, "A Comparison of Some Finite Element and Finite Difference Methods for a Simple Sloshing Problem," Technical Note, *AIAA J.*, Vol. 9, No. 10, 1971.

Y. P. Chugh and H. R. Hardy, "Application of Finite Element Analysis to Underground Storage of Natural Gas," *Trans. Am. Geophys. Union*, Vol. 51, No. 4, 1970.

A. Craggs, "An Acoustic Finite Element Approach for Studying Boundary Flexibility and Sound Transmission Between Irregular Enclosures," *J. Sound Vibration*, Vol. 30, No. 3, 1973.

C. S. Fang, W. Harrison and S. N. Wang, "Groundwater Flow in a Sandy Tidal Beach. II: Two-Dimensional Finite Element Analysis," *Water Resources Res.*, Vol. 8, No. 1, 1972.

M. Forthi, "Calcul Numérique des Fluides de Bingham et des Neutoniens Incompressibles parla Méthode des Eléments Finis," Ph.D. Thesis, Université de Paris VI, 1972.

P. W. France, C. J. Parekh, J. C. Peters, and C. Taylor, "Numerical Analysis of Free-Surface Seepage Problems," *Proc. ASCE, J. Irrig. Drainage Div.*, Vol. 97, No. IR1, Paper 7959, 1971.

G. Grotkop, "Die Berechnung von Flachwasserwellen nach der Methode der finiten Elemente," Ph.D. Thesis, Technische Universitat-Hanover, Hanover, 1972.

R.J. Guyan, B. H. Ujihara, and P. W. Welch, "Hydroelastic Analysis of Axisymmetric Systems by a Finite Element Method," *Proceedings of 2nd Conference on Matrix Methods in Structural Mechanics*, Wright-Patterson Air Force Base, Dayton, Ohio, 1968.

L. T. Issacs, "Finite Element Methods: Two-Dimensional Seepage with a Free Surface," *Bull.* 14, Dept. of Civil Eng., Queensland University, January 1971.

K. Iwata and K. Osakada, "Analysis of Hydrostatic Extrusion by the Finite Element Method," *Paper* 71-Proc-C for ASME Meeting, 1971.

I. Javandel and P. A. Witherspoon, "Method of Analyzing Transient Fluid Flow in Multi-layered aquifers," *Water Resources Res.*, Vol. 5, No. 4, 1969.

I. Javandel and P. A. Witherspoon, "Application of the Finite Element Method to Transient Flow in Porous Media," *Trans. Soc. Pet. Eng.*, Vol. 243, September 1968.

L. Loziuk, J. Anderson, and T. Belytschko, "Hydrothermal Analysis by the Finite Element Method," *Proc. ASCE*, Vol. 98, HY11, November 1972.

L. G. Napolitano, "Finite Element Methods in Fluid Dynamics," AGARD-VKI Lecture Series, *Advances in Numerical Fluid Dynamics*, Von Karman Institute for Fluid Dynamics, Brussels, March 1973.

S. P. Neuman and P. A. Witherspoon, "Analysis of Non-steady Flow with a Free Surface Using Finite Element Method," *Water Resources Res.*, Vol. 7, No. 3, 1971.

H. B. Nielson, "A Finite Element Method for Calculating a Two-Dimensional Open Channel Flow," *Basic Res. Progr. Rept.* 19, Technical University of Denmark, Copenhagen, August 1969.

J. T. Oden, "Finite-Element Analogue of Navier-Stokes Equation," *Proc. ASCE, J. Eng. Mech. Div.*, Vol. 96, No. EM4; August 1970.

K. Palit and R. T. Fenner, "Finite Element Analysis of Two-Dimensional Slow Non-Newtonian Flows," *Am. Inst. Chem. Eng. J.*, Vol. 18, No. 6, 1972.

A. K. Parkin, "Field Solutions for Turbulent Seepage Flow," *Proc. ASCE, J. Soil Mech. Found. Div.*, Vol. 97, No. SM1, 1971.

L. D. Pinson, "Evaluation of a Finite Element Analysis for the Longitudinal Vibrations of Liquid-Propellant Launch Vehicles," *NASA Rept.* TN D-5803, June 1970.

A. R. S. Ponter, "The Application of Dual-Minimum Theorems to the Finite Element Solution of Potential Problems with Special Reference to Seepage," *Int. J. Numer. Methods in Eng.*, Vol. 4, 1972.

R. S. Sandhu and E. L. Wilson, "Finite Element Analysis of Seepage in Elastic Media," *Proc. ASCE*, Vol. 95, No. EM3, 1969.

B. G. Secrest and W. W. Boyd, "The Finite Element Method Applied to Ideal Gaseous Mixtures," *Int. J. Numer. Methods Eng.*, Vol. 4, 1972.

I. M. Smith, R.V. Farraday, and B.A.O'Connor, "Rayleigh-Ritz and Galerkin Finite Elements for Diffusion-Convection Problems," *Water Resources Res.*, Vol. 9, No. 3, 1973.

A. O. Tay and G. de Vahl Davis, "Application of Finite Element Method to Convection Heat Transfer Between Parallel Planes," *Int. J. Heat Mass Transfer*, Vol. 14, No. 8, 1971.

C. Taylor and J. Davis, "Finite Element Numerical Modelling of Flow and Dispersion in Estuaries," International Association for Hydraulic Research, International Symposium on River Mechanics, 1973.

R. L. Taylor and C. B. Brown, "Darcy Flow Solution with a Free Surface," *Proc. ASCE, J. Hydraul. Div.*, Vol. 93, 1967.

E. G. Thompson and M. I. Haque, "A High-Order Finite Element for Completely Incompressible Creeping Flow," *Int. J. Numer. Methods Eng.*, Vol. 6, No. 3, 1973.

P. Tong, "Liquid Sloshing in an Elastic Container," A.F.O.S.R. 66-0943, June 1966.

R. E. Volker, "Non-Linear Flow in Porous Media by Finite Elements," *Proc. ASCE*, Vol. 95, No. HY6; November 1969.

C. Taylor, B. S. Patil and O. C. Zienkiewicz, "Harbor Oscillation: A Numerical Treatment for Undamped Natural Modes," *Proc. Instn. Civ. Engrs.* (London), Vol. 43, June, 1969, pp. 141-155. (Discussion: Vol. 46, June, 1970, pp. 203-211.)

S. T. Wu, "Unsteady MHD Duct Flow by the Finite Element Method," *Int. J. Numer. Methods Eng.*, Vol. 6, No. 1, 1973.

O. C. Zienkiewicz, P. Mayer, and Y. K. Cheung, "Solution of Anisotropic Seepage by Finite Elements," *Proc. ASCE*, Vol. 92, No. EM1, February 1966.

10

A SAMPLE COMPUTER CODE
AND OTHER PRACTICAL CONSIDERATIONS

10.1 INTRODUCTION

The major portion of this chapter is written for readers who have never programmed a digital computer to solve a continuum problem by the finite element method. Having an understanding of the underlying mathematics of the finite element method and knowing how to derive element equations for a given problem are not enough to solve the problem. It is also necessary to know how to translate the equations into computer instructions so that the element equations can be evaluated, assembled, and solved. In this regard, we discuss the details of writing a computer program to solve an example problem—the steady-state heat conduction equation in two dimensions. The program for this problem has been constructed without the embellishments or sophisticated coding techniques found in advanced finite element programs. Hence, if the reader is familiar with the FORTRAN language, he should be able to follow all of the coding with the aid of the many comment statements appearing throughout the program.

After a description of the program and a discussion of two sample solutions obtained by the program, we consider the problem of input data preparation for more advanced finite element programs. Instead of covering the details of the numerous ways to automate and make more efficient the handling of input data, we present an overview of this subject and then cite key references where detailed information can be found.

An important part of any finite element analysis is the evaluation of the integrals in the element equations. When these integrals cannot be evaluated in closed form, we must resort to approximate numerical methods. In Section 10.6 we summarize a number of useful numerical integration formulas for this purpose.

Though the subject of matrix equation solvers is beyond the scope of this work, in Section 10.7 we suggest references where pertinent information can be found on this subject.

Finally, we mention some general-purpose computer program packages that are available for structural analysis. These packages are especially useful to the analyst who does not have the time or inclination to develop his own program for structural problems.

10.2 SETTING UP A SIMPLE HEAT CONDUCTION PROBLEM

To illustrate the step-by-step procedure for developing a finite element computer code, we shall describe the FORTRAN computer program shown on pp. 384–399, which solves a simple continuum problem—the steady-state heat conduction problem for two-dimensional or axisymmetric three-dimensional bodies of arbitrary shape and inhomogeneous composition. The program code is not necessarily the most general or the most efficient, but it serves to point out all of the essential features.

10.2.1 Governing differential equations

We begin with a statement of the problem and the equations that must be coded. All of the equations are given in Chapter 7, but we will repeat some of them here for convenience. The steady-state temperature distribution in a two-dimensional body satisfies the thermal energy equation

$$\frac{\partial}{\partial x}\left(k_x \frac{\partial T}{\partial x}\right) + \frac{\partial}{\partial y}\left(k_y \frac{\partial T}{\partial y}\right) + \tilde{Q} = 0 \tag{7.18}$$

with boundary conditions

$$T = T(x, y) \quad \text{on } C_1 \tag{7.19}$$

$$k_x \frac{\partial T}{\partial x} n_x + k_y \frac{\partial T}{\partial y} n_y = q \quad \text{on } C_2 \tag{7.20}$$

$$k_x \frac{\partial T}{\partial x} n_x + k_y \frac{\partial T}{\partial y} n_y + h(T - T_\infty) = 0 \quad \text{on } C_3 \tag{7.21}$$

and, for a three-dimensional axisymmetric solid, the temperature distribution satisfies

$$\frac{\partial}{\partial r}\left(k_r r \frac{\partial T}{\partial r}\right) + \frac{\partial}{\partial z}\left(k_z r \frac{\partial T}{\partial z}\right) + r\tilde{Q} = 0 \tag{7.28}$$

with boundary conditions

$$T = T(r, z) \quad \text{on } S_1 \tag{7.29}$$

$$k_r r \frac{\partial T}{\partial r} n_r + k_z r \frac{\partial T}{\partial z} n_z + rq = 0 \quad \text{on } S_2 \tag{7.30}$$

$$k_r r \frac{\partial T}{\partial r} n_r + k_z r \frac{\partial T}{\partial z} n_z + rh(T - T_\infty) = 0 \quad \text{on } S_3 \tag{7.31}$$

We assume that $k_r = k_z = k(r, z)$ and $\tilde{Q}$, q, and h are known, specified functions of the coordinates. Figure 10.1 shows the coordinate systems for

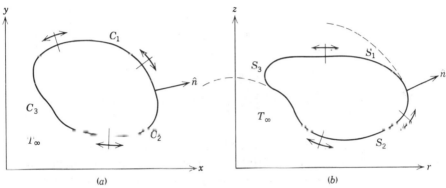

Figure 10.1. Definition of coordinate systems. (*a*) Coordinate system for the two-dimensional problem. (*b*) Coordinate system for axisymmetric three-dimensional problem.

the general two-dimensional body and the axisymmetric three-dimensional body. The problem that we want to solve may be summarized as follows:

Given: (1) The geometry of the body, (2) the data describing the physical properties of the heat-conducting medium, and (3) the thermal loading conditions (boundary conditions).

Find: (1) Temperature distribution throughout the body, and (2) heat flux at locations of interest.

10.2.2 The finite element equations

We recognize that the heat conduction equation is a special case of the general quasi-harmonic equation discussed in Section 7.3; hence the finite element formulation for the problem is already available. To simplify the discussion here, we shall consider only the computer code that implements the formulation for linear triangular elements. For this case, the element equations are given explicitly in Section 7.2.4. To designate clearly the identity of each of the matrices of the element equations, we shall change slightly the notation used in Chapter 7, but the definitions of the parameters remain the same. The element equations are as follows:

$$\overset{3\times3}{[K_T]}\overset{3\times1}{\{T\}^{(e)}} = \overset{3\times1}{\{K_{\tilde{Q}}\}} - \overset{3\times1}{\{K_q\}} - \overset{3\times3}{[K_h]}\overset{3\times1}{\{T\}^{(e)}} + \overset{3\times1}{\{K_{T_\infty}\}} \tag{10.1}$$

where

$$K_{T_{ij}} = \frac{k}{4\Delta}(b_i b_j + c_i c_j), \quad i,j = 1, 2, 3 \tag{10.2}$$

(These are terms of the thermal stiffness matrix)

$$\{T\}^{(e)} = \begin{Bmatrix} T_1 \\ T_2 \\ T_3 \end{Bmatrix} \tag{10.3}$$

$$\{K_{\tilde{Q}}\} = \frac{\Delta}{12}\begin{bmatrix} 2 & 1 & 1 \\ 1 & 2 & 1 \\ 1 & 1 & 2 \end{bmatrix}\begin{Bmatrix} \tilde{Q}_1 \\ \tilde{Q}_2 \\ \tilde{Q}_3 \end{Bmatrix} = \begin{matrix} \text{influence matrix for} \\ \text{internal heat generation} \end{matrix} \tag{10.4}$$

The remaining terms in equation 10.1 exist only if the element boundary is part of the solution region boundary, and external heat flux or boundary convection is specified on that boundary segment. If i and j are numbers designating the nodes that lie on the boundary, then

$$[K_h] = l_{ij}h\begin{bmatrix} \frac{1}{3} & \frac{1}{6} & 0 \\ \frac{1}{6} & \frac{1}{3} & 0 \\ 0 & 0 & 0 \end{bmatrix} = \begin{matrix} \text{influence matrix for} \\ \text{boundary heat convection} \end{matrix}$$

$$\{K_q\} = \begin{Bmatrix} Q_1 \\ Q_2 \\ Q_3 \end{Bmatrix} = \begin{matrix} \text{influence matrix for} \\ \text{externally specified heat} \\ \text{flux at interior nodes;} \\ Q_i \text{ is a heat flux allocated} \\ \text{to node } i \text{ of an element} \end{matrix} \tag{10.5}$$

or

$$\{K_q\} = \frac{q}{2}\begin{Bmatrix} l_{ij} \\ l_{ij} \\ 0 \end{Bmatrix} = \begin{matrix} \text{influence matrix for} \\ \text{external heat flux when} \\ \text{specified on boundary} \end{matrix} \tag{10.6}$$

$$\{K_{T_\infty}\} = \frac{hT_\infty}{2}\begin{Bmatrix} l_{ij} \\ l_{ij} \\ 0 \end{Bmatrix} = \begin{matrix} \text{influence matrix for} \\ \text{ambient temperatures} \\ \text{near convecting boundaries} \end{matrix} \tag{10.7}$$

For convenience in this program, we assume that k and $\tilde{Q}$ are constant within each element, that is, $\tilde{Q} = \tilde{Q}_1 = \tilde{Q}_2 = \tilde{Q}_3$. This assumption does not preclude the variation of k and $\tilde{Q}$ throughout the solution region, however, since k and $\tilde{Q}$ may be assigned different values for each element and the values need not be continuous from one element to another. Thus inhomogeneous materials may be treated.

As we noted in Section 7.2.4, many similarities exist between the axisymmetric problem and the two-dimensional plane problem. Because of these similarities, only slight program modifications are necessary to handle one problem or the other. The axisymmetrix problem differs from the plane problem in that all integrations have to be carried out over the *volumes* of the elements. To simplify these integrations as much as possible, we assume that the products (rk) and $(r\tilde{Q})$ remain constant within each element and take on the values $(\bar{r}k)$ and $(\bar{r}\tilde{Q})$, where $\bar{r} = \frac{1}{3}(r_1 + r_2 + r_3)$. This assumption introduces only a small error if none of the elements is close to the z axis, that is, $r \approx 0$. If we use the subscript A to designate the axisymmetric problem and P to designate the planar problem, the element matrices for the axisymmetric case in terms of the planar case become

$$k_A = \bar{r}k_P \tag{10.8}$$

$$k_{T_{Aij}} = 2\pi\bar{r}^2 k_{T_{Pij}} \tag{10.9}$$

The heat source matrix $\{K_{\tilde{Q}}\}$ requires additional integration. To evaluate the terms

$$K_{\tilde{Q}_i} = \int_{vol} r\tilde{Q}L_i \, d(\text{vol})$$

we set $r = \bar{r}$ to be consistent with the definition of the planar conductivity, k. Then we have

$$K_{\tilde{Q}_i} = \bar{r}\tilde{Q} \int_{r,z} L_i 2\pi r \, dr \, dz$$

or, substituting

$$L_i = \frac{1}{2\Delta}(a_i + b_i r + c_i z)$$

$$K_{\tilde{Q}_i} = \frac{\bar{r}Q\pi}{\Delta}\left[\underbrace{\int a_i r \, dr \, dz}_{\substack{\text{area of}\\\text{element}}} + \underbrace{\int b_i r^2 \, dr \, dz}_{\substack{\text{area of}\\\text{element}}} + \underbrace{\int c_i rz \, dr \, dz}_{\substack{\text{area of}\\\text{element}}}\right]$$

And, performing the integrations according to the formulas given in Zienkiewicz [1, p. 506], we can write

$$K_{\tilde{Q}_i} = \bar{r}\tilde{Q}\pi\left[a_i\bar{r} + b_i\bar{r}^2 + \frac{b_i}{12}(r_1'^2 + r_2'^2 + r_3'^2) \right.$$
$$\left. + c_i\bar{r}\bar{z} + \frac{c_i}{12}(r_1'z_1' + r_2'z_2' + r_3'z_3') \right] \tag{10.10}$$

where

$$r_i' = r_i - \bar{r}$$

and

$$z_i' = z_i - \bar{z}, \qquad \bar{z} = \tfrac{1}{3}(z_1 + z_2 + z_3)$$

For the surface integrals, the length l_{ij} becomes an area given by

$$l_{ij_A} = 2\pi\bar{r}_{ij}l_{ij_P} \tag{10.11}$$

where $\bar{r}_{ij} = \tfrac{1}{2}(r_i + r_j)$.

10.2.3   The overall program logic

All of the element equations are now available for coding. Writing the program to find the temperature distribution and heat flux for a given problem is essentially a three-step process. First, we must prepare the "front" of the program to (1) set up the necessary matrix arrays; (2) read the description of the triangular finite element mesh; and (3) accept all the input data, including the material properties of the elements and all the thermal boundary conditions. Second, the body of the program must evaluate the matrix equations for each element and then add the appropriate equations to the master matrix equations as each element is considered.† Finally, the master matrix equations must be modified to account for the boundary conditions, and then solved for the unknown nodal temperatures and heat fluxes. The flow chart of Figure 10.2 depicts these steps in greater detail.

Usually a finite element program is set up as a series of independent modules linked by a small driving program. Each module has a specific function such as data generation and input, calculation of element equations,

---

† Since all of the elements have been derived in terms of the global coordinate system, a transformation from local to global coordinates is unnecessary here.

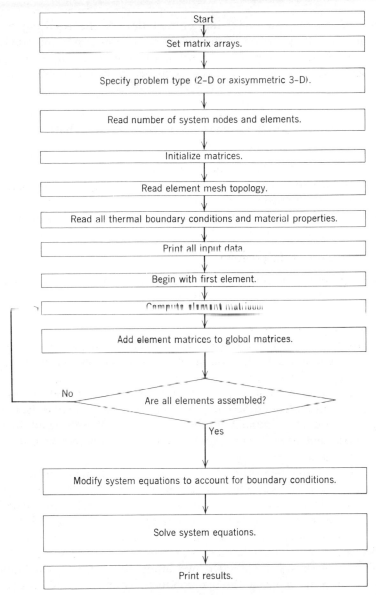

Figure 10.2. Flow chart for a computer program to solve steady-state heat conduction problems by the finite element method.

solution of system equations, and presentation of the resulting output. For simplicity, the program in the following section is not constructed in this way. Only the system equation-solving routine is set up as a module. The other steps are simply executed in sequence in the main program.

10.3 THE COMPUTER PROGRAM AND ITS EXPLANATION

Now we examine in detail the computer program (pp. 384–399) that performs each of the steps just mentioned. Throughout this discussion, we shall refer to the FORTRAN listing given in Section 10.3.5. The listing contains numerous COMMENT statements to explain the various operations.

10.3.1 Structure of the program

The program begins with a series of DIMENSION statements, which set up the arrays needed in the calculations. As indicated in statements 0002–0010, storage has been allocated for problems with up to 300 nodes; however, larger problems can be considered simply by increasing the dimensions of these matrices. The limit of problem size is dictated by the core storage of the available machine. Next, the program calls for a declaration of the type of problem to be solved (either two-dimensional or axisymmetric); then the appropriate problem label is printed (statements 0011–0016). Before proceeding to input the data describing the finite element mesh and the thermal loading conditions, we initialize all the matrix arrays by setting all the terms in the arrays equal to zero. This removes extraneous numbers which could be troublesome later.

Statements 0040–0045 read into the program the node numbers and the coordinates of the nodes for the complete finite element mesh. Also, the system topology, the element numbers, and the numbers of the three nodes associated with each element are read. At this point, the program could be improved by including an automatic mesh generator instead of using the present user-specified mesh. We shall elaborate on this concept in Section 10.5.

Beginning with statement 0046, we input all of the thermal loading conditions and the values of thermal conductivity and internal heat generation for each element. Since the solution obtained by the program depends intimately on this body of data, we ask the program to print out all data that have been input. This enables the analyst to check for input data errors. An improved version of the program would have a stop option to allow checking before proceeding. Statement 0096 marks the completion of these steps; the program is now ready to begin work on a particular thermal problem.

The loop to begin calculating the various element matrices starts with statement 0097. Most of the notation in this portion of the program closely follows that used in the equations of Section 10.2.2. Once the element matrices $[K_T]$ and $\{K_{\tilde{Q}}\}$ are computed for one element, they are assembled into the master matrices by the code in statements 0157–0163. This section of the program merits the reader's close inspection because it illustrates precisely how the assembly process is implemented on a computer. Since each element in the triangular mesh has three nodes, the local node numbers are I\$ = 1, 2, 3. The global node numbers for the element are recovered from the parameter NODE(K, I\$), which was read as input data for the element; that is, for element K, we read in the node numbers N(1) = NODE(K, 1), N(2) = NODE(K, 2) and N(3) = NODE(K, 3). Then the code in statements 0157–0163 loads the terms of the element matrices into their proper locations in the system matrices. Each time that a term of an element matrix is placed in a location in the system matrix where another term has already been placed, it is added to whatever value is there. The reader may want to see Section 2.3.2 for a review of this general assembly procedure.

After all of the elements have been processed in this way, the assembled system equations are ready to be modified to account for the boundary phenomena. This is done by statements 0168–0189. The boundary matrix $[K_h]$ for heat convection is added to the thermal stiffness matrix $[K_T]$ during this process, and the remaining heat fluxes are summed and stored as the column vector $\{RHS\}$, where

$$\{RHS\} = \{K_{\tilde{Q}}\} - \{K_q\} + \{K_{T_\infty}\}$$

Thus, at the conclusion of statement 0189, the system equations have the form

$$\overset{NN \times NN}{[K_T^*]} \ \overset{NN \times 1}{\{T\}} = \overset{NN \times 1}{\{RHS\}} \equiv \overset{NN \times 1}{\{R\}} \tag{10.12}$$

where

$$[K_T^*] = [K_T] + [K_h]$$

Not all of the components of the array $\{R\}$ are known because the heat fluxes at nodes where temperature is specified are unknown; that is, at each node i, we know either T_i or K_{q_i}. If no external heat flux is specified at node i, $q \equiv 0$ and $K_{q_i} = 0$. Hence, in the array $\{T\}$ we have a mixture of known and unknown nodal temperatures, and in the array $\{R\}$ a corresponding mixture of known and unknown resultant nodal heat fluxes. One way to sort out this mixture of parameters is to reorder the matrices [2]. If we do this, we can partition the equations as follows (see Section 2.3.4):

$$\begin{bmatrix} [K_T^*]_{11} & [K_T^*]_{12} \\ [K_T^*]_{12}^T & [K_T^*]_{22} \end{bmatrix} \begin{Bmatrix} \{T\}_1 \\ \{T\}_2 \end{Bmatrix} = \begin{Bmatrix} \{R\}_1 \\ \{R\}_2 \end{Bmatrix} \tag{10.13}$$

where $\{R\}_1$ and $\{T\}_2$ are known, and $\{T\}_1$ and $\{R\}_2$ are unknown. Effectively, we have segregated the known and unknown nodal temperatures and heat fluxes. Then we can solve directly for $\{T_1\}$ by solving the set of equations

$$[K_T^*]_{11}\{T\}_1 = \{R\}_1 - [K_T^*]_{12}\{T\}_2 \tag{10.14}$$

When $\{T_1\}$ has been found, we can calculate the unknown nodal heat fluxes if desired by evaluating the remaining equations:

$$\{R\}_2 = \{K_T^*]_{21}\{T\}_1 + [K_T^*]_{22}\{T\}_2 \tag{10.15}$$

Conceptually this procedure is simple, but in practice it poses an extra difficulty. The difficulty arises because the original (as assembled) equations — equations 10.12 — must be reordered, and this requires some additional programming effort while usually increasing the bandwidth of the final matrix. The subroutine SOLMIX (SOLVEMIX) called for by statement 0190 effectively implements this procedure. Note that within SOLMIX we call subroutine SOLVE, which solves equation 10.14 for $\{T\}_1$ by Gaussian elimination.†

A practical way to avoid the extra programming and to preserve the original size and bandwidths of $[K_T^*]$ is to follow the procedure discussed in Section 2.3.4. This procedure was originally suggested by Felippa and Clough [2] and later elaborated on by Desai and Abel [3].

Having outlined the functions of the various parts of the program, we turn now to a description of how the program is used in actual problem-solving situations. The first step is to describe the geometry of the solution domain and then represent it by a finite element model. The second step is to prepare the input data containing the model description.

10.3.2 Preparation of the finite element model (preliminary work)

To construct a finite element model for a problem, we must first draw to scale on a sheet of graph paper a cross section of the body to be analyzed.‡ After the geometry has been sketched, we pencil in node points along the boundary and in the interior of the body. Node points should be placed in

---

† The subroutines SOLMIX and SOLVE were written by Professor J. F. Booker of Cornell University.
‡ Wherever possible, we take advantage of symmetry to reduce the size of the solution region, the amout of labor required to prepare input data, and the computation time (problem size).

regions wherever values of temperature or heat flux are sought or specified. Nodes should be more closely spaced in regions where the temperature is expected to vary rapidly. Solution accuracy improves as the number of nodes increases; however, computation time is directly proportional to the cube of the number of nodes. Next, we begin connecting node points with straight lines to form a network of triangles throughout the region.

Only one rule governs the construction of a triangular mesh from a set of node points. This rule is that each node must be a vertex of the triangles it touches, that is, nodes may not be placed along the sides of any triangle. Figure 10.3 illustrates the point by considering triangles formed from a

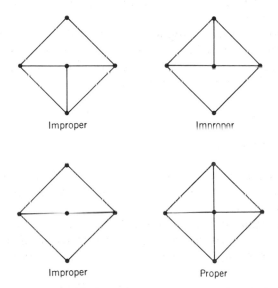

Improper   Improper

Improper   Proper

Figure 10.3.   Possible ways to form triangles from four mesh points.

group of five typical nodes. Generally, triangles that approach the equilateral state are better than long narrow triangles, which should be avoided if possible. When working with bodies composed of different materials, the parting lines of the materials should be represented by the boundaries of triangles. In fact, no difficulty arises if every triangle is chosen to be a different material; thus, almost any heterogeneous material can be handled easily.

The easiest way to treat regions with holes is to represent the holes by triangles and then imagine these elements to be a material whose thermal conductivity is smaller by three or four orders of magnitude than that of the surrounding elements. If the holes are fluid coolant passages, no elements are used to represent them. Instead, values of T_∞ are assigned to each hole and convective heat transfer coefficients are specified for the nodes on the hole boundaries.

Once the solution region has been properly triangulated, we are ready to establish the system topology. This is done by numbering the nodes consecutively, starting from 1. The elements must also be numbered consecutively starting from 1. Except for the requirement that nodes and elements must be consecutively numbered, the numbering scheme is arbitrary.† Once the numbers have been assigned, they must remain fixed because interpretation of the final output relies on the numbering scheme. The node numbers and element numbers should be placed on the previous sketch for later reference. They will comprise part of the input data to be discussed in the next section.

As mentioned, the present program capacity is limited to 300 nodes, but this capacity can be extended to the limit of central core storage. To enlarge the capacity, we simply need to change the dimension statements. For many problems, 100 nodes or fewer are sufficient to obtain the desired information.‡

10.3.3   Preparation of input data

In this section, we describe the preparation of input data sheets for a given problem. Since the coding language is FORTRAN IV, special attention must be given to format.

The data are arranged in eight different classifications, and they must be entered in the order listed below. Any consistent set of units can be used. One convenient set of engineering units is suggested in the nomenclature list.

Classification	Type of Data
I	Problem identification
II	Number of nodes and number of elements
III	Node numbers and nodal coordinates
IV	System topology (element number and associated node numbers)

† If special node-numbering schemes are used, it is possible to minimize the bandwidth of the overall system matrix. This program has not been designed to take advantage of matrix bandedness; thus the user need not be concerned about his method for assigning node numbers.
‡ As a matter of interest, a system with 100 nodes requires less than 0.4 minute of computation time on the GMR-IBM 360/65 system.

V	Node numbers and specified nodal temperatures
VI	Thermal conductivity and internal heat generation for each element
VII	Node numbers and specified *external* heat flux values
VIII	Definition of convective heat transfer boundary conditions

A detailed description of each of these classifications follows.

Classification I
Format: (I10)
Card columns used: 1–10
Number of cards in this group: 1
Explanation: This data card identifies the type of problem to be analyzed.
Place in column 10 of the first card the integer:
1   if the problem is two-dimensional;
2   if the problem is three-dimensional axisymmetric.

Classification II
Format: (2I10)
Card columns used: 1–10, 11–20
Number of cards in this group: 1
Explanation: This data card must contain two integers. The first is the number of nodes;† the second, the number of elements. Note that integers read with an I format must be right-adjusted.

Classification III
Format: (I10, 2F10.0)
Card columns used: 1–10, 11–20, 21–30
Number of cards in this group: equal to the number of nodes in the model
Explanation: Data cards in this set contain a list of the nodal coordinates. Before the coding sheets can be filled out for these cards, it is necessary to establish a scaled coordinate system on the scaled drawing of the part. Placement of the coordinate system is arbitrary for the two-dimensional problem, but for the axisymmetric problem, there is one rule: "The z axis must be parallel to the axis of symmetry." In either case, the

† Each time that a new problem with a different number of nodes is considered, one dimension statement in the main program must be changed. The matrices TM and B must be doubly dimensioned, as indicated in the first comment of the program listing (see Section 10.3.5).

units on both axes must be the same. Any consistent set of units can be used. There must be one data card for each node in the solution region. Each card contains a node number, the $\begin{Bmatrix} x \\ r \end{Bmatrix}$ coordinate and the $\begin{Bmatrix} y \\ z \end{Bmatrix}$ coordinate of that node, respectively. The cards in this set may be arranged in any order.

Classification IV

Format: (4I10)

Card columns used: 1–10, 11–20, 21–30, 31–40

Number of cards in this group: equal to the number of elements in the model

Explanation: System topology (the record of which node numbers belong to each element number) is specified in this data set. There must be one data card for each element in the solution region. Each card contains an element number and the three node numbers associated with that element. The order of the node numbers on each card is important. Referring to the sketch of the finite element model, we place node numbers on a card from left to right in the order in which we would encounter them if we started at any node of the element and walked around its boundary in a counterclockwise direction.

Classification V

Format: (6X, A4, I10, F10.0)

Card columns used: 7–10, 11–20, 21–30

Number of cards in this group: equal to the number of nodes where temperature is specified plus one STØP card

Explanation: This data set allows us to enter the temperature boundary conditions. There must be one card for each node where temperature is specified. Temperature may be specified at interior or exterior nodes. A typical card contains first a node number and then the value of the temperature at that node. The last card in this data set must contain only the word STØP, written in columns 7–10. This card signals that there are no more data of this type. For a valid solution, at least one nodal temperature must be specified unless convection occurs at all parts of the boundary, in which case only the STØP card is used.

Classification VI
  Format: (I10, 2F10.0)
  Card columns used: 1–10, 11–20, 21–30
  Number of cards in this group: equal to the number of elements in the
                   solution region
  Explanation: The thermal conductivity and a value of internal heat
        generation must be specified for each element. If there is no
        internal heat source in an element, the value 0.0 must be
        entered. The first number on a card is the integer represent-
        ing the element number. The next two numbers are the
        thermal conductivity and the amount of internal heat
        generation, respectively.

Classification VII
  Format: (6X, A4, I10, F10.0)
  Card columns used: 7–10, 11–20, 21–30
  Number of cards in this group: equal to the number of system nodes
                   minus the number of nodes where tem-
                   perature is specified plus one (TOP card)
  Explanation. A finite element model of a heat conduction problem re-
        quires a complete specification of the thermal loading
        conditions. This means that at every node in the solution
        region we must specify either a temperature or an *external
        heat flux.*† By "external heat flux," we mean one that the
        user imposes. When no external heat flux occurs at a node,
        the value 0.0 must be entered. *External heat flux must not be
        specified at any node where temperature is specified.* Numbers
        on a card are arranged as follows: (1) the integer denoting
        the node number, and (2) the signed‡ value of heat flux at the
        node.
        In practice, external heat fluxes are usually distributed or
        applied over some boundary area. For the finite element
        model, however, these fluxes must be represented by equiva-
        lent fluxes concentrated at the nodes. For boundary nodes,
        equivalent nodal fluxes are found by integrating the
        distributed flux over the half-boundaries nearest the node
        in question and then assigning this value of integrated flux
        to that node. Applied heat fluxes are something the pro-
        grammer must specify even if they are zero (as is usually the

---

† Heat flow out of the body is taken as positive.
‡ The program has an internal checking feature which insists on this condition. If this condition is not met, the error message is "SINGULAR MATRIX."

case). They should not be confused with boundary convection, which occurs when a convective heat transfer
coefficient and an ambient temperature are specified for a
node. The last card in this set must contain the word STØP,
written in columns 7–10.

Classification VIII

  Format: (6X, A4, 2I10, F10.0)

  Card columns used: 7–10, 11–20, 21–30, 31–40

  Number of cards in this group: equal to the number of boundary segments where heat convection takes place
  place plus one STØP card

  Explanation: These cards make possible the specifications of boundary
  heat convection. There must be one card for each boundary
  segment where convection occurs. The first two numbers on
  a card are the node numbers, which define the boundary
  segment end points. The third number is the value of the film
  coefficient over the segment, and the fourth is the ambient
  temperature. The film coefficient and ambient temperature
  may be different for each boundary segment if necessary.
  There is no provision in this program to account directly
  for radiation boundary conditions, but they can be approximated by specifying an equivalent film coefficient. Again, the
  last card in this set must contain the word STØP, written in
  columns 7–10. If no boundary convection occurs, only the
  STØP card appears in this set.

10.3.4   Description of output

The first line of output is a statement declaring the type of problem—two-
dimensional or three-dimensional axisymmetric. Next, all input data are
printed and labeled for easy identification. To ensure the validity of the
solution, the printed input data should be carefully checked against the
intended input. If there is an error in the specification of nodal temperatures
or external heat fluxes, the program will terminate with an error message.

The complete solution appears as a numbered list of nodal temperatures
and heat fluxes. Integers appearing at the far left of the list designate the node
numbers. Nodal heat flux must be interpreted as the *net* external heat flux
crossing the half-boundaries nearest the node.

The final line of output gives the average body temperature, found by
adding all nodal temperatures and dividing by the number of nodes. This
is useful in estimating the body's heat content.

The program contains no embellishments such as plotting routines to perform special functions. If the user wants to add these to meet his particular needs, he can do so without difficulty because the program listing is replete with guiding comments.

10.3.5   Definition of symbols and FORTRAN listing

Symbol†	*Description*
NCASE	integer which specifies the type of problem to be solved:
	NCASE = 1 2-D plane problem
	NCASE = 2 3-D axisymmetric problem
NN	number of nodes
NE	number of elements
XC(I), YC(I)	global coordinates of node I
NØDE(J,I)	J = 1, 2, ..., NE; I = 1, 2, 3 node numbers associated with element J
NT,NTS(I)	node number where temperature is specified
TN T	specified nodal temperature
NNST	number of nodes with specified temperature
TK(J)	thermal conductivity of element J
QQ(J)	value of internal heat generation for element J
NQ,NQS(I)	node number where external heat flux is specified
QNQ,Q(I)	specified external nodal heat flux
NNQS	number of nodes where external heat flux is specified
NC1(I),NC2(I)	node number pairs defining boundary segments where convection occurs
H(I)	convective heat transfer coefficient
TINF	ambient temperature
NNHC	number of boundary segments where convection occurs
XC$(I),YC$(I)	local coordinates of node I
T M$	element thermal stiffness matrix
TM	system thermal stiffness matrix
DEL	area of a triangular element
FQQ$	element influence matrix for internal heat generation
FQQ	system influence matrix for internal heat generation
SOLMIX	subroutine for partitioning system matrix
SOLVE	subroutine for solving a system of simultaneous linear equations by Gaussian elimination

† Symbols are listed in the same order as they appear in the main program.

```
FORTRAN IV G1          RELEASE 2.0                MAIN

      C         THIS PROGRAM WAS WRITTEN BY K. H. HUEBNER OF THE
      C         MECHANICAL RESEARCH DEPT. OF THE G. M. RESEARCH LABORATORIES.
      C         IT CAN BE USED FOR THE THERMAL ANALYSIS OF TWO AND AXISYMMETRIC
      C         THREE DIMENSIONAL BODIES OF IRREGULAR SHAPE AND HETEROGENEOUS
      C         COMPOSITION. THE FINITE ELEMENT METHOD IS USED TO OBTAIN THE
      C         APPROXIMATE NUMERICAL SOLUTION.
0001            DATA STOP/'STOP'/
      C$$$$$$$$$$$$$$$$$$$$$$$$$$$$$$$$$$$$$$$$$$$$$$$$$$$$$$$$$$$$$$$$$$$$$$$$$$$$$$$$
      C$$$$$$$$$$$$$$$$$$$$$$$$$$$$$$$$$$$$$$$$$$$$$$$$$$$$$$$$$$$$$$$$$$$$$$$$$$$$$$$$
      C         IMPORTANT NOTE
      C         THE MATRICES TM AND B MUST HAVE DIMENSIONS NNXNN, WHERE NN IS THE
      C         NUMBER OF NODES IN THE FINITE ELEMENT MODEL
0002            DIMENSION TM(12,12),B(12,12) ←─── NN = 12 FOR EXAMPLE PROBLEM
      C$$$$$$$$$$$$$$$$$$$$$$$$$$$$$$$$$$$$$$$$$$$$$$$$$$$$$$$$$$$$$$$$$$$$$$$$$$$$$$$$
      C$$$$$$$$$$$$$$$$$$$$$$$$$$$$$$$$$$$$$$$$$$$$$$$$$$$$$$$$$$$$$$$$$$$$$$$$$$$$$$$$
0003            DIMENSION XC(300),YC(300),NODE(300,3),NTS(300)
0004            DIMENSION T(300),TK(300),QQ(300),NQS(300),Q(300)
0005            DIMENSION NC1(300),NC2(300),H(300),TINF(300)
0006            DIMENSION TM$(3,3),FQQ$(3),N(3),FQQ(300)
0007            DIMENSION BL(300),Z(300),C(300)
0008            DIMENSION RHS(300),RHST(300),RHSQ(300)
0009            DIMENSION RP$(3),ZP$(3),RBL(300)
0010            DIMENSION XC$(3),YC$(3)
      C
      C         SPECIFY WHETHER PROBLEM IS TWO DIMENSIONAL( NCASE=1 )
      C         OR AXI-SYMMETRIC ( NCASE=2 )
0011            READ(5,500)NCASE
0012            IF(NCASE.EQ.1) GO TO 5
```

```
0013              WRITE(6,2015)
0014              GO TO 6
0015              WRITE(6,2020)
0016    6         CONTINUE
        C
        C     READ THE NUMBER OF SYSTEM NODES AND ELEMENTS
0017              READ(5,1005)NN,NE
        C
        C     INITIALIZE PARAMETERS
0018              DO 50 I=1,NN
0019              RHS(I)=0.0
0020              TK(I)=0.0
0021              QQ(I)=0.0
0022              Q(I)=0.0
0023              XC(I)=0.0
0024              YC(I)=0.0
0025              NQS(I)=0
0026              T(I)=0.0
0027              NTS(I)=0
0028              H(I)=0.0
0029              TINF(I)=0.0
0030              FQQ(I)=0.0
0031              BL(I)=0.0
0032              Z(I)=0.0
0033              C(I)=0.0
0034              RHST(I)=0.0
0035              RBL(I)=0.0
0036              DO 50 J=1,NN
0037              TM(I,J)=0.0
0038              B(I,J)=0.0
```

```
FORTRAN IV G1  RELEASE 2.0                    MAIN

0039       50 CONTINUE
C
C     READ NODE NUMBERS AND NODAL COORDINATES
0040          DO 100 J=1,NN
C
0041          READ(5,1006)I,XC(I),YC(I)
0042     100 CONTINUE
C
C     READ SYSTEM TOPOLOGY(ELEMENT NO. AND NODE NUMBERS
C     IN COUNTER-CLOCKWISE FASHION STARTING AT ANY NODE)
0043          DO 105 I=1,NE
0044          READ(5,1010)J,NODE(J,1),NODE(J,2),NODE(J,3)
0045     105 CONTINUE
C
C     CYCLE FOR EACH NODE HAVING SPECIFIED TEMPERATURE
0046          DO 110 I=1,NA
C
C     READ SPECIFIED TEMPERATURES AND NODE NUMBERS INTO A LIST
0047          READ(5,1015)WORD,NT,TNT
0048          IF (WORD.EQ.STOP)GOTO 120
CC49          NTS(I)=NT
0050          T(NTS(I))=TNT
0051     110 CONTINUE
C
C     COUNT NODES HAVING SPECIFIED TEMPERATURE
0052     120 NNST=I-1
C
```

```
C
C     READ THE VALUE OF THERMAL CONDUCTIVITY AND THE VALUE OF INTERNAL
C     HEAT GENERATION FOR EACH ELEMENT
0053        DO 125 I=1,NE
0054        READ(5,1020)J,TK(J),GQ(J)
0055  125 CONTINUE
C
C     READ NODE NUMBERS AND HEAT FLUX VALUES FOR
C     NODES WHERE HEAT FLUX IS SPECIFIED
0056        DO 130 I=1,NN
0057        READ(5,1025)WORD,NQ,QNQ
0058        IF(WORD.EQ.STOP)GO TO 135
0059        NQS(I)=NQ
0060        Q(NQS(I))=QNQ
0061  130 CONTINUE
C
C     COUNT EXTERNAL NODES WHERE HEAT FLUX IS SPECIFIED
0062  135 NNQS=I-1
C
C     READ PAIRS OF NODE NUMBERS THAT DEFINE BOUNDARY
C     SEGMENTS WHERE A CONVECTIVE HEAT TRANSFER
C     COEFFICIENT,H,AND AN AMBIENT TEMPERATURE,TINF,ARE
C     SPECIFIED
0063        DO 140 I=1,NN
0064        READ(5,1030)WORD,NC1(I),NC2(I),H(I),TINF(I)
0065        IF(WORD.EQ.STOP)GO TO 145
0066  140 CONTINUE
C
C     COUNT NO.OF BOUNDARY SEGMENTS WHERE HEAT CONVECTION OCCURS
0067  145 NNHC=I-1
```

```
FORTRAN IV G1        RELEASE 2.0              MAIN

                    C
                    C     PRINT ALL DATA THAT HAS BEEN READ IN
0068                      WRITE(6,1035)NN,NE
0069                      WRITE(6,1040)
0070                      DO 150 I=1,NN
0071                      WRITE(6,1045)I,XC(I),YC(I)
0072                150   CONTINUE
0073                      WRITE(6,1050)
0074                      DO 155 I=1,NE
0075                      WRITE(6,1055)I,NODE(I,1),NODE(I,2),NODE(I,3)
0076                155   CONTINUE
0077                      WRITE(6,1060)
0078                      DO 160 I=1,NNST
0079                      WRITE(6,1065)I,NTS(I),T(NTS(I))
0080                160   CONTINUE
0081                      WRITE(6,1070)
0082                      DO 165 I=1,NE
0083                      WRITE(6,1075)I,TK(I),QQ(I)
0084                165   CONTINUE
0085                      WRITE(6,1080)
0086                      DO 170 I=1,NNQS
0087                      WRITE(6,1085)I,NQS(I),Q(NQS(I))
0088                170   CONTINUE
0089                      WRITE(6,1090)
0090                      IF(NNHC.EQ.0) GO TO 176
0091                      DO 175 I=1,NNHC
0092                      WRITE(6,1095)I,NC1(I),NC2(I),H(I),TINF(I)
0093                175   CONTINUE
```

```
0094          GO TO 177
0095      176 WRITE(6,1056)
0096      177 CONTINUE
      C
      C CYCLE FOR EACH ELEMENT AND FORM SYSTEM MATRICES
      C
0097          DO 300 K=1,NE
      C K IS NOW THE ELEMENT NUMBER
0098          N1=NODE(K,1)
0099          N2=NODE(K,2)
0100          N3=NODE(K,3)
0101          XC$(1)=XC(NODE(K,1))
0102          XC$(2)=XC(NODE(K,2))
0103          XC$(3)=XC(NODE(K,3))
0104          YC$(1)=YC(NODE(K,1))
0105          YC$(2)=YC(NODE(K,2))
0106          YC$(3)=YC(NODE(K,3))
0107          ZBAR=(YC$(1)+YC$(2)+YC$(3))/3.
0108          RBAR=(XC$(1)+XC$(2)+XC$(3))/3.
0109          A=1.0
0110          IF(NCASE.EQ.2)A=RBAR
0111          AA=1.0
0112          IF(NCASE.EQ.2)AA=2.*3.1415926*(RBAR)
0113          A1=XC$(2)*YC$(3)-XC$(3)*YC$(2)
0114          A2=XC$(3)*YC$(1)-XC$(1)*YC$(3)
0115          A3=XC$(1)*YC$(2)-XC$(2)*YC$(1)
0116          B1=YC$(2)-YC$(3)
0117          B2=YC$(3)-YC$(1)
0118          B3=YC$(1)-YC$(2)
0119          C1=XC$(3)-XC$(2)
0120          C2=XC$(1)-XC$(3)
```

389

```
0121              C3=XC$(2)-XC$(1)
0122              DFL=ABS(0.5*(XC$(1)*(YC$(2)-YC$(3))+
             1XC$(2)*(YC$(3)-YC$(1))+
             2XC$(3)*(YC$(1)-YC$(2))))
       C
       C      FORM INFLUENCE MATRIX FOR TEMPERATURE
0123              CONST=(TK(K)*A/(4.0*DFL))*AA
0124              TM$(1,1)=(B1**2+C1**2)*CONST
0125              TM$(1,2)=(B1*B2+C1*C2)*CONST
0126              TM$(1,3)=(B1*B3+C1*C3)*CONST
0127              TM$(2,1)=TM$(1,2)
0128              TM$(2,2)=(B2**2+C2**2)*CONST
0129              TM$(2,3)=(B2*B3+C2*C3)*CONST
0130              TM$(3,1)=TM$(1,3)
0131              TM$(3,2)=TM$(2,3)
0132              TM$(3,3)=(B3**2+C3**2)*CONST
       C
       C      FORM INFLUENCE MATRIX FOR HEAT GENERATION
0133              QQ$1=QQ(K)
0134              QQ$2=QQ(K)
0135              QQ$3=QQ(K)
0136              CONSTQ=DFL/12.0
0137              FQQ$(1)=(2.*QQ$1+QQ$2+QQ$3)*CONSTQ
0138              FQQ$(2)=(QQ$1+2.*QQ$2+QQ$3)*CONSTQ
0139              FQQ$(3)=(QQ$1+QQ$2+2.*QQ$3)*CONSTQ
0140              IF(NCASE.EQ.1)GO TO 185
0141              RP$(1)=RBAR-XC$(1)
0142              RP$(2)=RBAR-XC$(2)
0143              RP$(3)=RBAR-XC$(3)
0144              ZP$(1)=ZBAR-YC$(1)
```

```
0145          ZP$(2)=ZBAR-YC$(2)
0146          ZP$(3)=ZBAR-YC$(3)
0147          FACT1=(RP$(1)**2+RP$(2)**2+RP$(3)**2)/12.0
0148          FACT2=(RP$(1)*ZP$(1)+RP$(2)*ZP$(2)+RP$(3)*ZP$(3))/12.0
0149          FACT3=RBAR*QQ(K)*3.14159
0150          FQQ$(1)=FACT3*(A1*RBAR+B1*(RBAR**2)
             1+R1*FACT1+C1*RBAR*ZBAR+C1*FACT2)
0151          FQQ$(2)=FACT3*(A2*RBAR+R2*(RBAR**2)+B2*FACT1
             1+C2*RBAR*ZBAR+C2*FACT2)
0152          FQQ$(3)=FACT3*(A3*RBAR+B3*(RBAR**2)+B3*FACT1
             1+C3*RBAR*ZBAR+C3*FACT2)
0153      185 CONTINUE
0154          N(1)=N1
0155          N(2)=N2
0156          N(3)=N3
      C
      C   ASSEMBLE SYSTEM EQUATIONS WITHOUT REGARDING
      C   BOUNDARY CONDITIONS
0157          DO 200 I$=1,3
0158          I=N(I$)
0159          FQQ(I)=FQQ(I)+FQQ$(I$)
0160          DO 200 J$=1,3
0161          J=N(J$)
0162          TM(I,J)=TM(I,J)+TM$(I$,J$)
0163      200 CONTINUE
      C   RECYCLE FOR NEXT ELEMENT
0164      300 CONTINUE
0165          DO 305 I=1,NN
```

391

```
0166              RHS(I)=FQQ(I)
0167          305 CONTINUE
      C
      C     MODIFY SYSTEM EQUATIONS TO INCLUDE BOUNDARY
      C     PHENOMENA
      C     CHECK TO SEE IF ANY BOUNDARY FLUX IS SPECIFIED
0168              IF(NNQS.EQ.0)GO TO 310
      C
      C     INSERT BOUNDARY HEAT FLUX
0169              DO 310 I=1,NNQS
0170              RHS(NQS(I))=RHS(NQS(I))-Q(NQS(I))
0171          310 CONTINUE
0172              IF(NNHC.EQ.0)GO TO 320
      C
      C     ACCOUNT FOR CONVECTION ON BOUNDARY
0173              DO 315 I=1,NNHC
0174              BL(I)=SQRT((XC(NC1(I))-XC(NC2(I)))**2
                 1+(YC(NC1(I))-YC(NC2(I)))**2)
0175              RBL(I)=2.*3.14159*((0.5*(XC(NC1(I))+XC(NC2(I)))))**2
0176              IF(NCASE.EQ.2)BL(I)=RBL(I)*BL(I)
0177              CTINF=H(I)*TINF(I)*BL(I)/2.0
0178              RHS(NC1(I))=RHS(NC1(I))+CTINF
0179              RHST(NC1(I))=RHST(NC1(I))+CTINF
0180              RHS(NC2(I))=RHS(NC2(I))+CTINF
0181              RHST(NC2(I))=RHST(NC2(I))+CTINF
0182          315 CONTINUE
```

```
0183            DO 320 I=1,NNHC
0184            CONSTC=H(I)*BL(I)
0185            TM(NC1(I),NC1(I))=TM(NC1(  ,NC1(I))+CONSTC/3.0
0186            TM(NC1(I),NC2(I))=TM(NC1(  ,NC2(I))+CONSTC/6.0
0187            TM(NC2(I),NC1(I))=TM(NC2(I ,NC1(I))+CONSTC/3.0
0188            TM(NC2(I),NC1(I))=TM(NC2(  ,NC1(I))+CONSTC/6.0
0189        320 CONTINUE
            C
            C
0190        C   NOW THE SYSTEM EQUATION OF THE FORM AX=C IS READY
0191        C   TO BE SOLVED FOR THE UNKNOWN TEMPERATURES
0192            CALL SOLMIX(NN,TM,T,RHS,NMET,NTS,NNQS,B,Z,C)
0193            WRITE(6,2000)
0194            DO 325 I=1,NN
0195            Q(I)=FQQ(I)-RHS(I)+RHST(I)
0196            WRITE(6,2005)I,T(I),Q(I)
0197        325 CONTINUE
0198            TTOT=0.0
0199            DO 330 I=1,NN
0200        330 TTOT=TTOT+T(I)
C199            TNN=NN
0200            TAVG=TTOT/TNN
0201            WRITE(6,2010)TAVG
            C
            C
0202        C   FORMAT STATEMENTS
0203        500 FORMAT(I10)
0204       1005 FORMAT(2I10)
0205       1006 FORMAT(I10,2F10.0)
           1010 FORMAT(4I10)
```

```
0206   1015 FORMAT(6X,A4,I10,F10.0)
0207   1016 FORMAT(6X,A4,I10)
0208   1020 FORMAT(I10,2F10.0)
0209   1025 FORMAT(6X,A4,I10,F10.0)
0210   1030 FORMAT(6X,A4,2I10,2F10.0)
0211   1035 FORMAT(5X,'NO. OF NODES=',I3,//,5X,'NO.OF ELEMENTS=',I3,//)
0212   1040 FORMAT(5X,'SUMMARY OF NODAL COORDINATES',//,
              17X,'I',12X,'X(I)',12X,'Y(I)',//)
0213   1045 FORMAT(5X,I3,2(7X,F10.3))
0214   1050 FORMAT(5X,'LISTING OF SYSTEM TOPOLOGY',//,5X,
              1'ELEMENT NUMBER',20X,'NODE NUMBERS',//)
0215   1055 FORMAT(5X,I3,10X,3(5X,I3))
0216   1060 FORMAT(7X,'NODES WHERE TEMPERATURE IS SPECIFIED',
              1//,5X,'I',5X,'NODE',5X,'TEMPERATURE',//)
0217   1065 FORMAT(2X,2(4X,I3),3X,F12.3)
0218   1070 FORMAT(5X,'SPECIFIED THERMAL CONDUCTIVITY AND INTERNAL HEAT GENERA
              1TION',//,2X,'ELEMENT NO.',5X,'THERMAL CONDUCTIVITY',5X,'INTERNAL H
              2EAT GENERATION',//)
0219   1075 FORMAT(5X,I3,2(10X,F12.3))
0220   1080 FORMAT(5X,'
              1          //,5X,'I',5X,'NODE',10X,'HEAT FLUX',//)  NODES WHERE HEAT FLUX IS SPECIFIED',
0221   1085 FORMAT(5X,I3,3X,I3,10X,F12.3)
0222   1090 FORMAT(5X,'BOUNDARY SEGMENTS WHERE CONVECTIVE HEAT FLUX OCCURS',
              1//,5X,'I',10X,'NODE NO. PAIRS',10X,'H',10X,'AMBIENT TEMP.',//)
0223   1095 FORMAT(5X,3(3X,I3),2(5X,F12.3))
0224   1096 FORMAT(5X,'  NO BOUNDARY HEAT CONVECTION IS SPECIFIED',//)
0225   2000 FORMAT(5X,'NODE NO.',9X,'T(I)',10X,'Q(I)',//)
0226   2005 FORMAT(9X,I3,2(5X,E15.5),//)
0227   2010 FORMAT(5X,'AVERAGE BODY TEMPERATURE=',E15.5,//)
```

```
0228   2015 FORMAT(1H1,5X,'THIS IS A 3-D AXISYMMETRIC PROBLEM',//)
0229   2020 FORMAT(1H1,5X,'THIS IS A 2-D PROBLEM',//)
0230        STOP
0231        END
```

```
FORTRAN IV G LEVEL    19              SOLMIX

0001          SUBROUTINE SOLMIX( N,A,X,Y,NX,LISTX,NY,LISTY,B,Z,C)
0002          DIMENSION A(N,N),X(N),Y(N),LISTX(N),LISTY(N)
0003          DIMENSION B(NY,NY),Z(NY),C(NY)
       C  SUBROUTINE SOLMIX SOLVES THE MIXED PROBLEM
       C          A(I,J)*X(J) = Y(I)          I,J = 1,N
       C  WHERE LISTX(I) AND LISTY(I) ARE MUTUALLY EXCLUSIVE LISTS
       C  WHERE LISTX(I) OF LENGTH NX CONTAINS INDICES OF KNOWN X VALUES
       C  WHERE LISTY(I) OF LENGTH NY CONTAINS INDICES OF KNOWN Y VALUES
       C  SUCH THAT NX + NY = N
       C
       C  CHECK FOR IMPROPER PROBLEM
0004          IF (N.EQ.(NX+NY)) GO TO 100
0005          WRITE (6,111) N,NX,NY
0006   111    FORMAT(10X,'N=',I2,10X,'NX=',I2,10X,'NY=',I2)
0007          STOP
0008   100    CONTINUE
       C
       C  CHECK FOR STANDARD PROBLEM WITH ALL Y VALUES KNOWN
0009          IF (NX.NE.0) GO TO 120
0010          CALL SOLVE(N,A,Y,X)
0011          RETURN
0012   120    CONTINUE
       C
       C  CHECK FOR STANDARD PROBLEM WITH ALL X VALUES KNOWN
0013          IF (NY.NE.0) GO TO 140
0014          DO 130 I = 1,N
0015          Y(I) = 0.
0016          DO 130 J = 1,N
0017   130    Y(I) = Y(I) + A(I,J)*X(J)
```

```
0018          RETURN
0019   140 CONTINUE
       C
       C FORM STANDARD PROBLEM IN NY UNKNOWN X VALUES
       C          B(I,J)*Z(J) = C(I)          I,J = 1,NY
              DO 200 I = 1,NY
              DO 200 J = 1,NY
0020          DO 200 I = 1,NY
0021          DO 200 J = 1,NY
0022   200 B(I,J) = A(LISTY(I),LISTY(J))
0023          DO 300 I = 1,NY
0024          C(I) = Y(LISTY(I))
0025          DO 300 J = 1,NX
0026   300 C(I) = C(I) - A(LISTY(I),LISTX(J))*X(LISTX(J))
       C FIND UNKNOWN X(I)
0027          CALL SOLVE(NY,B,C,Z)
0028          DO 400 I = 1,NY
0029   400 X(LISTY(I)) = Z(I)
       C FIND UNKNOWN Y(I)
0030          DO 800 I = 1,NX
0031          C(I) = 0.
0032          DO 600 J = 1,NY
0033   600 C(I) = C(I) + A(LISTX(I),LISTY(J))*X(LISTY(J))
0034          DO 700 J = 1,NX
0035   700 C(I) = C(I) + A(LISTX(I),LISTX(J))*X(LISTX(J))
0036   800 Y(LISTX(I)) = C(I)
0037          RETURN
0038          END
```

```
0001          SUBROUTINE SOLVE(N,A,C,X)
0002          DIMENSION A(N,N),X(N),C(N)
       C SUBROUTINE SOLVES BY GAUSSIAN ELIMINATION THE N LINEAR EQUATIONS
       C A(I,J)*X(J) = C(I)   (SUM ON J = 1,N)   (FOR I = 1,N)
       C MORE POWERFUL VERSIONS WOULD INCLUDE
       C BUFFERS TO AVOID CHANGING A(I,J) AND C(I)
       C
       C SELECT KTH ROW AS 'PIVOT'
0003          DO 400 K = 1,N
       C FIND LARGEST |A(I,K)| FOR I = K,N
0004          BIG = ABS(A(K,K))
0005          IBIG = K
0006          DO 100 I = K,N
0007          SIZE = ABS(A(I,K))
0008          IF (SIZE.LT.BIG) GO TO 100
0009          BIG = SIZE
0010          IBIG = I
0011      100 CONTINUE
       C SWAP ROWS SO |A(K,K)| IS BIGGEST
0012          IF (K.EQ.IBIG) GO TO 280
0013          DO 200 J = K,N
0014          ABIG = A(IBIG,J)
0015          A(IBIG,J) = A(K,J)
0016      200 A(K,J) = ABIG
0017          CBIG = C(IBIG)
0018          C(IBIG) = C(K)
0019          C(K) = CBIG
0020      280 CONTINUE
       C CHECK FOR NULL PIVOT
```

```
0021          IF (A(K,K).EQ.0.) GO TO 600
      C DECOUPLE SYSTEM BY SUCCESSIVE SUBTRACTION
      C OF FRACTIONS OF K-TH ROW FROM ALL OTHERS
0022          DO 400 I = 1, N
0023          IF (I.EQ.K) GO TO 400
0024          RATIO = A(I,K)/A(K,K)
0025          DO 300 J = K, N
0026  300     A(I,J) = A(I,J) -RATIO*A(K,J)
0027          C(I)  = C(I)  -RATIO*C(K)
0028  400 CONTINUE
      C SOLVE DECOUPLED SYSTEM
0029          DO 500 K = 1, N
0030  500     X(K) = C(K)/A(K,K)
0031          RETURN
      C ARRANGE ABORT
0032  600     WRITE (6,666)
0033  666     FORMAT(10X,'SINGULAR MATRIX')
0034          DO 700 I = 1, N
0035  700     X(I) = 0.
0036          RETURN
0037          END
```

399

10.4   EXAMPLE PROBLEMS WITH INPUT AND OUTPUT

Example 1.   One-Dimensional Heat Flow in a Bar

The problem of heat flow along an insulated bar of unit thickness is a simple example to illustrate the use of this program. One end of the bar is held at 1000°F, while the other end is maintained at 0°F. What are the temperature distribution and heat flux in the bar?

Solution. The bar is shown in Figure 10.4.

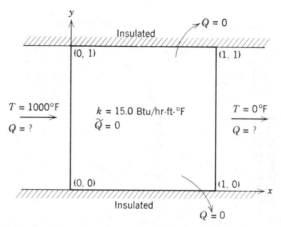

Figure 10.4.   One-dimensional heat flow problem.

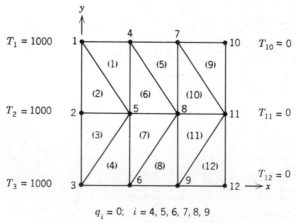

$$q_i = 0; \quad i = 4, 5, 6, 7, 8, 9$$

Figure 10.5.   Finite element mesh for the problem of Figure 10.4.

Although we can write the solution to this problem at once,† let us use the program to obtain a finite element solution. Suppose that we decide to use the finite element model given in Figure 10.5. The specified nodal temperatures and heat fluxes are as illustrated. There is no internal heat generation or boundary heat convection.

After numbering the nodes and elements as indicated, we prepare the input coding sheets as shown on pp. 402–404. The fact that the number of nodes equals the number of elements is fortuitous here.

Inspection of the output pp. 405–406 shows that the finite element solution for this linear problem is exact. This result is to be expected because, in this case, the linear approximating functions used in the finite element model exactly represent actual behavior.

Example 2. Convection from a Blunt Fin

Consider the problem of heat convection from a blunt fin whose dimensions are the same as those of the bar in Example 1. Suppose that the fin is attached to a wall (maintained at 1000°F) while it is exposed to an environment whose ambient temperature is 1500°F. Heat transfer is expected from the environment to the wall through the fin. What is the temperature distribution in the fin and the resultant heat flux?

Solution. The conditions of the problem are shown schematically in Figure 10.6.

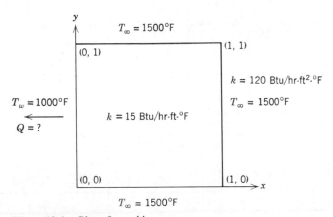

Figure 10.6. Blunt fin problem.

† The exact solution is as follows: $T(x) = 1000(1 - x)$, $Q = AkT = 15,000$ Btu/hr.

Coding form. Column guide (character positions): Name (1), Operation (8–14), Operand (16–30), Comments (35–71).

Name	Operation	Operand			Comments
	1 2		1 2		
	1	0.0		1.000	
	2	0.0		0.500	
	3	0.0		0.0	
	4	0.333		1.0	
	5	0.333		0.500	
	6	0.333		0.0	
	7	0.667		1.000	
	8	0.667		0.500	
	9	0.667		0.0	
	10	1.0		1.000	
	11	1.0		0.500	
	12	1.0		0.0	
	1		1	5	4
	2		1	2	5
	3		2	3	5
	4		3	6	5
	5		4	8	7
	6		4	5	8
	7		5	6	8
	8		6	9	8
	9		7	1 1	1 0
	10		7	8	1 1

402

Name		Operation			Operand			Comments						
1	8	10	14	16	20	22	30	35	40	45	50	55	60	65	71
	1 1	1 1			8		5		1 1					
	1 2	1 2			9		1 2		1 1					
					1	1 0 0 0 .								
					2	1 0 0 0 .								
					3	1 0 0 0 .								
					1 0	0 . 0								
					1 1	0 . 0								
					1 2	0 . 0								
	S T O P													
		1	1 5 .			0 . 0								
		2	1 5 .			0 . 0								
		3	1 5 .			0 . 0								
		4	1 5 .			0 . 0								
		5	1 5 .			0 . 0								
		6	1 5 .			0 . 0								
		7	1 5 .			0 . 0								
		8	1 5 .			0 . 0								
		9	1 5 .			0 . 0								
		1 0	1 5 .			0 . 0								
		1 1	1 5 .			0 . 0								
		1 2	1 5 .			0 . 0								
					4	0 . 0								
					5	0 . 0								
					6	0 . 0								

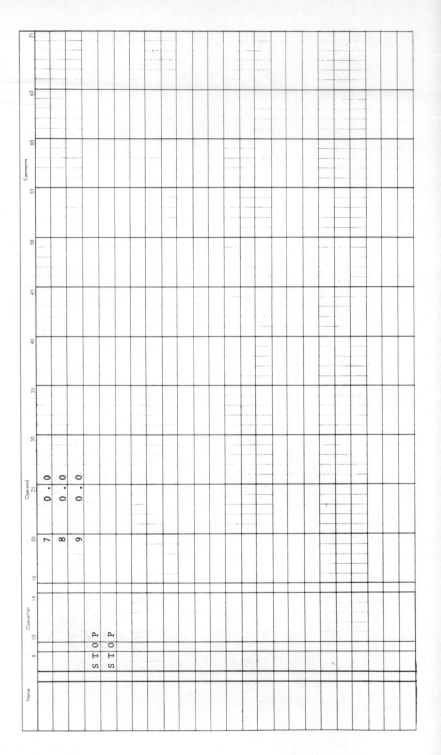

THIS IS A 2-D PROBLEM

NO. OF NODES= 12

NO. OF ELEMENTS= 12

SUMMARY OF NODAL COORDINATES

I	X(I)	Y(I)
1	0.0	1.000
2	0.0	0.500
3	0.0	0.0
4	0.333	1.000
5	0.333	0.500
6	0.333	0.0
7	0.667	1.000
8	0.667	0.500
9	0.667	0.0
10	1.000	1.000
11	1.000	0.500
12	1.000	0.0

LISTING OF SYSTEM TOPOLOGY

ELEMENT NUMBER   NODE NUMBERS

1	1	5	4
2	1	2	5
3	2	3	5
4	3	6	5
5	4	8	7
6	4	5	8
7	5	6	8
8	6	9	8
9	7	11	10
10	7	8	11
11	8	9	11
12	9	12	11

NODES WHERE TEMPERATURE IS SPECIFIED

I	NODE	TEMPERATURE
1	1	1000.000
2	2	1000.000
3	3	1000.000
4	10	0.0
5	11	0.0
6	12	0.0

SPECIFIED THERMAL CONDUCTIVITY AND INTERNAL HEAT GENERATION

ELEMENT NO.	THERMAL CONDUCTIVITY	INTERNAL HEAT GENERATION
1	15.000	0.0
2	15.000	0.0
3	15.000	0.0
4	15.000	0.0
5	15.000	0.0
6	15.000	0.0
7	15.000	0.0
8	15.000	0.0
9	15.000	0.0
10	15.000	0.0
11	15.000	0.0
12	15.000	0.0

NODES WHERE HEAT FLUX IS SPECIFIED

I	NODE	HEAT FLUX
1	4	0.0
2	5	0.0
3	6	0.0
4	7	0.0
5	8	0.0
6	9	0.0

BOUNDARY SEGMENTS WHERE CONVECTIVE HEAT FLUX OCCURS

I	NODE NO. PAIRS	H	AMBIENT TEMP.

NO BOUNDARY HEAT CONVECTION IS SPECIFIED

NODE NO.	T(I)	Q(I)
1	0.10000E 04	-0.37500E 04
2	0.10000E 04	-0.75000E 04
3	0.10000E 04	-0.37500E 04
4	0.66700E 03	0.0
5	0.66700E 03	0.0
6	0.66700E 03	0.0
7	0.33300E 03	0.0
8	0.33300E 03	0.0
9	0.33300E 03	0.0
10	0.0	0.37500E 04
11	0.0	0.75000E 04
12	0.0	0.37500E 04

AVERAGE BODY TEMPERATURE= 0.50000E 03

An approximate solution could be obtained for this problem by assuming that the temperature in the fin depends only on x. Under this assumption, the governing differential equation is

$$\frac{d^2T}{dx^2} = \frac{2h}{k}(T - T_\infty) = \alpha^2(T - T_\infty), \qquad \alpha^2 = \frac{2h}{k} \qquad (10.16a)$$

with boundary conditions

$$T(0) = T_w \qquad (10.16b)$$

$$-k \left.\frac{dT}{dx}\right|_{x=1} = h[T(1) - T_\infty] \qquad (10.16c)$$

The solution to equations 10.16 is

$$\frac{T - T_\infty}{T_w - T_\infty} = \frac{\cosh \alpha(1 - x) + (\alpha/2)\sinh \alpha(1 - x)}{\cosh \alpha + (\alpha/2)\sinh \alpha} \qquad (10.17)$$

Unfortunately, for the blunt fin in this problem, the one-dimensional assumption is not good; however, we can use it for purposes of comparison. To obtain a two-dimensional finite element solution to this problem, let us use the same model as for Example 1. Only the boundary conditions will be different. Thus the finite element model for this problem has the form sketched in Figure 10.7. The coded input data and the resulting output are

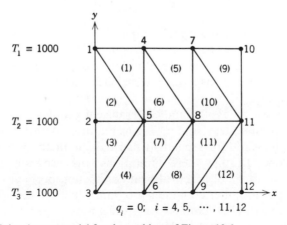

Figure 10.7.   Finite element model for the problem of Figure 10.6.

given on pp. 409–414. Figure 10.8 shows how the finite element solution compares with the approximate one-dimensional solution for the temperature distribution. Note that this simple finite element model gives an indicated of surface temperatures as well as interior temperatures.

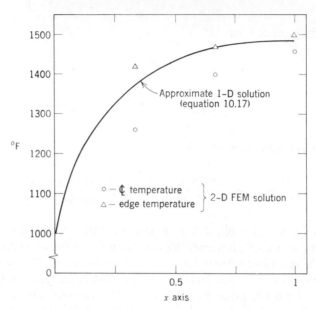

Figure 10.8. Comparison of the finite element solution and an approximate one-dimensional solution for the temperature distribution in a blunt fin.

    These example problems, though simple, illustrate all the essential features of the program, and they demonstrate its reliability. Diversity, flexibility, and simplicity are the program's most useful features. Problems with nonuniform thermal conductivity, distributed internal heat generation, boundary heat convection, and arbitrary specification of boundary or interior temperatures and heat fluxes are all easily handled. Transient problems requiring successive solutions in time steps, and problems requiring iterative solutions such as those with temperature-dependent thermal conductivity, are beyond the program's scope. Despite these limitations, the program can solve an extremely broad range of practical heat conduction problems.

    An example of a similar computer program which solves two- and three-dimensional transient heat conduction problems by using linear triangular or tetrahedral elements is given by Heuser [4].

Operation (10)	(14)	(16)	(20)	Operand (25)	(30)	(40)
1 2			1 2			
1	0.	.0		1 .	0 0 0	
2	0.	.0		0 .	5 0 0	
3	0.	.0		0 .	0	
4		.3 3 3		1 .	0 0 0	
5		.3 3 3		0 .	5 0 0	
6		.3 3 3		0 .	0	
7		.6 6 7		1 .	0 0 0	
8		.6 6 7		0 .	5 0 0	
9		.6 6 7		0 .	0	
1 0	1.	.0		1 .	0 0 0	
1 1	1.	.0		0 .	5 0 0	
1 2	1.	.0		0 .	0	
1			1		5	4
2			1		2	5
3			2		3	5
4			3		6	5
5			4		8	7
6			4		5	3
7			5		6	3
8			6		9	3
9			7		1 1	1 0
1 0			7		8	1

409

410

Name	Operation		Operand		Comments
	1 1		8	9	1 1
	1 2		9	1 2	1 1
			1	1 0 0 0 .	
			2	1 0 0 0 .	
			3	1 0 0 0 .	
S T O P					
	1	1 5 .		0 . 0	
	2	1 5 .		0 . 0	
	3	1 5 .		0 . 0	
	4	1 5 .		0 . 0	
	5	1 5 .		0 . 0	
	6	1 5 .		0 . 0	
	7	1 5 .		0 . 0	
	8	1 5 .		0 . 0	
	9	1 5 .		0 . 0	
	1 0	1 5 .		0 . 0	
	1 1	1 5 .	4	0 . 0	
	1 2	1 5 .	5	0 . 0	
			6	0 . 0	
			7	0 . 0	
			8	0 . 0	
			9	0 . 0	

Name	Operation		Operand						Comments
			10	0.0					
			11	0.0					
			12	0.0					
	STOP								
			3		6	120.		1500.	
			6		9	120.		1500.	
			9		12	120.		1500.	
			12		11	120.		1500.	
			11		10	120.		1500.	
			10		7	120.		1500.	
			7		4	120.		1500.	
			4		1	120.		1500.	
	STOP								

411

```
THIS IS A 2-D PROBLEM

NO. OF NODES= 12

NO.OF ELEMENTS= 12

SUMMARY OF NODAL COORDINATES

   I                X(I)                Y(I)


   1                0.0                 1.000
   2                0.0                 0.500
   3                0.0                 0.0
   4                0.333               1.000
   5                0.333               0.500
   6                0.333               0.0
   7                0.667               1.000
   8                0.667               0.500
   9                0.667               0.0
  10                1.000               1.000
  11                1.000               0.500
  12                1.000               0.0
LISTING OF SYSTEM TOPOLOGY

ELEMENT NUMBER                             NODE NUMBERS


   1                1        5        4
   2                1        2        5
   3                2        3        5
   4                3        6        5
   5                4        8        7
   6                4        5        8
   7                5        6        8
   8                6        9        8
   9                7       11       10
  10                7        8       11
  11                8        9       11
  12                9       12       11
 NODES WHERE TEMPERATURE IS SPECIFIED

 I      NODE        TEMPERATURE


   1       1         1000.000
   2       2         1000.000
   3       3         1000.000
```

SPECIFIED THERMAL CONDUCTIVITY AND INTERNAL HEAT GENERATION

ELEMENT NO.	THERMAL CONDUCTIVITY	INTERNAL HEAT GENERATION
1	15.000	0.0
2	15.000	0.0
3	15.000	0.0
4	15.000	0.0
5	15.000	0.0
6	15.000	0.0
7	15.000	0.0
8	15.000	0.0
9	15.000	0.0
10	15.000	0.0
11	15.000	0.0
12	15.000	0.0

NODES WHERE HEAT FLUX IS SPECIFIED

I	NODE	HEAT FLUX
1	4	0.0
2	5	0.0
3	6	0.0
4	7	0.0
5	8	0.0
6	9	0.0
7	10	0.0
8	11	0.0
9	12	0.0

BOUNDARY SEGMENTS WHERE CONVECTIVE HEAT FLUX OCCURS

I	NODE NO. PAIRS		H	AMBIENT TEMP.
1	3	6	120.000	1500.000
2	6	9	120.000	1500.000
3	9	12	120.000	1500.000
4	12	11	120.000	1500.000
5	11	10	120.000	1500.000
6	10	7	120.000	1500.000
7	7	4	120.000	1500.000
8	4	1	120.000	1500.000

NODE NO. I	T(I)	Q(I)
1	0.10000E 04	0.11924E 05
2	0.10000F 04	0.60469F 04
3	0.10000F 04	0.11924E 05
4	0.14202E 04	0.0
5	C.12685F 04	0.0
6	0.14202E 04	0.0
7	0.14775E 04	0.0
8	0.14025F 04	0.0
9	0.14775F 04	0.0
10	0.15010E 04	0.0
11	0.14696F 04	0.0
12	0.15010E 04	0.0

AVERAGE BODY TEMPERATURE= 0.13282E 04

10.5 AUTOMATIC MESH GENERATION

A cumbersome aspect of obtaining a finite element solution to a problem is the preparation of the input data. Most of the input data consist of a description of the element mesh topology. We must, by some means, provide the computer program with the node numbers and the coordinates of node points along with element numbers and the node numbers associated with each element. When our finite element mesh contains hundreds or even thousands of nodes, this task takes on major proportions. If all of these data had to be prepared by hand and input via cards, as was done for the sample program in Section 10.3, the job would be very tedious and time-consuming indeed. And, if the input data contain errors, these are, of course, reflected in an erroneous solution which is simply a waste of expensive computer time.

Because of this situation, many serious users of the finite element method have developed or adopted automatic or semiautomatic means of generating the system topology. All of these are labor-saving devices, but none is completely general. They are usually designed to produce meshes containing only one kind of element. For example, they could be intended for the automatic triangulation of any multiply connected planar domain, or they might generate only isoparametric quadrilateral elements in a solution domain. Consequently, many different special-purpose mesh generators are currently being used.

A description and survey of the variety of mesh generation methods and their algorithms are far beyond our purpose here. Instead, we merely suggest a number of references where various schemes are presented, and we give an example of what some mesh generation schemes can do. Samples of the different kinds of approaches to developing mesh generation schemes can be found in refs 5–10. Buell and Bush [11] provide a survey of many of the useful mesh generation schemes, and their reference list is a good source of additional information on the subject.

Mesh generation always involves some engineering judgment to decide on the effective placement of the nodes and elements within the solution region. Hence an automatic mesh generation scheme usually requires the user to specify some information about the mesh he wants. Often he selects zones in which a certain node-placement density is desired, and he specifies some of the important boundary node locations. Then the mesh generator places additional nodes on the zone boundaries and within the zones and assembles a consistent network of elements from them. The only other data that the user needs to specify are those defining the connectivity of the zones. Figure 10.9 illustrates the general idea.

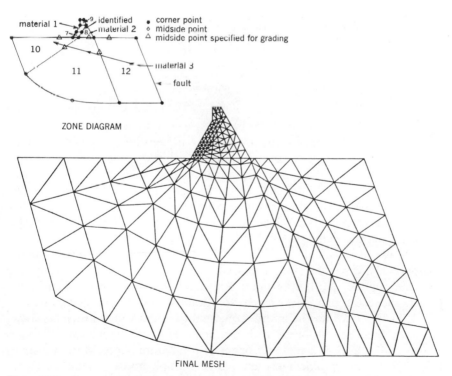

Figure 10.9. Examples of automatic mesh generation [7]. (*a*) Dam on an earth foundation.

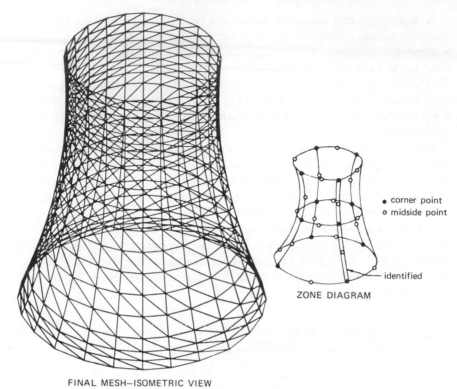

• corner point
○ midside point

— identified

ZONE DIAGRAM

FINAL MESH—ISOMETRIC VIEW

Figure 10.9 *(continued)*.   (*b*)   Hyperbolic cooling tower.

When a problem calls for the use of three-dimensional elements (such as tetrahedra or rectangular prisms), automatic mesh generation is almost mandatory because the spatial visualization of a body divided into such elements is appallingly difficult. Unfortunately, few general schemes have been developed for three-dimensional problems.

Computer graphics facilities are also an invaluable aid to the finite element analyst. After the element mesh has been generated (either by hand or automatically), it is most helpful to have the mesh graphically displayed by the computer. Then a quick visual check is all that is needed to detect unwanted errors. Computer plotting of the output data is also most useful in interpreting the results of a finite element solution. In our simple heat conduction problems, we saw that the output consisted of nodal values of temperature and heat flux. For problems with hundreds of nodes, tabulation of these results could result in pages and pages of printed numbers. Moreover, in a more complicated problem where we have multiple nodal variables (such as the velocity components and pressure in a fluid flow problem) or in transient problems involving time histories, the output data become even more voluminous. Interpretation of these output data is easier if computer-driven plotters can be used to produce isometric solution surfaces or contour plots.

10.6 NUMERICAL INTEGRATION FORMULAS

Often the evaluation of the element equations for a particular problem involves the evaluation of integrals over an element. We face the problem of evaluating

$$J(x) = \int f(x)\, dx \qquad (10.18)$$

$$J(x, y) = \iint f(x, y)\, dx\, dy \qquad (10.19)$$

$$J(x, y, z) = \iiint f(x, y, z)\, dx\, dy\, dz \qquad (10.20)$$

where f is a known function, and the region of integration is defined by the element boundaries. As we have seen, it is sometimes possible to obtain closed-form exact expressions for these integrals when natural coordinates are used. But, when the form of f or the shape of the element does not permit closed-form integration, we must resort to numerical integration. A number of effective formulas are available for this task, but we shall mention here just a few of them.

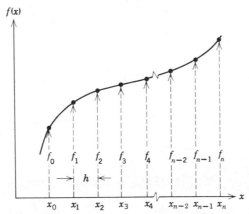

Figure 10.10. Numerical integration of $f(x)$ via the Newton–Cotes formulas. The sample points are equally spaced along the x axis.

We consider first the one-dimensional case—equation 10.18. A class of numerical integration formulas known as *Newton-Cotes (closed) formulas* assumes that the function f is evaluated at equally spaced points along the x axis (Figure 10.10). These integration formulas are derived by integrating exactly a Lagrangian interpolation polynomial of order n fit to $n + 1$ values of $f(x)$. The first eight formulas in the infinite series that can be derived are as follows [12, pp. 885–887]:

$n = 1$:

$$\int_{x_0}^{x_1} f(x)\, dx = \frac{h}{2} (f_0 + f_1) - \frac{h^3}{12} f^{(2)}(\xi) \tag{10.21a}$$

$n = 2$:

$$\int_{x_0}^{x_1} f(x)\, dx = \frac{h}{3} (f_0 + 4f_1 + f_2) - \frac{h^5}{90} f^{(4)}(\xi) \tag{10.21b}$$

$n = 3$:

$$\int_{x_0}^{x_3} f(x)\, dx = \frac{3h}{8} (f_0 + 3f_1 + 3f_2 + f_3) - \frac{3h^5}{80} f^{(4)}(\xi) \tag{10.21c}$$

$n = 4$:

$$\int_{x_0}^{x_4} f(x)\, dx = \frac{2h}{45} (7f_0 + 32f_1 + 12f_2 + 32f_3 + 7f_4) - \frac{8h^7}{945} f^{(6)}(\xi) \tag{10.21d}$$

$n = 5$:

$$\int_{x_0}^{x_5} f(x)\, dx = \frac{5h}{288} (19f_0 + 75f_1 + 50f_2 + 50f_3 + 75f_4 + 19f_5)$$

$$- \frac{275h^7}{12,096} f^{(6)}(\xi) \tag{10.21e}$$

$n = 6$:

$$\int_{x_0}^{x_6} f(x)\, dx = \frac{h}{140} (41f_0 + 216f_1 + 27f_2 + 272f_3 + 27f_4 + 216f_5 + 41f_6)$$

$$- \frac{9h^9}{1400} f^{(8)}(\xi) \tag{10.21f}$$

$n = 7$:

$$\int_{x_0}^{x_7} f(x)\, dx = \frac{7h}{17,280} (751f_0 + 3577f_1 + 1323f_2 + 2989f_3 + 2989f_4$$

$$+ 1323f_5 + 3577f_6 + 751f_7) - \frac{8183h^9}{518,400} f^{(8)}(\xi) \tag{10.21g}$$

$n = 8$:

$$\int_{x_0}^{x_8} f(x)\, dx = \frac{4h}{14{,}175}\, (989f_0 + 5888f_1 - 928f_2 + 10{,}496f_3 - 4540f_4$$

$$+ 10{,}496f_5 - 928f_6 + 5888f_7 + 989f_8) - \frac{2368h^{11}}{467{,}775}\, f^{(10)}(\xi) \quad (10.21h)$$

where $x_0 < \xi < x_n$ and $f^{(k)}$ designates the kth derivative of f.
All of the formulas in this series can be written generally as

$$\int_{x_a}^{x_b} f(x)\, dx = \sum_{i=0}^{n} W_i^x f(x_i) + E_x^n \qquad (10.22)$$

where W_i^x are the weight coefficients and E_x^n is the error, which is of the order of h^n. As an example, for $n = 4$, we have

$$W_0^x = \frac{14h}{45}, \qquad W_1^x = \frac{64h}{45}, \qquad W_2^x = \frac{24h}{45}$$

$$W_3^x = \frac{64h}{45}, \qquad W_4^x = \frac{14h}{45}$$

$$E_x^4 = \frac{-8h^7}{945}\, f^{(6)}(\xi)$$

The Newton-Cotes formulas for one-dimensional integration can be easily extended to evaluate the multiple integrals of equations 10.19 and 10.20 when the integration is taken over a rectangle or a right prism. We can develop formulas for the multiple-dimensional integrals by applying successively the formulas for one-dimensional integration. Suppose that we select equally spaced integration points (n in the x direction, l in the y direction, and m in the z direction). Then, for the double integral, equation 10.19, we can write—from equation 10.22—

$$J(x, y) = \int_{y_a}^{y_b} \int_{x_a}^{x_b} f(x_i, y)\, dx\, dy = \int_{y_a}^{y_b} \left[\sum_{i=0}^{n} W_i^x f(x_i, y) + E_x^n \right] dy$$

$$= \sum_{j=0}^{l} W_j^y \left[\sum_{i=0}^{n} W_i^x f(x_i, y_j) + E_x^n \right] + E_y^l$$

$$= \sum_{j=0}^{l} \sum_{i=0}^{n} W_j^y W_i^x f(x_i, y_j) + E_{xy}^{ln} \qquad (10.23)$$

where E_{xy}^{ln} is the combined integration error, and W_j^y are the weight coefficients for the y direction of integration. Similarly, for the triple integral, equation 10.20, we can write

$$J(x, y, z) = \int_{z_a}^{z_b} \int_{y_a}^{y_b} \int_{x_a}^{x_b} f(x, y, z)$$

$$= \sum_{k=0}^{m} \sum_{j=0}^{l} \sum_{i=0}^{n} W_k^z W_j^y W_i^x f(x_i, y_j, z_k) + E_{xyz}^{mln} \qquad (10.24)$$

Table 10.1. Abscissas and weight factors for Legendre-Gauss integration (from Davis and Rabinowitz [13])

$$\int_{-1}^{1} f(x)\, dx \approx \sum_{i=1}^{n} W_i^x f(x_i)$$

Abscissas $= \pm x_i$ (zeros of Legendre polynomials)
Weight factors $= W_i^x$

$\pm x_i$			W_i^x		
		n = 2			
0.57735	02691	89626	1.00000	00000	00000
		n = 3			
0.00000	00000	00000	0.88888	88888	88889
0.77459	66692	41483	0.55555	55555	55556
		n = 4			
0.33998	10435	84856	0.65214	51548	62546
0.86113	63115	94053	0.34785	48451	37454
		n = 5			
0.00000	00000	00000	0.56888	88888	88889
0.53846	93101	05683	0.47862	86704	99366
0.90617	98459	38664	0.23692	68850	56189
		n = 6			
0.23861	91860	83197	0.46791	39345	72691
0.66120	93864	66265	0.36076	15730	48139
0.93246	95142	03152	0.17132	44923	79170
		n = 7			
0.00000	00000	00000	0.41795	91836	73469
0.40584	51513	77397	0.38183	00505	05119
0.74153	11855	99394	0.27970	53914	89277
0.94910	79123	42759	0.12948	49661	68870
		n = 8			
0.18343	46424	95650	0.36268	37833	78362
0.52553	24099	16329	0.31370	66458	77887
0.79666	64774	13627	0.22238	10344	53374
0.96028	98564	97536	0.10122	85362	90376

Table 10.2. Numerical integration formulas for triangles
(from Felippa [14], after Hammer and Stroud [15])

No.	Order	Figure	Rem.	Points	Triangular Coordinates	Weights W_k
1	Linear		$R = 0(h^2)$	a	1/3, 1/3, 1/3	1
2	Quadratic		$R = 0(h^3)$	a b c	1/2, 1/2, 0 0, 1/2, 1/2 1/2, 0, 1/2	1/3 1/3 1/3
3	Cubic		$R = 0(h^4)$	a b c d	1/3, 1/3, 1/3 11/15, 2/15, 11/15 2/15, 11/15, 2/15 2/15, 2/15, 11/15	$-27/48$ 25/48
4	Cubic		$R = 0(h^4)$	a b c d e f g	1/3, 1/3, 1/3 1/2, 1/2, 0 0, 1/2, 1/2 1/2, 0, 1/2 1, 0, 0 0, 1, 1 0, 0, 1	27/60 8/60 3/60
5	Quintic		$R = 0(h^6)$	a b c d e f g	1/3, 1/3, 1/3 $\alpha_1, \beta_1, \beta_1$ $\beta_1, \alpha_1, \beta_1$ $\beta_1, \beta_2, \beta_2$ $\alpha_2, \beta_2, \beta_2$ $\beta_2, \alpha_2, \beta_2$ $\beta_2, \beta_2, \alpha_2$	0.225 0.13239415 0.12593918

With:
$\alpha_1 = 0.05961587$
$\beta_1 = 0.47014206$
$\alpha_2 = 0.79742699$
$\beta_2 = 0.10128651$

Notes:
1. Use of the cubic formula (3) is not recommended since the negative weight may cause severe cancellation error.
2. When the function $f(L_i)$ is unbounded on a certain portion of the boundary (for instance, on the symmetry axis for axisymmetric ring elements), an expression like (5) with only internal integration points should be used.

Another class of numerical integration formulas, known as *Gaussian type formulas*, does not rely on equally spaced integration, but rather uses abscissas which are the zeros of the particular interpolation polynomial employed. For a given number of abscissas, this presumably leads to better accuracy. If Legendre polynomials are used, the resulting formulas are known as Legendre-Gauss formulas. Integration in one dimension is then accomplished by

$$J(x) = \int_{-1}^{1} f(x)\, dx \approx \sum_{i=1}^{n} W_i f(x_i) \tag{10.25}$$

where the weights W_i and the abscissas x_i are given in Table 10.1. Multiple integration can be handled in the same way as indicated in equations 10.23 and 10.24.

Similar integration formulas have been developed for evaluating area and volume integrals when the regions of integration are triangles or tetrahedra and the integrands are expressed in terms of natural coordinates. For triangles, for example, we have

$$J(L_1, L_2, L_3) = \int_{0}^{1} \int_{0}^{1-L_1} f(L_1, L_2, L_3)\, dL_1\, dL_2 = \Delta \sum_{i=1}^{n} W_i f(L_i) + R \tag{10.26}$$

where Δ is the area of the triangle and L_i are the values of the natural coordinates of the integration points. Table 10.2 gives the values of the coordinates of the integration points and the weights. Equation 10.26 is exact for all polynomials f of order up to $n - 1$. If f is not a polynomial, the error R is of order h^n, where h is a characteristic length of the triangle. Corresponding formulas for tetrahedral volumes may be found in Hammer et al. [16].

10.7 SOME LITERATURE ON EQUATION SOLVERS

A chapter on coding techniques and other practical considerations associated with the finite element method would be remiss if it did not mention the subject of equation-solving procedures. Every finite element analysis ultimately reduces to solving a set of linear or nonlinear algebraic equations of the standard form

$$[K]\{x\} = \{R\}$$

Improving a given finite element analysis often means increasing the number of nodes and the number of degrees of freedom in the solution region and, hence, enlarging the system matrix $[K]$. Consequently, highly accurate finite element analyses usually involve the solution of a large number of equations, and what the analyst desires is an efficient scheme for solving these large-order systems.

Matrix equation-solving schemes can be generally classified as *direct* or *iterative* [17,18]. Gaussian elimination and Cholesky decomposition are examples of direct schemes, while the Gauss-Seidel and the Given methods are two of the many iterative schemes available. The equation solvers that are best for finite element system equations are those that take advantage of the symmetry, sparseness, and bandedness properties that usually characterize $[K]$. By utilizing these properties, computer storage and solution time can be minimized.

Because of the multitude of techniques available for matrix equation solving, it is impractical to attempt any discussion of the subject in this book. Readers who are interested in the details of the various techniques should consult refs. 17–28.

10.8   LARGE-SCALE COMPUTER PROGRAMS

One of the first questions a potential user of the finite element method asks is: "Must I develop my own computer program, or can I adapt one of the existing general-purpose programs?" The answer depends on the nature of the problem that the user is considering.

In new fields of application, such as fluid mechanics, general-purpose programs are not yet available. Some universities and major industries have on-going research to develop finite element computer programs for solving general fluids problems, but at the present time only a few special-purpose programs have emerged. As time goes on, more general programs can be expected because research work in this area is quite active.

For solid mechanics problems, on the other hand, the situation is quite different. The analyst, if he so chooses, does not have to do his own development work because he can call on a number of existing programs. Table 10.3 lists a number of large general-purpose programs for structural analysis. Some of these are available at nominal cost from the government-sponsored COSMIC library at the University of Georgia. The extensive documentation on these programs makes it relatively easy to learn how to use them. Also, the nature of the finite element method is such that potent and sophisticated programs can be employed by persons who are not experts in all the underlying mathematics. However, it should be borne in mind that to use a versatile general-purpose program to solve a specialized problem is often far more costly (in computer time) than to write a program expressly for solving the specialized problem. Therefore, if a particular type of problem is to be solved repeatedly, the analyst should consider writing his own program for the job.

The book by Fenves et al. [30] provides a good starting point for becoming familiar with several large-scale structural analysis programs. In it, several authors have given critical reviews of ASKA, NASTRAN, DAISEY, STARDYNE, STRUDL, and MARC. These reviews are especially helpful to the potential user because they provide overviews on program capability and suitability, as well as examples of analyses performed by the programs.

Using the large-scale programs requires access to high-speed digital computers capable of processing and solving large systems of equations. Analysts employing the finite element method often rely on computers such as the

Table 10.3.   A sample listing of finite element computer programs
(from Huebner [29])

Program	Developer
ANSYS	Swanson Analysis Systems Inc.

ASKA
Automatic System for Kinematic
  Analysis
	Professors J. H. Argyris and
	  H. A. Kamel

ASTRA Advanced Structural Analyzer	The Boeing Co.
BEST 3 Bending Evaluation of Structures	Structural Dynamics Research Corp.
DAISY	Professor H. A. Kamel for Lockheed Missile & Space Co.

ELAS and ELAS8[a]
General-Purpose Computer Programs
  for Equilibrium Problems for Linear
  Structures
	Jet Propulsion Lab.

FESS/FINESSE Systems of Programs	Civil Engineering Dept., Univ. of Wales, O. C. Zienkiewicz, Head
FORMAT Fortran Matrix Abstraction Technique	McDonnell Douglas Corp.
MAGIC Matrix Analysis Via Generative and Interpretive Computations	Bell Aerosystems
MARC Nonlinear Finite Element Analysis Program	Marc Analysis Research Corp.

CDC 6600, Honeywell 6080, IBM 370/165, RCA Spectra, and Univac 1100. If only small, special-purpose programs are needed, smaller computers can, of course, be used. For those who do not have their own computer facilities, outside computer service is easy to find. Many service bureaus can arrange remote terminal access to their computers on a time-sharing basis.

It is obvious from the proliferation of literature and computer programs that the finite element method is one of the newest and most powerful tools of the engineering analyst. The method requires sophisticated computers, but it is not an impractical or excessively elegant approach to real engineering problems. More and more industries are beginning to realize that the finite element method can be an invaluable aid in their product design and develop-

Table 10.3. (*continued*)

Program	Developer
NASTRAN[a] NASA Structural Analysis	Computer Science Corp., The Martin Co., MacNeal-Schwendler Corp.
SAFE Structural Analysis by Finite Elements	Gulf General Atomic, Inc.
SAMIS[a] Structural Analysis and Matrix Interpretive System	Philco Div., Ford Motor Co.
SAP A General Structural Analysis Program	Professor E. L. Wilson, Univ. of California, Berkeley
SBO-38	Martin Marietta Corp.
SLADE A Computer Program for the Static Analysis of Thin Shells	Sandia Lab.
STARDYNE	Mechanics Research Inc.
STARS 2 Shell Theory Automated for Rotational Structures	Grumman Aircraft Engineering Corp.
STRESS	M.I.T.
STRUDL (ICES) A Structural Design Language in the Integrated Civil Engineering System	M.I.T. Engineering Computer International

[a] Copies of these programs can be obtained from COSMIC Computer Center, Barrow Hall, University of Georgia, Athens, Ga. 30601.

ment work. A recent paper by Hamann [31] from one of the major automotive industries gives an extensive and convincing description of how the method can be effectively used in industry to solve complex structural analysis problems. In the future, we can expect that complex fluid mechanics problems will also yield to routine finite element analysis in an industrial environment.

10.9   CLOSURE

The sample computer program presented in this final chapter illustrates how a real problem is actually solved by the finite element method. The program was kept simple and straightforward, so that none of the essential features would be obscured.

Once the reader has gained an understanding of the finite element method fundamentals presented in the preceding chapters and has learned how these fundamentals are implemented on a digital computer, he has at his command a powerful engineering analysis tool.

REFERENCES

1.   O. C. Zienkiewicz, *The Finite Element Method in Engineering Science*, McGraw-Hill Book Company, London, 1971.
2.   C. A. Felippa and R. W. Clough, "The Finite Element Method in Solid Mechanics," *Numerical Solution of Field Problems in Continuum Physics*, SIAM-AMS Proceedings, Vol. 2, American Mathematical Society, Providence, R.I., 1970.
3.   C. S. Desai and J. F. Abel, *Introduction to the Finite Element Method: A Numerical Method for Engineering Analysis*, Van Nostrand-Reinhold, New York, 1972.
4.   J. Heuser, "Finite Element Method for Thermal Analysis," *NASA Rept.* TN D-7274, Goddard Space Flight Center, Greenbelt, Md., November 1973.
5.   W. D. Barfield, "Numerical Method for Generating Orthogonal Curvilinear Meshes," *J. Comput. Phys.*, Vol. 5, 1970.
6.   C. O. Frederick, Y. C. Wong, and F. W. Edge, "Two-Dimensional Automatic Mesh Generation for Structural Analysis," *Int. J. Numer. Methods Eng.*, Vol. 2, No. 1, 1970.
7.   O. C. Zienkiewicz and D. V. Phillips, "An Automatic Mesh Generation Scheme for Plane and Curved Surfaces by 'Isoparametric' Coordinates," *Int. J. Numer. Methods Eng.*, Vol. 3, No. 4, 1971.
8.   G. M. McNeice and P. V. Marcal, "Optimization of Finite Element Grids Based on Minimum Potential Energy," Brown Report, Brown University, 1971.
9.   J. Fukuda and J. Suhara, "Automatic Mesh Generation for Finite Element Analysis," in *Advances in Computational Methods in Structural Mechanics and Design*, J. T. Oden, R. W. Clough, and Y. Yammamoto (eds.), University of Alabama Press, Huntsville, Ala., 1972.

10. E. G. Sewell, "Automatic Generation of Triangulations for Piecewise Polynomial Approximation," *Rept.* CSD TR 83, Computer Sci. Dept., Purdue University, Lafayette, Ind., 1973.

11. W. R. Buell and B. A. Bush, "Mesh Generation—A Survey," *ASME Paper* 72-WA/DE-2.

12. M. Abramowitz and I. A. Stegun (ed.), *Handbook of Mathematical Functions*, National Bureau of Standards Applied Mathematics Series 55, June 1964.

13. P. Davis and P. Rabinowitz, "Abscissas and Weights for Gaussian Quadratures of High Order," *J. Res. Natl. Bur. Stand.*, Vol. 56, RP2645, 1956.

14. C. A. Felippa, "Refined Finite Element Analysis of Linear and Nonlinear Two-Dimensional Structures," *Struct. Mater. Res. Rept.* PB 178418, University of California at Berkeley, October 1966.

15. P. C. Hammer and A. H. Stroud, "Numerical Evaluation of Multiple Integrals," *Math. Tables Aids Comput.*, Vol. 12, 1958.

16. P. C. Hammer, O. P. Marlowe, and A. H. Stroud, "Numerical Integration over Simplexes and Cones," *Math. Tables Aids Comput.*, Vol. 10, 1957.

17. L. Fox, *Introduction to Numerical Linear Analysis*, Oxford University Press, London, 1964.

18. G. E. Forsythe, *Solving Linear Algebraic Equations Can Be Interesting*, Oxford University Press, London, 1964.

19. G. M. Gere and W. Weaver, Jr., *Matrix Algebra for Engineers*, Van Nostrand-Reinhold, New York, 1965.

20. A. S. Householder, *Principles of Numerical Analysis*, McGraw-Hill Book Company, New York, 1953.

21. A. Ralstan and H. S. Wilf (ed.), *Mathematical Methods for Digital Computers*, John Wiley and Sons, New York, 1960.

22. R. S. Varga, *Matrix Iterative Analysis*, Prentice-Hall, Englewood Cliffs, N.J., 1963.

23. M. R. Hestenes and E. Stiefel, "Methods of Conjugate Gradients for Solving Linear Systems," *J. Res. Natl. Bur. Stand.*, Vol. 49, RP2379, 1952.

24. B. M. Irons, "A Frontal Solution Program for Finite Element Analysis," *Int. J. Numer. Methods Eng.*, Vol. 2, No. 1, 1970.

25. H. G. Jensen and G. A. Parks, "Efficient Solutions for Linear Matrix Equations," *Proc. ASCE*, Vol. 96, No. ST-1, January 1970.

26. S. Klein, "The Cholesky Equation Solver," *Aerospace Rept.* ATR-70(S9990)-2, January 1970.

27. R. J. Melosh and R. M. Bamford, "Efficient Solution of Load-Deflection Equations," *Proc. ASCE*, Vol. 95, No. ST-4, April 1969.

28. I. Fried, "A Gradient Computational Procedure for the Solution of Large Problems Arising from the Finite Element Discretization Method," *Int. J. Numer. Methods Eng.*, Vol. 2, No. 4, 1970.

29. K. H. Huebner, "Finite Element Method—Stress Analysis and Much More," *Mach. Des.*, January 10, 1974.

30. S. J. Fenves, N. Perrone, A. R. Robinson, and W. C. Schnobrich (eds.), *Numerical and Computer Methods in Structural Mechanics*, Part II, Academic Press, New York, 1973.

31. W. C. Hamann, "Interfacing Finite Element Methods with Product Design Engineering," *ASME Paper* 73-DE-24, Design Engineering Conference and Show, Philadelphia, Pa., April 9–12, 1973.

APPENDIX A

MATRICES

In the 1850's Cayley introduced matrix notation as an effective shorthand scheme for dealing with systems of linear algebraic equations. Without this convenient notation, the labor of writing out every term of a large system of simultaneous equations would be enough to discourage anyone.

Matrix notation is intimately associated with the finite element method because ultimately the application of finite element techniques leads to sets of simultaneous equations. In many ways matrix notation is like a parachute. To conveniently carry and store the parachute we keep it folded, but to use it we must unfold it. Similarly, to conveniently discuss or manipulate ordered sets of numbers or functions we represent them compactly with matrix notation. But at the calculation stage the notation must be unraveled by hand or by computer so that individual terms can be treated explicitly.

Here† we present the fundamentals of matrix notation and matrix algebra necessary for the finite element method.

A.1 DEFINITIONS

Suppose that we have a rectangular array of numbers or functions arranged as follows:

$$\begin{bmatrix} a_{11} & a_{12} & a_{13} & \cdots & a_{1n} \\ a_{21} & a_{22} & a_{23} & \cdots & a_{2n} \\ \vdots & & & & \\ a_{m1} & a_{m2} & a_{m3} & \cdots & a_{mn} \end{bmatrix}$$

Such an array is called an $m \times n$ *matrix* when certain laws of manipulation, to be given later, are specified. The entries in the matrix are themselves called

† A more extensive treatment can be found in standard reference books such as *Linear Algebra* by G. Hadley, Addison-Wesley Publishing Company, Reading, Mass., 1961.

elements, and these elements may be real or complex.† The elements appearing in any given horizontal line comprise a row of the matrix, while those appearing in any vertical line comprise a column. Each element of a matrix carries a double subscript label. The first subscript indicates its row position, and the second subscript its column position. Thus a_{ij} is the element in the ith row and jth column. The matrix we illustrated has m rows and n columns. If $m = n$, the matrix is a *square matrix* of *order n*.

In the application of finite element techniques, we also encounter row matrices ($m = 1$) and column matrices ($n = 1$). A matrix with one row and n columns is called a row matrix, or row vector. A matrix with m rows and one column is called a column matrix, or column vector. The convention that we use to distinguish the different types of matrices is as follows:

We denote a *rectangular $m \times n$ matrix A* by enclosing A in brackets: $[A]$. For the special case where $m = 1$ *and* $n = 1$, $[A]$ is simply a scalar.

If $m = 1$ and $n > 1$, $[A]$ is a *row matrix*, and we denote it as $\lfloor A \rfloor$.

If $m > 1$ and $n = 1$, $[A]$ is a *column matrix*, and we denote it as $\{A\}$.

A.2 SPECIAL TYPES OF SQUARE MATRICES

1. A *null matrix* is one whose elements are all zero.

2. An *identity matrix* denoted as $[I]$ is one whose main diagonal elements are unity. Main diagonal elements have equal row and column subscripts. Thus the main diagonal runs from the upper left corner to the lower right corner. If the elements of an identity matrix are symbolized as δ_{ij}, then

$$\delta_{ij} = \begin{cases} 1, & i = j \\ 0, & i \neq j \end{cases}$$

3. A *diagonal matrix* has zero elements everywhere except on its main diagonal. Thus we recognize that an identity matrix is a special form of a diagonal matrix.

4. A *symmetric matrix* is one whose elements satisfy the condition $a_{ij} = a_{ji}$. If corresponding rows and columns of a symmetric matrix are interchanged, the matrix remains unchanged.

A.3 MATRIX OPERATIONS

The definitions that we have given thus far describe certain kinds of arrays of numbers or matrices. Before we can hope to manipulate these matrices, we

† In this text we assume that the elements of matrices are real numbers or functions. However, the results we present here hold also for matrices with complex elements.

first must define some rules for manipulations. Hence we consider next some definitions for matrix operations.

Two matrices are *equal* to one another only if they are identical; that is, the elements of one matrix must be the same as the corresponding elements of the other matrix. For example, if $[A] = [B]$, then $a_{ij} = b_{ij}$ for all i and j. The *sum* of $[A]$ and $[B]$ is defined only when $[B]$ has the same number of rows and the same number of columns as $[A]$. When these conditions hold, the addition of $[A]$ and $[B]$ results in a matrix, $[C]$, whose elements are the sum of the corresponding elements of $[A]$ and $[B]$. Thus, if $[C] = [A] + [B]$, then $c_{ij} = a_{ij} + b_{ij}$. Since the addition of matrices reduces to the addition of elements, matrix addition follows the associative and commutative laws of real numbers. Therefore

$$[A] + [B] = [B] + [A]$$

and

$$[A] + ([B] + [C]) = ([A] + [B]) + [C] = [A] + [B] + [C]$$

Multiplication of matrix $[A]$ *by a scalar* β results in a matrix $[B]$ whose elements are $b_{ij} - \beta a_{ij}$.

When we write $[B] = \beta[A]$, we imply that every element of $[A]$ is multiplied by the constant β. The rule for *matrix subtraction* now follows from the other rules we have established, namely,

$$[A] + (-1)[B] = [A] - [B]$$

Thus, if $[C] = [A] - [B]$, then $c_{ij} = a_{ij} - b_{ij}$. We can see from this rule that the restrictions on the number of rows and columns are the same for matrix addition and subtraction.

There are many feasible ways to define matrix multiplication. The definition that we give here is the most common because it leads to an efficient means for dealing with systems of simultaneously linear equations. Suppose that we have an $m \times n$ matrix $[A]$ and a $p \times q$ matrix $[B]$. If $n = p$, the matrix product $[A][B]$ is defined and is an $m \times q$ matrix $[C]$ whose elements are

$$c_{ij} = \sum_{k=1}^{n} a_{ik} b_{kj}, \qquad i = 1, 2, \ldots, n, \quad j = 1, 2, \ldots, q$$

If $n \neq p$, the matrix product is undefined.

In the matrix product $[A][B]$, matrix $[A]$ is the premultiplier and matrix $[B]$ is the postmultiplier. Matrix multiplication is defined only when the number of columns of the premultiplier equals the number of rows of the postmultiplier. The following equation illustrates the procedure for matrix multiplication:

$$
\underset{2 \times 3}{\begin{bmatrix} a_{11} & a_{12} & a_{13} \\ a_{21} & a_{22} & a_{23} \end{bmatrix}} \underset{3 \times 3}{\begin{bmatrix} b_{11} & b_{12} & b_{13} \\ b_{21} & b_{22} & b_{23} \\ b_{31} & b_{32} & b_{33} \end{bmatrix}}
$$

$$
= \begin{bmatrix} (a_{11}b_{11} + a_{12}b_{21} + a_{13}b_{31}), & (a_{11}b_{12} + a_{12}b_{22} + a_{13}b_{32}), \\ (a_{21}b_{11} + a_{22}b_{21} + a_{23}b_{31}), & (a_{21}b_{12} + a_{22}b_{22} + a_{23}b_{32}), \end{bmatrix}
$$

$$
\begin{bmatrix} (a_{11}b_{13} + a_{12}b_{23} + a_{13}b_{33}) \\ (a_{21}b_{13} + a_{22}b_{23} + a_{23}b_{33}) \end{bmatrix}
$$

In general, matrix multiplication does not follow all the rules for the multiplication of real numbers. When the matrix products are defined, the associated law and the distributive law hold, but the commutative law does not. Thus

$$([A][B])[C] = [A]([B][C]) = [A][B][C] \quad \text{(associative law)}$$

and

$$[A]([B] + [C]) = [A][B] + [A][C] \quad \text{(distributive law)}$$

but, generally, matrices do not commute, that is,

$$[A][B] \neq [B][A]$$

A.4   SPECIAL MATRIX PRODUCTS

1.   Product of a square matrix and a column matrix

Suppose, for example, that we have a system of three simultaneous linear equations such as

$$
\begin{aligned}
a_{11}x_1 + a_{12}x_2 + a_{13}x_3 &= b_1 \\
a_{21}x_1 + a_{22}x_2 + a_{23}x_3 &= b_2 \\
a_{31}x_1 + a_{32}x_2 + a_{33}x_3 &= b_3
\end{aligned}
\qquad \text{(A.1)}
$$

The a's and b's are known constants, while the x's are the unknowns. We have a convenient way to write equations A.1 if we recognize that a square matrix containing the a's premultiplying a column matrix containing the x's is equivalent to a column matrix containing the b's. In matrix notation equation A.1 is equivalent to the equation

$$\begin{bmatrix} a_{11} & a_{12} & a_{13} \\ a_{21} & a_{22} & a_{23} \\ a_{31} & a_{32} & a_{33} \end{bmatrix} \begin{Bmatrix} x_1 \\ x_2 \\ x_3 \end{Bmatrix} = \begin{Bmatrix} b_1 \\ b_2 \\ b_3 \end{Bmatrix}$$

or

$$[A]\{x\} = \{b\} \tag{A.2}$$

Anyone would agree that equation A.2 is easier to write than equation A.1. Therefore the product of a square matrix and a column is a most useful way to symbolize simultaneous linear equations, and we use it often in this text.

2. Product of a row matrix and a square matrix

When a row matrix premultiplies a square matrix, the result is a row matrix of the same order as the initial row matrix. For example,

$$\overset{1 \times n}{\lfloor A \rfloor} \overset{n \times n}{[B]} = \overset{1 \times n}{[C]}$$

If $n = 2$,

$$\lfloor a_1 \quad a_2 \rfloor \begin{bmatrix} b_{11} & b_{12} \\ b_{21} & b_{22} \end{bmatrix} = \lfloor (a_1 b_{11} + a_2 b_{21})(a_1 b_{12} + a_2 b_{22}) \rfloor$$

3. Product of a row matrix and a column matrix

When a row matrix premultiplies a column matrix, the result is a matrix of order 1 or, in other words, a scalar. For example,

$$\overset{1 \times n}{\lfloor A \rfloor} \overset{n \times 1}{\{B\}} = \overset{1 \times 1}{C} = \sum_{i=1}^{n} a_i b_i$$

If $n = 2$,

$$\lfloor a_1 \quad a_2 \rfloor \begin{Bmatrix} b_1 \\ b_2 \end{Bmatrix} = a_1 b_1 + a_2 b_2$$

4. Product of the identity matrix and any other matrix

We can readily verify that for any matrix $[A]$

$$[I][A] = [A][I] = [A]$$

The dimensions of $[I]$ are adjusted so that the matrix products are defined.

A.5   MATRIX TRANSPOSE

The transpose of an $m \times n$ matrix $[A]$ is an $n \times m$ matrix $[A]'$, whose rows
are the columns of $[A]$. To form the transpose of a matrix, we simply inter-
change its rows and columns so that row i of the matrix becomes column i of
its transpose. Applying this definition, we find that the transpose of a row
matrix is a column matrix, and that a symmetric matrix is its own transpose.
Also, it is easy to show that the transpose of the product of two matrices is the
product of their transposes in reverse order, that is, $([A][B])' = [B]'[A]'$.

A.6   QUADRATIC FORMS

Although matrix notation is most often employed to deal with sets of linear
equations, it is also useful in symbolizing special nonlinear expressions called
quadratic forms. A function of n variables $x_1, \ldots, x_n$ in quadratic form is
defined as

$$F(x_1, \ldots, x_n) = \sum_{i=1}^{n} \sum_{j=1}^{n} a_{ij} x_i x_j$$

$$= a_{11} x_1{}^2 + a_{12} x_1 x_2 + \cdots + a_{1n} x_1 x_n + \cdots + a_{21} x_2 x_1$$

$$+ \cdots + a_{2n} x_2 x_n + \cdots + a_{n1} x_n x_1 + \cdots + a_{nn} x_n{}^2$$

We encounter a quadratic form whenever we express the energy of a continuous system in a set of discretized coordinates of the system. A quadratic
form in one variable, say x_1, is simply $ax_1{}^2$. In two variables, x_1 and x_2, the
most general quadratic form is

$$F(x_1, x_2) = a_{11} x_1{}^2 + 2a_{12} x_1 x_2 + a_{22} x_2{}^2$$

Using matrix notation, we can write this as

$$F(x_1, x_2) = \lfloor x_1 \quad x_2 \rfloor \begin{bmatrix} a_{11} & a_{12} \\ a_{21} & a_{22} \end{bmatrix} \begin{Bmatrix} x_1 \\ x_2 \end{Bmatrix}$$

or

$$F(x_1, x_2) = \{x\}'[A]\{x\} = \lfloor x \rfloor [A]\{x\} \tag{A.3}$$

We wrote equation A.3 for a quadratic form of two variables, but the same
matrix symbolism holds also for a quadratic form of n variables.

Another useful observation is the following. If we write

$$\frac{1}{2} \frac{\partial F}{\partial x_i} = \frac{1}{2} \frac{\partial}{\partial x_i} (\lfloor x \rfloor [A]\{x\}) = b_i, \quad i = 1, 2, \ldots, n$$

we obtain a set of equations in the form of equations A.2, that is,

$$\begin{bmatrix} a_{11} & a_{12} & a_{13} & \cdots & a_{1n} \\ a_{21} & a_{22} & a_{23} & \cdots & a_{2n} \\ \vdots & \vdots & \vdots & \cdots & \vdots \\ a_{n1} & a_{n2} & a_{n3} & \cdots & a_{nn} \end{bmatrix} \begin{Bmatrix} x_1 \\ x_2 \\ \vdots \\ x_{n-1} \\ x_n \end{Bmatrix} = \begin{Bmatrix} b_1 \\ b_2 \\ \vdots \\ b_{n-1} \\ b_n \end{Bmatrix}$$

or $[A]\{x\} = \{b\}$, where $[A]$ is a *symmetric* matrix. We will use this result frequently.

A.7 MATRIX INVERSE

When we considered matrix multiplication, we avoided discussing *matrix division* because such an operation is *undefined*. However, there is a concept known as the *matrix inverse*, which is symbolized in much the same way as division is symbolized for real numbers. We know that for any real number $b \neq 0$ there exists an inverse b^{-1} such that $b^{-1}b = bb^{-1} = 1$. Under certain circumstances, matrices also possess this inverse property. If for a given square matrix $[A]$ there exists another square matrix $[A]^{-1}$ such that $[A]^{-1}[A] = [A][A]^{-1} = [I]$, then $[A]^{-1}$ is the inverse of $[A]$. Matrix $[A]$ is *nonsingular* when its inverse, $[A]^{-1}$, exists†.

With this definition and symbolism we now have a convenient way to write the solution of a system of equations of the form $[A]\{x\} = \{b\}$. If we premultiply both sides of the equation by $[A]^{-1}$, we obtain

$$[A]^{-1}[A]\{x\} = [A]^{-1}\{b\}$$

or

$$[I]\{x\} = [A]^{-1}\{b\}$$

which reduces to

$$\{x\} = [A]^{-1}\{b\} \tag{A.4}$$

A.8 MATRIX PARTITIONING

We introduced a matrix as an array of numbers that obeys certain laws of manipulation. Sometimes one subject of an array is of greater interest than

† A square matrix has an inverse when its determinant is nonzero. Determinants and the problem of computing a matrix inverse are beyond the scope of this brief discussion on matrix notation. A clear yet thorough treatment of these topics may be found in F. B. Hildebrand, *Methods of Applied Mathematics*, Prentice-Hall, Englewood Cliffs, N.J., 1952. Appendix E treats selected matrix inversion techniques well suited to digital computers.

another, and special notation is useful to highlight the various subsets. For this purpose, we use *matrix partitioning*. As an example of this technique, let us consider a 5×4 matrix $[A]$ that has been divided into sections as shown by the dashed lines:

$$[A] = \begin{bmatrix} a_{11} & a_{12} & a_{13} & a_{14} & a_{15} \\ a_{21} & a_{22} & a_{23} & a_{24} & a_{25} \\ a_{31} & a_{32} & a_{33} & a_{34} & a_{35} \\ \hdashline a_{41} & a_{42} & a_{43} & a_{44} & a_{45} \end{bmatrix}$$

We may now define submatrices as

$$[A_{11}] = \begin{bmatrix} a_{11} & a_{12} & a_{13} \\ a_{21} & a_{22} & a_{23} \\ a_{31} & a_{32} & a_{33} \end{bmatrix}, \qquad [A_{12}] = \begin{bmatrix} a_{14} & a_{15} \\ a_{24} & a_{25} \\ a_{34} & a_{35} \end{bmatrix}$$

$$[A_{21}] = \begin{bmatrix} a_{41} & a_{42} & a_{43} \end{bmatrix}, \qquad [A_{22}] = \begin{bmatrix} a_{44} & a_{45} \end{bmatrix}$$

and write $[A]$ in partitioned form as

$$[A] = \begin{bmatrix} [A_{11}] & [A_{12}] \\ [A_{21}] & [A_{22}] \end{bmatrix}$$

We could have partitioned $[A]$ in many other ways. The only requirement is that the partition lines be drawn straight and run the full length and width of the matrix.

The rules for the manipulation of partitioned matrices are the same as those for ordinary matrices, provided that all the submatrices involved are compatible for the particular operation to be performed. Consider matrix multiplication, for example. Suppose that we wish to postmultiply our partitioned 4×5 matrix $[A]$ by a 5×2 matrix $[B]$. Multiplication by submatrices, that is,

$$\overset{4 \times 5}{\begin{bmatrix} \overset{3 \times 3}{[A_{11}]} & \overset{3 \times 2}{[A_{12}]} \\ \underset{1 \times 3}{[A_{21}]} & \underset{1 \times 2}{[A_{22}]} \end{bmatrix}} \overset{5 \times 2}{\begin{bmatrix} [B_1] \\ [B_2] \end{bmatrix}} = \overset{4 \times 2}{\begin{bmatrix} [A_{11}][B_1] + [A_{12}][B_2] \\ [A_{21}][B_1] + [A_{22}][B_2] \end{bmatrix}}$$

is permissible only when $[B]$ is partitioned so that

$$[B] = \begin{bmatrix} b_{11} & b_{12} \\ b_{21} & b_{22} \\ b_{31} & b_{32} \\ \hdashline b_{41} & b_{42} \\ b_{51} & b_{52} \end{bmatrix} = \begin{bmatrix} [B_1] \\ [B_2] \end{bmatrix}$$

where

$$[B_1] = \begin{bmatrix} b_{11} & b_{12} \\ b_{21} & b_{22} \\ b_{31} & b_{32} \end{bmatrix}, \qquad [B_2] = \begin{bmatrix} b_{41} & b_{42} \\ b_{51} & b_{52} \end{bmatrix}$$

We note that in the multiplication of partitioned matrices the submatrices are treated as though they were scalar matrix elements. The submatrices are treated in the same way for the other matrix operations as well.

A.9   THE CALCULUS OF MATRICES

1. Differentiation of a matrix

If the elements of a matrix $[A]$ are themselves functions of n independent variables, $x_1, x_2, \ldots, x_n$, the matrix $[A]$ is called a matrix function of x_1, $x_2, \ldots, x_n$. The nth-order derivative of $[A]$ exists when the nth order derivative of each of its elements exists. We calculate the derivative of $[A]$ by differentiating each of its elements. Hence the element in the ith row and jth column of $\partial[A]/\partial x_k$ is $\partial(a_{ij})/\partial x_k$

2.   Integration of a matrix

The integral of a function matrix $[A]$ exists only when the integral of each element of the matrix exists. To calculate $\int[A]\,dx_k$ we must perform the integration on each of its elements. Hence the element in the ith row and jth column of $\int[A]\,dx_k$ is $\int a_{ij}\,dx_k$.

APPENDIX B

VARIATIONAL CALCULUS

B.1 INTRODUCTION

The calculus of variations is, in a sense, an extension of the calculus, although both were developed almost simultaneously. Here we review some of the concepts of differential calculus and variational calculus to show their parallel natures and to establish the terminology necessary for the variational approach to finite element analysis.

B.2 CALCULUS—THE MINIMA OF A FUNCTION

1. Definitions

Suppose we have a function F which depends on n independent variables, that is,

$$F = F(x_1, x_2, \ldots, x_n)$$

and suppose also that F has a minimum value at $(x_1^*, x_2^*, \ldots, x_n^*)$. Then, by definition, F has an *absolute minimum* at $(x_1^*, x_2^*, \ldots, x_n^*)$ if

$$\Delta F = F(x_1^* + \Delta x_1, \ldots, x_n^* + \Delta x_n) - F(x_1^*, \ldots, x_n^*) > 0$$

for arbitrary $\Delta x_1, \ldots, \Delta x_n$ consistent with the domain of definition for the variables $x_1, x_2, \ldots, x_n$. Also we say that F has a *local minimum* at $(x_1^*, x_2^*, \ldots, x_n^*)$ if $\Delta F > 0$ for arbitrary infinitely small changes $\Delta x_1, \ldots, \Delta x_n$ consistent again with the domain of definition. We shall confine our discussion to local extremals (minima or maxima) of functions and functionals.

Functions and functionals are said to have a *stationary value* at the points in the space of independent variables or independent functions, respectively, where the necessary conditions for local extremals are met. In the following

438

discussion we shall derive these necessary conditions for a number of different cases. Since stationary values are defined in terms of only necessary conditions for local extremals and not sufficient conditions, they may be neither maxima or minima.

2.   Functions of one variable

As an example in one dimension, consider the function $F = F(x_1)$ for $x_{1l} < x_1 < x_{1u}$. If we assume that F has a local minimum at x_1^*, then

$$\Delta F(x_1^*) = F(x_1^* + \Delta x_1) - F(x_1^*) > 0$$

Because Δx_1 is infinitely small, we can expand $F(x_1^* + \Delta x_1)$ about the value x_1^* and obtain

$$F(x_1^* + \Delta x_1) = F(x_1^*) + \frac{dF}{dx}\bigg|_{x_1 = x_1^*} \Delta x_1 + \frac{1}{2}\frac{d^2F}{dx_1^2}\bigg|_{x_1 = x_1^*} \Delta x_1^2 + \cdots$$

Thus

$$\Delta F(x_1^*) = \frac{dF}{dx_1}\bigg|_{x_1 = x_1^*} \Delta x_1 + \frac{d^2F}{dx_1^2}\bigg|_{x_1 = x_1^*} \frac{\Delta x_1^2}{2} + \cdots$$

Now it is obvious that, to have $\Delta F(x_1^*) > 0$, we must have

$$\frac{dF}{dx_1}\bigg|_{x_1 = x_1^*} = 0 \tag{B.1}$$

Equation B.1 is the necessary condition for $F(x_1)$ to have a local minimum at x_1^*.

3.   Functions of two or more variables

Now suppose that F is a function of two independent variables, that is, if F has a local minimum at (x_1^*, x_2^*) and $x_{1l} < x_1 < x_{1u}, x_{2l} < x_2 < x_{2u}$, then we must have

$$\Delta F = F(x_1^* + \Delta x_1, x_2^* + \Delta x_2) - F(x_1^*, x_2^*) > 0$$

Expanding $F(x_1^* + \Delta x_1, x_2^* + \Delta x_2)$ about x_1^*, x_2^*, we have, as before,

$$\Delta F = \frac{\partial F}{\partial x_1}\bigg|_{(x_1^*, x_2^*)} \Delta x_1 + \frac{\partial F}{\partial x_2}\bigg|_{(x_1^*, x_2^*)} \Delta x_2$$

$$+ \frac{1}{2}\left(\frac{\partial^2 F}{\partial x_1^2}\Delta x_1^2 + \frac{2\partial^2 F}{\partial x_1\,\partial x_2}\Delta x_1\,\Delta x_2 + \frac{\partial^2 F}{\partial x_2^2}\right)_{x_1 = x_1^*,\, x_2 = x_2^*} + \cdots$$

A necessary condition for $\Delta F > 0$ is that at (x_1^*, x_2^*)

$$\frac{\partial F}{\partial x_1} = 0, \qquad \frac{\partial F}{\partial x_2} = 0 \tag{B.2}$$

Following the same reasoning, we can show that for a function of n independent variables, $F(x_1, x_2, ..., x_n)$, a necessary condition for a local minimum is that each of its first partials must vanish at $(x_1^*, x_2^*, ..., x_n^*)$:

$$\frac{\partial F}{\partial x_1}, \frac{\partial F}{\partial x_2}, \quad ..., \quad \frac{\partial F}{\partial x_n} = 0 \tag{B.3}$$

Equations B.1, B.2, and B.3 represent only necessary conditions for a minimum. Sufficient conditions involve inequalities among the second partial derivatives and are beyond the scope of this discussion.

4. Constraints

Sometimes we face the problem of minimizing a function subject to certain specified constraint conditions. For example, suppose that we wish to minimize the function $F(x_1, x_2)$ subject to the condition that $g(x_1, x_2) = 0$, where g is a known function. This problem is handled by the *Lagrangian* multiplier method. Instead of minimizing $F(x_1, x_2)$ directly, we construct a new function

$$\tilde{F}(x_1, x_2, \lambda) = F(x_1, x_2) + \lambda g(x_1, x_2)$$

where λ is the Lagrangian multiplier. The new function $\tilde{F}$ can now be minimized in the usual fashion. If a local minimum of $\tilde{F}$ occurs at $(x_1^*, x_2^*, \lambda^*)$, then $F(x_1^*, x_2^*)$ is a local minimum of F subject to $g(x_1, x_2) = 0$. A necessary condition for $\tilde{F}$ to have a local minimum is that at $(x_1^*, x_2^*, \lambda^*)$

$$\frac{\partial \tilde{F}}{\partial x_1} = \frac{\partial \tilde{F}}{\partial x_2} = \frac{\partial \tilde{F}}{\partial \lambda} = 0$$

These three equations provide the conditions for determining x_1^*, x_2^*, λ^*.

This procedure immediately generalizes to the case of n independent variables $x_1, x_2, ..., x_n$ and $m < n$ constraint conditions $g(x_1, x_2, ..., x_n)$, $g_2(x_1, x_2, ..., x_n), ..., g_m(x_1, x_2, ..., x_n)$. The new function to be minimized is formed by writing

$$\tilde{F}(x_1, x_2, ..., x_n, \lambda_1, \lambda_2, ..., \lambda_m) = F(x_1, x_2, ..., x_n) + \sum_{i=1}^{m} \lambda_i g_i(x_1, x_2, ..., x_n)$$

Then a local minimum of F can be found by solving the $n + m$ equations

$$\frac{\partial \tilde{F}}{\partial x_i} = 0, \quad i = 1, 2, \ldots, n$$

$$\frac{\partial \tilde{F}}{\partial \lambda_j} = 0, \quad j = 1, 2, \ldots, m$$

for

$$(x_1^*, x_2^* \ldots, x_n^*, \lambda_1^*, \lambda_2^*, \ldots, \lambda_m^*)$$

B.3 VARIATIONAL CALCULUS—THE MINIMA OF FUNCTIONALS

1. Definitions

In contrast to the calculus, variational calculus is concerned primarily with the theory of maxima and minima, but the functions to be extremized are, in general, functions of functions, or *functionals*. A simple functional in terms of one independent variable would have the typical form

$$I(\phi) = \int_{x_1}^{x_2} F(x, \phi, \phi_x, \phi_{xx}) \, dx \tag{B.4}$$

where $\phi = \phi(x)$ and $\phi_x = \partial\phi/\partial x$, $\phi_{xx} = \partial^2\phi/\partial x^2$. Of course, more complicated functionals involving higher-order derivatives and more dependent variables are possible. The problem of variational calculus is to choose a function ϕ, called an extremal, so as to minimize or maximize (extremize) functionals like equation B.4. In other words, variational calculus is concerned with finding functions that extremize integrals whose integrands contain these functions.

Variational calculus has always been associated with realistic problems of continuum mechanics. Often the functionals whose extreme values are sought are expressions of some form of system energy. For example, many of the phenomena governing the elastic distortion of bodies can be deduced from a principle of minimum potential energy.

2. Functionals of one variable

In the following we shall derive a necessary condition to be satisfied by $\phi(x)$ to give, say, a minimum† value to $I(\phi)$, equation B.4. Often values of ϕ at the

† For the purpose of discussion, we restrict our attention to finding minima of a functional. The same arguments follow through if conditions for maxima are sought, and the results are the same.

end points of the interval $x_1 \leq x \leq x_2$ are given

$$\phi(x_1) = \phi_1, \qquad \phi(x_2) = \phi_2 \qquad \text{(B.5)}$$

but, as we shall see, other end conditions or boundary conditions are permissible. We seek a function $\phi(x)$ which satisfies equation B.5 and minimizes the functional of equation B.4. Suppose that we postulate the existence of a comparison function $\tilde{\phi}(x, \epsilon)$ which differs slightly from the extremizing function $\phi(x)$ and depends on a small parameter ϵ. We require that

$$\lim_{\epsilon \to 0} \tilde{\phi}(x, \epsilon) = \phi(x)$$

The comparison function can approach the extremizing function in a number of ways. Two particular classes of comparison functions are known as weak and strong variations. These are classified as follows:

Weak variation: $\displaystyle\lim_{\epsilon \to 0} \frac{d\tilde{\phi}(x, \epsilon)}{dx} = \frac{d\phi(x)}{dx}$

Strong variation: $\displaystyle\lim_{\epsilon \to 0} \frac{d\tilde{\phi}(x, \epsilon)}{dx} \neq \frac{d\phi(x)}{dx}$

We shall consider only comparison functions characterized by weak variations. Since ϵ is a small parameter, we may expand $\tilde{\phi}(x, \epsilon)$ about ϵ to obtain

$$\tilde{\phi}(x, \epsilon) = \phi(x) + \frac{\partial \tilde{\phi}}{\partial \epsilon}\bigg|_{\epsilon=0} \epsilon + (\text{terms of order } \epsilon^2)$$

We define the *first variation* in $\tilde{\phi}$ with respect to ϵ as

$$\delta\tilde{\phi} = \frac{\partial \tilde{\phi}}{\partial \epsilon}\bigg|_{\epsilon=0} \epsilon$$

For a general function Ω we would have

$$\delta\Omega = \frac{d\Omega}{d\epsilon}\bigg|_{\epsilon=0} \epsilon \qquad \text{(B.6)}$$

The notation $\delta\Omega$ is standard variational notation that represents the first-order change in Ω about $\epsilon = 0$ due to changing the variables that depend on ϵ. Let us consider, in the interval $x_1 \leq x \leq x_2$, a specific type of comparison function of the form

$$\tilde{\phi}(x, \epsilon) = \phi(x) + \epsilon\eta(x) \qquad \text{(B.7)}$$

The function $\eta(x)$ is any continuously differentiable function with the property $\eta(x_1) = \eta(x_2) = 0$. The difference between the comparison function $\tilde{\phi}$ and the minimizing function ϕ is defined as the first variation in ϕ. Hence

$$\delta\phi \equiv \epsilon\eta(x)$$

The value of the integral, equation B.4, along the curve $\tilde{\phi}(x, \epsilon)$ now becomes a function of ϵ, that is,

$$I(\epsilon) = \int_{x_1}^{x_2} F(x, \tilde{\phi}, \tilde{\phi}_x, \tilde{\phi}_{xx}) \, dx$$

$$= \int_{x_1}^{x_2} F(x, \phi + \epsilon\eta, \phi_x + \epsilon\eta_x, \phi_{xx} + \epsilon\eta_{xx}) \, dx \qquad (B.8)$$

Now, since $I(0)$ is the minimum value of I, we can be sure that ΔI, the change in I due to the change in the function ϕ, will be such that

$$\Delta I = I(\epsilon) - I(0) \geq 0 \qquad (B.9)$$

Since ϵ is infinitely small, we may expand $I(\epsilon)$ about $\epsilon = 0$ to obtain

$$I(\epsilon) = I(0) + \frac{dI}{d\epsilon}\bigg|_{\epsilon=0} \epsilon + \frac{1}{2}\frac{d^2 I}{d\epsilon^2}\bigg|_{\epsilon=0} \epsilon^2 + \cdots \qquad (B.10)$$

To satisfy equation B.9 in view of equation B.10, it is necessary that

$$\frac{dI}{d\epsilon}\bigg|_{\epsilon=0} = 0$$

since ϵ can be positive or negative. Thus $\delta I = 0$; that is, the first variation of I must vanish. From our previous definition we see that $\delta I = 0$ implies that I has a stationary value. The analogy between finding the minimum of a function via ordinary calculus and finding the minimum of a functional via variational calculus is now obvious. From equation B.8 we have

$$\frac{dI}{d\epsilon}\bigg|_{\epsilon=0} = \int_{x_1}^{x_2} \left(\frac{\partial F}{\partial \tilde{\phi}}\frac{\partial \tilde{\phi}}{\partial \epsilon} + \frac{\partial F}{\partial \tilde{\phi}_x}\frac{\partial \tilde{\phi}_x}{\partial \epsilon} + \frac{\partial F}{\partial \tilde{\phi}_{xx}}\frac{\partial \tilde{\phi}_{xx}}{\partial \epsilon} \right) dx = 0$$

$$= \int_{x_1}^{x_2} \left(\frac{\partial F}{\partial \phi}\eta + \frac{\partial F}{\partial \phi_x}\eta_x + \frac{\partial F}{\partial \phi_{xx}}\eta_{xx} \right) dx = 0 \qquad (B.11)$$

Frequently, the δ notation is used and equation B.11 is written as

$$\delta I = \int_{x_1}^{x_2} \left(\frac{\partial F}{\partial \phi}\delta\phi + \frac{\partial F}{\partial \phi_x}\delta\phi_x + \frac{\partial F}{\partial \phi_{xx}}\delta\phi_{xx} \right) dx = 0 \qquad (B.12)$$

This follows from the definition of the first variation of a functional. Before we proceed we shall establish several definitions. The change in F due to a change in the function ϕ is, as before,

$$\Delta F = F(x, \phi + \epsilon\eta, \phi_x + \epsilon\eta_x, \phi_{xx} + \epsilon\eta_{xx}) - F(x, \phi, \phi_x, \phi_{xx})$$

Expanding the right-hand side in powers of ϵ, we find

$$\Delta F = \frac{\partial F}{\partial \phi} \epsilon\eta + \frac{\partial F}{\partial \phi_x} \epsilon\eta_x + \frac{\partial F}{\partial \phi_{xx}} \epsilon\eta_{xx} + \text{(higher powers of } \epsilon\text{)}$$

By definition

$$\delta F \equiv \frac{\partial F}{\partial \phi} \epsilon\eta + \frac{\partial F}{\partial \phi_x} \epsilon\eta_x + \frac{\partial F}{\partial \phi_{xx}} \epsilon\eta_{xx}$$

The analogy between the first variation of a functional and the first differential of a function now becomes obvious. In fact, all the laws of differentiation of sums, products, quotients, powers, and so forth carry over directly to become laws of the first variation. For example,

$$\delta(B_1 + B_2) = \delta B_1 + \delta B_2$$

$$\delta(B_1 B_2) = B_1\,\delta B_2 + B_2\,\delta B_1$$

$$\delta\left(\frac{B_1}{B_2}\right) = \frac{B_2\,\delta B_1 - B_1\,\delta B_2}{B_2{}^2}$$

$$\delta B_1{}^n = n B_1{}^{n-1}\,\delta B_1$$

etc.

The first variation is also a commutative operator with both differentiation and integration if the integration limits are not to be varied. Thus

$$\delta \int F\,dx = \int \delta F\,dx, \qquad \delta\frac{d\phi}{dx} = \frac{d}{dx}\delta\phi$$

Now we return our attention to equation B.12. The necessary conditions for $\phi(x)$ to be a minimizing function can be discovered if we integrate each term by parts. The result is

$$\delta I = \int_{x_1}^{x_2} \left[\frac{\partial F}{\partial \phi} - \frac{d}{dx}\left(\frac{\partial F}{\partial \phi_x}\right) + \frac{d^2}{dx^2}\left(\frac{\partial F}{\partial \phi_{xx}}\right)\right] \delta\phi\,dx$$

$$+ \left[\frac{\partial F}{\partial \phi_x} - \frac{d}{dx}\left(\frac{\partial F}{\partial \phi_{xx}}\right)\right]\delta\phi\,\Bigg|_{x_1}^{x_2} + \left[\left(\frac{\partial F}{\partial \phi_{xx}}\,\delta\phi_x\right)\right]_{x_1}^{x_2} = 0 \qquad \text{(B.13)}$$

Because $\delta\phi$ and $\delta\phi_x$ are arbitrary admissible variations, equation B.13 can be satisfied only if the integrand and the remaining terms vanish. Thus the necessary conditions for $\phi(x)$ to minimize $I(\phi)$ are as follows:

$$\frac{\partial F}{\partial \phi} - \frac{d}{dx}\frac{\partial F}{\partial \phi_x} + \frac{d^2}{dx^2}\left(\frac{\partial F}{\partial \phi_{xx}}\right) = 0 \tag{B.14a}$$

$$\left[\frac{\partial F}{\partial \phi_x} - \frac{d}{dx}\left(\frac{\partial F}{\partial \phi_{xx}}\right)\right]\delta\phi \Bigg|_{x_1}^{x_2} = 0 \tag{B.14b}$$

$$\left[\left(\frac{\partial F}{\partial \phi_{xx}}\right)\delta\phi_x\right]_{x_1}^{x_2} = 0 \tag{B.14c}$$

Equation B.14a is called the *Euler-Lagrange equation* and is the differential equation that $\phi(x)$ must satisfy. The other two conditions, equations B.14b and B.14c, dictate the necessary boundary conditions. From equation B.14b we must have

$$\left[\frac{\partial F}{\partial \phi_x} - \frac{d}{dx}\left(\frac{\partial F}{\partial \phi_{xx}}\right)\right]_{x_1}^{x_2} = 0 \tag{B.15}$$

or

$$\phi(x_1) = \phi(x_2) = 0 \quad \text{since } \delta\phi \Bigg|_{x_1}^{x_2} = 0 \tag{B.16}$$

and from equation B.14c either

$$\frac{\partial F}{\partial \phi_{xx}}\Bigg|_{x_1}^{x_2} = 0 \tag{B.17}$$

or

$$\frac{\partial \phi}{\partial x}\Bigg|_{x_1}^{x} = 0 \tag{B.18}$$

Equations B.15 and B.17 are called *natural boundary conditions*, while equations B.16 and B.18 are *geometric boundary conditions*.

Since equation B.14a, the Euler-Lagrange equation, is an ordinary second-order, in general, nonlinear differential equation, two boundary conditions are needed for a well-posed problem. Any combination of natural and geometric boundary conditions that meets the conditions of equations B.14b and B.14c is appropriate.

3.   More general functionals

Proceeding in a similar manner, we may derive Euler-Lagrange equations and boundary conditions for other functionals. For example, if we have extremal functions of two independent variables and derivatives up to and including second order, the functional has the form

$$I(\phi) = \iint\limits_{R} F(x, y, \phi, \phi_x, \phi_y, \phi_{xx}, \phi_{xy}, \phi_{yy}) \, dx \, dy \qquad (B.19)$$

and the corresponding Euler-Lagrange equation is

$$\frac{\partial^2}{\partial x^2}\left(\frac{\partial F}{\partial \phi_{xx}}\right) + \frac{\partial^2}{\partial x\, \partial y}\left(\frac{\partial F}{\partial \phi_{xy}}\right) + \frac{\partial^2}{\partial y^2}\left(\frac{\partial F}{\partial \phi_{yy}}\right)$$

$$-\frac{\partial}{\partial x}\left(\frac{\partial F}{\partial \phi_x}\right) - \frac{\partial}{\partial y}\left(\frac{\partial F}{\partial \phi_y}\right) + \frac{\partial F}{\partial \phi} = 0 \quad (B.20)$$

In terms of one independent variable and nth-order derivatives, the functional has the form

$$I(\phi) = \int_{x_1}^{x_2} F(x, \phi, \phi_x, \phi_{xx}, \ldots, \phi_{xx}, \ldots) \, dx \qquad (B.21)$$

and the corresponding Euler-Lagrange equation is

$$\frac{\partial F}{\partial \phi} - \frac{d}{dx}\left(\frac{\partial F}{\partial \phi_x}\right) + \frac{d^2}{dx^2}\left(\frac{\partial F}{\partial \phi_{xx}}\right) - \cdots (-1)^n \frac{d^n}{dx^n}\left(\frac{\partial F}{\partial \phi_{xx}}\right) = 0 \quad (B.22)$$

Example

Suppose that we have a functional given by

$$I(\phi) = \iint\limits_{R} \left[\left(\frac{\partial \phi}{\partial x}\right)^2 + \left(\frac{\partial \phi}{\partial y}\right)^2 - 2\phi f(x, y)\right] dV$$

and we want to find the necessary condition that $\phi(x, y)$ must satisfy to give $I(\phi)$ a stationary value. From equation B.19 we recognize that

$$F(x, y, \phi, \phi_x, \phi_y) = \left(\frac{\partial \phi}{\partial x}\right)^2 + \left(\frac{\phi}{\partial y}\right)^2 - 2\phi f$$

and when we apply equation B.20 we find that ϕ must satisfy

$$\frac{\partial^2 \phi}{\partial x^2} + \frac{\partial^2 \phi}{\partial y^2} + f(x, y) = 0$$

This we recognize as Poisson's equation in two dimensions.

It is important to emphasize that the solution of an Euler-Lagrange equation may not yield a function ϕ that minimizes a given functional because the Euler-Lagrange equation expresses only a necessary and not a sufficient condition for a minimum. In the general case we must verify whether the ϕ found by solving the Euler-Lagrange equation yields a minimum, and this involves computation of second variations.† However, in many situations we can use some physical considerations to tell whether the ϕ so obtained actually minimizes the functional.

We have considered functionals of only one function and several independent variables, but this procedure may be used to derive necessary conditions for the minima of functionals of more than one function. For such a case several Euler-Lagrange equations would result—one for each independent function in the functional.

4. Constraint conditions

The procedure for minimizing a functional subject to constraining functionals is similar to that for minimizing a function subject to constraining functions. Suppose that we seek the minimum of the functional

$$I(\phi) = \int_{x_1}^{x_2} F(x, \phi, \phi_x, \phi_{xx})\, dx \tag{B.23}$$

subject to the auxiliary condition that $\phi(x)$ also satisfies

$$J = \int_{x_1}^{x_2} G(x, \phi, \phi_x, \phi_{xx})\, dx \tag{B.24}$$

where J is a specified constant. As before, we introduce a Lagrange multiplier λ and construct the new integral

$$\tilde{I}(\phi) = I(\phi) + \lambda J$$

Now the $\phi(x)$ that minimizes $I(\phi)$ and satisfies equation B.24 must also satisfy $\delta \tilde{I} = 0$. From equation B.14a it follows that the minimizing function satisfies the Euler-Lagrange equation

$$\frac{\partial H}{\partial \phi} - \frac{d}{dx}\frac{\partial H}{\partial \phi_x} + \frac{d^2}{dx^2}\left(\frac{\partial H}{\partial \phi_{xx}}\right) = 0$$

where $H = I + \lambda J = \tilde{I}$.

---

† For a concise review of the concept of second variations, the reader is referred to P. N. Berg, "Calculus of Variations," Chapter 16 in *Handbook of Engineering Mechanics*, edited by W. Flügge, McGraw-Hill Book Company, New York, 1962.

If we are dealing with functionals of more than one independent function, there may be several constraint conditions. These are handled in the same way by forming an augmented functional. Each constraint is multiplied by its corresponding Lagrange multiplier and added to the original functionals. Then the necessary conditions for a minimum apply to the new functional, just as they would have applied to the original functional with no imposed constraints.

APPENDIX C

BASIC EQUATIONS FROM
LINEAR ELASTICITY THEORY

C.1 INTRODUCTION

Here we summarize some of the concepts and basic equations of linear elasticity theory. Except for the general definitions of stress and strain, the equations given without derivation or proof apply to homogeneous isotropic materials unless otherwise noted. Because most finite element analyses are carried out in a Cartesian coordinate system, only this system is used in the following. Other special coordinate systems are sometimes employed in particular applications, and these are examined in the text where they appear.

This summary is intended only as a quick reference when the theory of finite element analysis is applied to solid mechanics problems in Part 2 of this book. Hence the reader who questions the origin or limitations of a particular equation is encouraged to consult one of the many comprehensive texts on elasticity theory, for example, refs. 1–6.

A natural extension of the study of thin-plate bending is the study of thin-shell distortion. Because of the many complications and the different levels of theoretical approach to shell analysis, we present no summary equations for this case. Instead, the special considerations needed for the finite element analysis of shell problems are given in the body of the text. Readers interested in shell theories, both classical and higher order, may consult standard text books such as those of Kraus [9] and Flügge [10].

C.2 STRESS COMPONENTS

A state of stress exists in a body acted upon by external forces. If these external forces act over the surface of the body, they are called *surface forces*; if they are distributed throughout the volume of the body, they are called *body forces*.

449

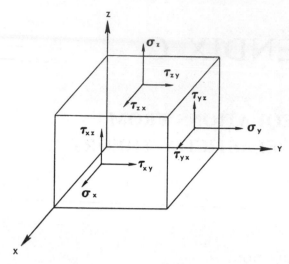

Figure C.1.   Definition of the components of the Cartesian stress tensor.

Hydrostatic pressure over an area is an example of a possible surface force, while gravitational forces or inertia forces due to accelerating a body are examples of body forces.

Figure C.1 shows the notation used to define the state of stress in a three-dimensional body. The components of normal stresses are denoted as σ_x, σ_y, σ_y; the components of shear stress area, as τ_{xy}, τ_{yz}, τ_{zx}. Normal stress carries a single subscript to indicate that the stress acts on a plane normal to the axis in the subscript direction. The first letter of the double subscript on shear stress indicates that the plane on which the stress acts is normal to the axis in the subscript direction; the second letter designates the coordinate direction in which the stress acts.

We have designated only three components of shear stress because only three are independent. By considering the equilibrium of an elemental volume and neglecting the vanishing body force as the volume diminishes in size, we can show that $\tau_{xy} = \tau_{yx}$, $\tau_{zx} = \tau_{xz}$, $\tau_{yz} = \tau_{zy}$. Hence the six quantities $\sigma_x, \sigma_y, \sigma_z, \tau_{xy}, \tau_{xz}$, and τ_{yz} completely describe the state of stress at a point.

C.3 STRAIN COMPONENTS

When an elastic body is subjected to a state of stress, we shall assume that the particles of the body move only a small amount and the body in its deformed state remains perfectly elastic, so that its deformation disappears when the stress is removed. If the displacement field in the deformed body is represented by the three components u, v, and w parallel to the coordinate axes x, y, and z, respectively, the *strain* at a point in the body may be expressed as

$$\epsilon_x = \frac{\partial u}{\partial x}, \qquad \epsilon_y = \frac{\partial v}{\partial y}, \qquad \epsilon_z = \frac{\partial w}{\partial z}$$

$$\gamma_{xy} = \frac{\partial u}{\partial y} + \frac{\partial v}{\partial x}, \qquad \gamma_{xz} = \frac{\partial u}{\partial z} + \frac{\partial w}{\partial x}, \qquad \gamma_{yz} = \frac{\partial v}{\partial z} + \frac{\partial w}{\partial y} \tag{C.1}$$

The strain ϵ_x, for example, is defined as the unit elongation of the body at a point, the shearing strain γ_{xy} as the distortion of the angle between the x-z and y-z planes, and so forth.

The relationships between strain and displacement expressed by equation C.1 are approximate, but are consistent with the assumption of small deformations usually encountered in engineering situations. More exact strain-displacement relations contain higher-order terms [2].

A body acted upon by external forces may, in general, experience deformation characterized by strain, and it may translate and rotate. The rotation of an elemental volume of the body can be expressed in terms of its displacements in the following manner:

$$\omega_x = \omega_{zy} = \frac{1}{2}\left(\frac{\partial w}{\partial y} - \frac{\partial v}{\partial z}\right)$$

$$\omega_y = \omega_{xz} = \frac{1}{2}\left(\frac{\partial u}{\partial z} - \frac{\partial w}{\partial x}\right)$$

$$\omega_z = \omega_{yz} = \frac{1}{2}\left(\frac{\partial v}{\partial x} - \frac{\partial u}{\partial y}\right) \tag{C.2}$$

where ω_x is the rotation about the x axis, and so forth. Sometimes it is convenient to define a quantity called the *dilatation* at a point in the body. This is related to strains and displacements as follows:

$$\psi = \epsilon_x + \epsilon_y + \epsilon_z = \frac{\partial u}{\partial x} + \frac{\partial v}{\partial y} + \frac{\partial w}{\partial z} \tag{C.3}$$

C.4   GENERALIZED HOOKE'S LAW (CONSTITUTIVE EQUATIONS)

The six components of stress are related to the six components of strain through a proportionality matrix $[C]$ containing 36 terms for a general anisotropic material:

$$
\begin{Bmatrix} \sigma_x \\ \sigma_y \\ \sigma_z \\ \tau_{xy} \\ \tau_{yz} \\ \tau_{zx} \end{Bmatrix} = \begin{bmatrix} C_{11} & \cdots\cdots\cdots\cdots & C_{16} \\ \cdot & & \cdot \\ \cdot & & \cdot \\ \cdot & & \cdot \\ \cdot & & \cdot \\ C_{61} & \cdots\cdots\cdots\cdots & C_{66} \end{bmatrix} \begin{Bmatrix} \epsilon_x \\ \epsilon_y \\ \epsilon_z \\ \gamma_{xy} \\ \gamma_{yz} \\ \gamma_{zx} \end{Bmatrix} \tag{C.4}
$$

Because the matrix $[C]$ is symmetric, only 21 different coefficients can be identified. Using matrix notation, we may rewrite equation C.4 as

$$\{\sigma\} = [C]\{\epsilon\}$$

and by inversion the strains may be expressed as

$$\{\epsilon\} = [C]^{-1}\{\sigma\} = [D]\{\sigma\} \tag{C.5}$$

The matrix $[C]$ is termed the *material stiffness matrix*, while its inverse $[D]$ is the *material flexibility matrix*. These relations between stress and strain take on less complex forms if we assume that the elastic body is *homogeneous* and *isotropic*. By "homogeneous" we mean that any elemental volume of the body possesses the same specific physical properties as any other elemental volume of the body; by "isotropic" we mean that the physical properties are the same in all directions. For homogeneous isotropic materials only two physical constants are required to express all the coefficients in Hooke's law; these are Young's modulus, E, and Poisson's ratio, v. In terms of these constants the matrices in Hooke's law are as follows:

$$
[C] = \frac{E}{(1+v)(1-2v)}
\begin{bmatrix}
1-v & v & v & 0 & 0 & 0 \\
v & 1-v & v & 0 & 0 & 0 \\
v & v & 1-v & 0 & 0 & 0 \\
0 & 0 & 0 & \dfrac{1-2v}{2} & 0 & 0 \\
0 & 0 & 0 & 0 & \dfrac{1-2v}{2} & 0 \\
0 & 0 & 0 & 0 & 0 & \dfrac{1-2v}{2}
\end{bmatrix}
$$

$$\tag{C.6}$$

$$[D] = [C]^{-1} = \frac{1}{E} \begin{bmatrix} 1 & -v & -v & 0 & 0 & 0 \\ -v & 1 & -v & 0 & 0 & 0 \\ -v & -v & 1 & 0 & 0 & 0 \\ 0 & 0 & 0 & 2(1+v) & 0 & 0 \\ 0 & 0 & 0 & 0 & 2(1+v) & 0 \\ 0 & 0 & 0 & 0 & 0 & 2(1+v) \end{bmatrix} \quad \text{(C.7)}$$

Sometimes these matrices are given in terms of Lame's constants, λ and μ, which are related to E and v as follows:

$$E = \frac{\mu(3\lambda + 2\mu)}{\lambda + \mu} \quad \text{(C.8a)}$$

$$v = \frac{\lambda}{2(\lambda + \mu)} \quad \text{(C.8b)}$$

or

$$\lambda = \frac{Ev}{(1 + v)(1 - 2v)} \quad \text{(C.9a)}$$

$$\mu = \frac{E}{2(1 + v)} \quad \text{(C.9b)}$$

In addition to these constants, two other constants characterizing a material's behavior under pure shear and pure volumetric distortion often appear in the engineering literature.

The constant G, known as the *modulus of rigidity*, relates shearing stress and shearing strain and is defined in terms of E and v as

$$G = \frac{E}{2(1 + v)} = \mu \quad \text{(C.10a)}$$

The *modulus of volume expansion*, K, is defined as

$$K = \frac{E}{3(1 - 2v)} \quad \text{(C.10b)}$$

The constant K relates unit volume expansion (or contraction) to volumetric stresses such as hydrostatic pressure.

In general, the 21 individual coefficients in equation C.4 or any subset of these for special cases such as homogeneous isotropic solids must be determined by carefully controlled experiment. For general anisotropic materials, determination of the constitutive properties is a most arduous task.

When a body deforms, certain constraint conditions must be satisfied. These conditions, known as the equilibrium and compatibility conditions, impose force and geometric constraints, respectively, on the deformations. We will now discuss the forms of these conditions.

C.5 STATIC EQUILIBRIUM EQUATIONS

The equilibrium of an elastic body in a state of stress is governed by three partial differential equations for the stress components. These equations may be derived by writing force balances for an elemental volume of the material acted upon by body forces X, Y, and Z. Forces acting on the element are calculated by assuming that the sides of the element have infinitesimal area and by multiplying the stresses at the centroid of the element by the area of the sides. Summing forces in the three coordinate directions gives

$$\frac{\partial \sigma_x}{\partial x} + \frac{\partial \tau_{xy}}{\partial y} + \frac{\partial \tau_{xz}}{\partial z} + X = 0$$

$$\frac{\partial \sigma_y}{\partial y} + \frac{\partial \tau_{xy}}{\partial x} + \frac{\partial \tau_{yz}}{\partial z} + Y = 0 \qquad \text{(C.11)}$$

$$\frac{\partial \sigma_z}{\partial z} + \frac{\partial \tau_{xz}}{\partial x} + \frac{\partial \tau_{yz}}{\partial y} + Z = 0$$

A balance of moments about the three coordinate directions shows that $\tau_{xy} = \tau_{yx}$, and so on. Boundary conditions for the equilibrium equations, equations C.11, may be found by considering the external surface forces acting on the body. If the components of these unit surface forces are denoted as $\bar{\sigma}_x$, $\bar{\sigma}_y$, and $\bar{\sigma}_z$, we find that at any point *on the surface* of the body we have

$$\begin{aligned}
\bar{\sigma}_x &= \sigma_x l + \tau_{xy} m + \tau_{xz} n \\
\bar{\sigma}_y &= \sigma_y l + \tau_{yz} m + \tau_{xy} n \\
\bar{\sigma}_z &= \sigma_z l + \tau_{xz} m + \tau_{yz} n
\end{aligned} \qquad \text{(C.12)}$$

where l, m, and n are the direction cosines of the outward normal to the surface at the point of interest.

Given a set of boundary conditions in the form of equations C.12, it is impossible to obtain a unique solution to these equations for the six stress components because we have six unknowns and only three equations. This apparent difficulty can be remedied by taking into account the displacement field u, v, w and its relation to strain in the body.

C.6 COMPATIBILITY CONDITIONS

The definition of strain in terms of derivatives of displacement (given in equation C.1) lead to other relations between strain and displacement that must be satisfied. These relations, known as *compatibility conditions*, can be derived by differentiation of equations C.1. For example, from the first of these equations we have

$$\frac{\partial^2 \epsilon_x}{\partial y^2} = \frac{\partial^3 u}{\partial x \, \partial y^2}, \qquad \frac{\partial^2 \epsilon_y}{\partial x^2} = \frac{\partial^3 v}{\partial x^2 \, \partial y}$$

$$\frac{\partial^2 \gamma_{xy}}{\partial x \, \partial y} = \frac{\partial^3 u}{\partial x \, \partial y^2} + \frac{\partial^3 v}{\partial x^2 \, \partial y}$$

From these equations it is obvious that

$$\frac{\partial^2 \epsilon_x}{\partial y^2} + \frac{\partial^2 \epsilon_y}{\partial x^2} = \frac{\partial^2 \gamma_{xy}}{\partial x \, \partial y}$$

Continuing in this manner, we arrive at the following six differential relations, the compatibility conditions between components of stress and strain:

$$
\left.
\begin{aligned}
\frac{\partial^2 \epsilon_x}{\partial y \, \partial z} &= \frac{\partial}{\partial x}\left(-\frac{\partial \gamma_{yz}}{\partial x} + \frac{\partial \gamma_{zx}}{\partial y} + \frac{\partial \gamma_{xy}}{\partial z} \right) \\[2mm]
\frac{\partial^2 \epsilon_y}{\partial z \, \partial x} &= \frac{\partial}{\partial y}\left(-\frac{\partial \gamma_{zx}}{\partial y} + \frac{\partial \gamma_{yx}}{\partial z} + \frac{\partial \gamma_{yz}}{\partial x} \right) \\[2mm]
\frac{\partial^2 \epsilon_z}{\partial x \, \partial y} &= \frac{\partial}{\partial z}\left(-\frac{\partial \gamma_{xy}}{\partial z} + \frac{\partial \gamma_{yz}}{\partial x} + \frac{\partial \gamma_{zx}}{\partial y} \right) \\[2mm]
2\frac{\partial^2 \gamma_{xy}}{\partial x \, \partial y} &= \frac{\partial^2 \epsilon_x}{\partial y^2} + \frac{\partial^2 \epsilon_y}{\partial x^2} \\[2mm]
2\frac{\partial^2 \gamma_{yz}}{\partial y \, \partial z} &= \frac{\partial^2 \epsilon_y}{\partial z^2} + \frac{\partial^2 \epsilon_z}{\partial y^2} \\[2mm]
2\frac{\partial^2 \gamma_{zx}}{\partial z \, \partial x} &= \frac{\partial^2 \epsilon_z}{\partial x^2} + \frac{\partial^2 \epsilon_x}{\partial z^2}
\end{aligned}
\right\}
\qquad \text{(C.13)}
$$

The compatibility conditions may also be expressed in terms of stress components if we use Hook's law. The results, known as the Beltrami-Michell equations, are as follows:

$$
\left.
\begin{aligned}
\nabla^2 \sigma_x + \frac{1}{1+\sigma}\frac{\partial^2 \theta}{\partial x^2} &= -\frac{\sigma}{1-\sigma}\nabla \cdot \mathbf{F}^* - 2\frac{\partial X^*}{\partial x} \\[2mm]
\nabla^2 \sigma_y + \frac{1}{1+\sigma}\frac{\partial^2 \theta}{\partial y^2} &= -\frac{\sigma}{1-\sigma}\nabla \cdot \mathbf{F}^* - 2\frac{\partial Y^*}{\partial y} \\[2mm]
\nabla^2 \sigma_z + \frac{1}{1+\sigma}\frac{\partial^2 \theta}{\partial z^2} &= -\frac{\sigma}{1-\sigma}\nabla \cdot \mathbf{F}^* - 2\frac{\partial Z^*}{\partial z} \\[2mm]
\nabla^2 \tau_{yz} + \frac{1}{1+\sigma}\frac{\partial^2 \theta}{\partial y\,\partial z} &= -\left(\frac{\partial Y^*}{\partial z}+\frac{\partial Z^*}{\partial y}\right) \\[2mm]
\nabla^2 \tau_{zx} + \frac{1}{1+\sigma}\frac{\partial^2 \theta}{\partial z\,\partial x} &= -\left(\frac{\partial Z^*}{\partial x}+\frac{\partial X^*}{\partial z}\right) \\[2mm]
\nabla^2 \tau_{xy} + \frac{1}{1+\sigma}\frac{\partial^2 \theta}{\partial x\,\partial y} &= -\left(\frac{\partial X^*}{\partial y}+\frac{\partial Y^*}{\partial x}\right)
\end{aligned}
\right\} \quad \text{(C.14)}
$$

where

$$
\nabla = \frac{\partial}{\partial x}\,\hat{\imath} + \frac{\partial}{\partial y}\,\hat{\jmath} + \frac{\partial}{\partial z}\,\hat{k} \quad \text{and} \quad \nabla^2 = \nabla \cdot \nabla
$$

is the Laplacian, $\nabla^2 = \partial^2/\partial x^2 + \partial^2/\partial y^2 + \partial^2/\partial z^2$; θ is the stress invariant, $\theta = \sigma_x + \sigma_y + \sigma_z$; and $\mathbf{F}^* = X^*\hat{\imath} + Y\hat{\jmath} + Z^*\hat{k}$. Equations C.13 or C.14 are generally sufficient for determining the components of strain or stress, respectively.

C.7   DIFFERENTIAL EQUATIONS FOR DISPLACEMENTS

If we return to the static equilibrium equations for stress, equations C.11, and their boundary conditions, equation C.12, we may derive a set of three differential equations for the three components of displacement. The derivation involves the use of Hooke's law to replace the stress components with strain components; then the strain components must be expressed in terms of derivatives of displacements via the definitions of equations C.1. The result is

$$
\begin{aligned}
\nabla^2 u + \frac{1}{1-2v}\frac{\partial}{\partial x}\left(\frac{\partial u}{\partial x}+\frac{\partial v}{\partial y}+\frac{\partial w}{\partial z}\right) + \frac{X^*}{\mu} &= 0 \\[2mm]
\nabla^2 v + \frac{1}{1-2v}\frac{\partial}{\partial y}\left(\frac{\partial u}{\partial x}+\frac{\partial v}{\partial y}+\frac{\partial w}{\partial z}\right) + \frac{Y^*}{\mu} &= 0 \\[2mm]
\nabla^2 w + \frac{1}{1-2v}\frac{\partial}{\partial z}\left(\frac{\partial u}{\partial x}+\frac{\partial v}{\partial y}+\frac{\partial w}{\partial z}\right) + \frac{Z^*}{\mu} &= 0
\end{aligned}
\qquad \text{(C.15)}
$$

On the surface the boundary conditions take the form

$$
T_x^* = \lambda\psi l + \mu\left[2\frac{\partial u}{\partial x}l + \left(\frac{\partial u}{\partial y} + \frac{\partial v}{\partial x}\right)m + \left(\frac{\partial u}{\partial z} + \frac{\partial w}{\partial x}\right)n\right]
$$

$$
T_y^* = \lambda\psi m + \mu\left[2\frac{\partial v}{\partial y}m + \left(\frac{\partial v}{\partial z} + \frac{\partial w}{\partial y}\right)n + \left(\frac{\partial v}{\partial x} + \frac{\partial u}{\partial y}\right)l\right] \quad\text{(C.16)}
$$

$$
T_z^* = \lambda\psi n + \mu\left[2\frac{\partial w}{\partial z}n + \left(\frac{\partial w}{\partial x} + \frac{\partial u}{\partial z}\right)l + \left(\frac{\partial w}{\partial z} + \frac{\partial v}{\partial z}\right)m\right]
$$

where, as before, ψ is the strain invariant, given by

$$
\psi = \epsilon_x + \epsilon_y + \epsilon_z = \frac{\partial u}{\partial x} + \frac{\partial v}{\partial y} + \frac{\partial w}{\partial z}
$$

C.8 VARIATIONAL PRINCIPLES

In the preceding sections of this appendix, we reviewed the differential statements of equilibrium and compatibility for linear elastic bodies subjected to stress. Now we turn our attention to the *integral* statements that must hold for these bodies. Hence we essentially shift our emphasis from relations that hold for any point in the system to relations that hold for the entire system. These integral relations are the different variational principles that apply to solid mechanics problems. All of them can be derived from the principle of virtual work, but we shall simply state them here without proof. Much of finite element analysis in solid mechanics is based on one variational principle or another. We shall now discuss the three most widely used variational principles—two are actually minimum principles, while the other is a stationary principle.

1. Minimum potential energy principle (principle of virtual displacement)

Consider an elastic body of a given shape, deformed by the action of body forces and surface tractions. The potential energy of such a body is defined as the energy of deformation of the body (the strain energy) minus the work done on the body by the external forces. The theorem of minimum potential energy may be stated as follows [2]:

"The displacement (u, v, w) which satisfies the differential equations of equilibrium, as well as the conditions at the bounding surface, yields a smaller value for the potential energy ... than any other displacement which satisfies the same conditions at the bounding surface."

Hence, if $\Pi(u, v, w)$ is the potential energy, $U_p(u, v, w)$ is the strain energy, and $V_p(u, v, w)$ is the work done by the applied loads during displacement changes, then, according to the minimum principle, we have at equilibrium

$$\delta\Pi(u, v, w) = \delta[U_p(u, v, w) - V_p(u, v, w)]$$
$$= \delta U_p(u, v, w) - \delta V_p(u, v, w) = 0 \qquad (C.17)$$

In equation C.17 we tacitly assume that the variation is taken with respect to the displacements while all the other parameters are held fixed.

The strain energy of a linear elastic body is defined as

$$U_p(u, v, w) = \frac{1}{2} \iiint_V \lfloor\epsilon\rfloor\{\sigma\}\, dV$$

and with Hooke's law

$$U_p(u, v, w) = \frac{1}{2} \iiint_V \lfloor\epsilon\rfloor[C]\{\epsilon\}\, dV \qquad (C.18)$$

where V is the volume of the body. From equation (C.1)

$$\{\epsilon\} = \begin{Bmatrix} \epsilon_x \\ \epsilon_y \\ \epsilon_z \\ \gamma_{xy} \\ \gamma_{xz} \\ \gamma_{yz} \end{Bmatrix} = \begin{bmatrix} \dfrac{\partial}{\partial x} & 0 & 0 \\ 0 & \dfrac{\partial}{\partial y} & 0 \\ 0 & 0 & \dfrac{\partial}{\partial z} \\ \dfrac{\partial}{\partial y} & \dfrac{\partial}{\partial x} & 0 \\ \dfrac{\partial}{\partial z} & 0 & \dfrac{\partial}{\partial x} \\ 0 & \dfrac{\partial}{\partial z} & \dfrac{\partial}{\partial y} \end{bmatrix} \begin{Bmatrix} u \\ v \\ w \end{Bmatrix}$$

or

$$\{\epsilon\} = [B]\{\delta\} \qquad (C.19)$$

Substituting equation C.19 into equation C.18 gives the strain energy in terms of the displacement field, that is,

$$U_p(u, v, w) = \frac{1}{2} \iiint_V \lfloor\delta\rfloor[B]^T[C][B]\{\delta\}\, dV$$

If initial strains $\{\epsilon_0^*\}^*$ are present,† the strain energy becomes

$$U_P(u, v, w) = \frac{1}{2} \iiint_V [\lfloor\tilde{\delta}\rfloor[B]^T[C][B]\{\delta\} - 2\lfloor\tilde{\delta}\rfloor[B]^T[C]\{\epsilon_0^*\}]\, dV \quad (C.20)$$

The work done by the external forces is given by

$$V_P(u, v, w) = \iiint_V (X_u^* + Y_v^* + Z_w^*)\, dV$$

$$+ \iint_{S_1} (T_x^* u + T_y^* v + T_z^* w)\, dS_1$$

$$= \iiint_V \lfloor F^*\rfloor\{\delta\}\, dV + \iint_{S_1} \lfloor T^*\rfloor\{\delta\}\, dS_1 \quad (C.21)$$

where

$$\lfloor F^*\rfloor = \lfloor X^*, Y^*, Z^*\rfloor, \quad \lfloor T^*\rfloor = \lfloor T_x^* \ \ T_y^* \ \ T_z^*\rfloor, \quad \lfloor \delta\rfloor = \lfloor u \ \ v \ \ w\rfloor$$

The body force components $\lfloor F^*\rfloor$ and the surface tractions $\lfloor T^*\rfloor$ are the given external forces, and S_1 is the portion of the surface of the body on which the tractions are prescribed. Combining these equations, we can write the general potential energy functionals as

$$\Pi(u, v, w) = \frac{1}{2} \iiint_V [\lfloor\tilde{\delta}\rfloor[B]^T[C][B]\{\delta\} - 2\lfloor\tilde{\delta}\rfloor[B]^T[C]\{\epsilon_0^*\}]\, dV$$

$$- \iiint_V \lfloor F^*\rfloor\{\delta\}\, dV + \iint_{S_1} \lfloor T^*\rfloor\{\delta\}\, dS_1$$

The displacement field u, v, w that minimizes Π and satisfies all the boundary conditions is the equilibrium displacement field. When we use the principle of minimum potential energy in finite element analysis, we assume the form of the displacement field in each element and then use the functional Π to derive the element equations as explained in Chapter 3. This approach is called the *displacement method* or the *stiffness method*, and the element equations that result are the approximate equilibrium equations. The compatibility conditions are identically satisfied.

† Throughout this appendix, quantities denoted by a superscript asterisk are known or are given a priori.

2. Minimum complementary energy principle (principle of virtual stress)

Just as the potential energy principle corresponds to the equilibrium condition in an elastic body, the complementary energy principle corresponds to the compatibility condition. The principle states that, in an elastic body, "the state of stress that satisfies the stress-strain relations in the interior and all the prescribed displacement boundary conditions also minimizes the system's complementary energy."

Hence, if $\Pi_c(\sigma_x, \sigma_y, \ldots, \tau_{zx})$ is the complementary energy, $U_c(\sigma_x, \sigma_y, \ldots, \tau_{zx})$ is the complementary stress energy, and V_c is the work done by the applied loads during stress changes, then, according to the minimum complementary energy principle,

$$\begin{aligned} \delta\Pi_c &= \delta(U_c - V_c) \\ &= \delta U_c - \delta V_c = 0 \end{aligned} \tag{C.22}$$

Here the variation is taken with respect to the stress components rather than the displacement components. The complementary stress energy is defined as

$$U_c(\sigma_x, \sigma_y, \ldots, \tau_{zx}) = \frac{1}{2} \iiint_V \lfloor \sigma \rfloor [D]\{\sigma\}\, dV$$

and with initial strains $\{\epsilon_0\}$

$$U_c = \frac{1}{2} \iiint_V \left[\lfloor \sigma \rfloor [D]\{\sigma\} + 2\lfloor \sigma \rfloor\{\epsilon_0^*\} \right] dV \tag{C.23}$$

The complementary work of the external loads is given by

$$\begin{aligned} V_c &= \iint_{S_2} (T_x^* \bar{u} + T_y^* \bar{v} + T_z^* \bar{w})\, dS \\ &= \iint_{S_2} \lfloor T^* \rfloor \{\bar{\delta}\}\, dS_2 \end{aligned} \tag{C.24}$$

where $\bar{u}, \bar{v}, \bar{w}$ are the prescribed displacement components, and S_2 is the part of the bounding surface where they are prescribed. Hence the complementary energy functional becomes

$$\Pi_c(\sigma_x, \sigma_y, \ldots, \tau_{zx}) = \frac{1}{2} \iiint_V \left[\lfloor \sigma \rfloor [D]\{\sigma\} + 2\lfloor \sigma \rfloor\{\epsilon_0^*\} \right] dV - \iint_{S_2} \lfloor T^* \rfloor\{\bar{\delta}\}\, dS \tag{C.25}$$

When we use the complementary energy principle in finite element analysis, we assume the form of the stress field in each element and proceed as explained in Chapter 3. This approach is called the *force method* or the *flexibility method*, and the element equations that result are the approximate compatibility equations. The equilibrium equations are identically satisfied.

3.   Reissner's principle [7]

In contrast to the potential energy functional, which allows variations of displacement, and the complementary energy functional, which allows variations of stress, the Reissner functional allows variations of both displacement and stress. Another contrasting feature of the Reissner principle is that it does not evolve naturally from the concept of virtual work, but rather is a contrived principle which may be obtained either from the potential energy or the complementary energy theorem. Consequently, *Reissner's principle* embodies aspects of both the equilibrium and the compatibility conditions.

Reissner's principle states that

$$\delta \Pi_R(u, v, w, \sigma_x, \sigma_y, \ldots, \tau_{xz}) = 0 \qquad (C.26)$$

where the functional Π_R for a linear elastic material is given as

$$\Pi_R = \iiint_V \left[\lfloor \sigma \rfloor \{c\} - \tfrac{1}{2} \lfloor \sigma \rfloor [D] \{\sigma\} - \lfloor \delta \rfloor \{F^*\} \right] dV$$

$$- \iint_{S_1} \lfloor \delta \rfloor \{T^*\} \Big|_{\substack{\text{on sur-}\\ \text{face } S_1}} dS_1 - \iint_{S_2} \lfloor \tilde{\delta} - \delta^* \rfloor \{\bar{T}\} \Big|_{\substack{\text{on sur-}\\ \text{face } S_2}} dS_2 \qquad (C.27)$$

The notation is the same as used previously except that $\lfloor \bar{T} \rfloor = \lfloor T_x \quad T_y \quad T_z \rfloor$ are nonprescribed surface tractions on portion S_2 of the bounding surface. The variation of Π_R with respect to $\{\tilde{\delta}\}$ and $\{\sigma\}$ yields neither a maximum nor a minimum but only a stationary value.

To use Reissner's principle in finite element analysis, we must assume the form of both the displacement and the stress fields within each element. Application of Reissner's principle is not extensive; it appears most often in the analysis of plate and shell problems.

4.   Hamilton's minimum principle

Another variational principle sometimes used to study the dynamic behavior of elastic structures is *Hamilton's principle*. This differs from the preceding three principles in that the variation of the functional is taken with respect to only one independent variable, namely, time. According to Hamilton's principle, the first variation of the Lagrangian function, L, must vanish, that is,

$$\delta \int_{t_0}^{t_1} L \, dt = 0 \qquad (C.28)$$

where

$$L = E_k - U_s - W_P \qquad (C.29)$$

and E_k = the total kinetic energy of the body,
  U_s = the internal strain energy,
  W_P = the work done by the applied loads when displacement is varied.
  If we define $\{\dot{\delta}\}$ as the three-component velocities at a point and ρ as the mass density of the material, equation C.29 may be written in matrix form as

$$L = \frac{1}{2} \iiint_V [\rho \lfloor \dot{\delta} \rfloor \{\dot{\delta}\} - \lfloor \epsilon \rfloor [C]\{\epsilon\} + 2\lfloor \delta \rfloor \{F^*\}] \, dV + \iint_{S_1} \lfloor \delta \rfloor \{T^*\} \, dS_1$$

$$(C.30)$$

C.9 PLANE STRAIN AND PLANE STRESS

Two of the many important specializations of three-dimensional linear elasticity theory are the cases of plane strain and plane stress. Both of these physically distinct situations can be described in terms of two independent coordinates, say x and y. In plane strain the component of displacement normal to the x-y plane is zero, while in plane stress the component of stress normal to the x-y plane is zero. For these two cases the foregoing equations of elasticity theory simplify to the following.

Plane strain

  Stress components:

$$\lfloor \sigma \rfloor = \lfloor \sigma_x \quad \sigma_y \quad \tau_{xy} \rfloor$$
$$\sigma_z = v(\sigma_x + \sigma_y) \qquad (C.31)$$

  Strain components:

$$\lfloor \epsilon \rfloor = \lfloor \epsilon_x \quad \epsilon_y \quad \gamma_{xy} \rfloor \qquad (C.32)$$

  Hooke's law:

$$\{\sigma\} = [C]\{\epsilon\} \qquad (C.33)$$

$$[C] = \frac{E}{(1 + v)(1 - 2v)} \begin{bmatrix} 1 - v & v & 0 \\ v & 1 - v & 0 \\ 0 & 0 & \dfrac{1 - 2v}{2} \end{bmatrix} \qquad (C.34)$$

Static equilibrium:

$$\frac{\partial \sigma_x}{\partial x} + \frac{\partial \tau_{xy}}{\partial y} + F_x = 0$$

$$\frac{\partial \sigma_y}{\partial y} + \frac{\partial \tau_{xy}}{\partial x} + F_y = 0 \tag{C.35}$$

$$\frac{\partial \sigma_z}{\partial z} + F_z = 0$$

Compatibility:

$$\frac{\partial^2 \epsilon_x}{\partial y^2} + \frac{\partial^2 \epsilon_y}{\partial x^2} = 2 \frac{\partial^2 \gamma_{xy}}{\partial x \, \partial y} \tag{C.36}$$

Differential equations for displacements:

$$\nabla^2 u + \frac{1}{1 - 2v} \frac{\partial}{\partial x} \left(\frac{\partial u}{\partial x} + \frac{\partial v}{\partial y} \right) + \frac{F_x}{u} = 0$$

$$\nabla^2 v + \frac{1}{1 - 2v} \frac{\partial}{\partial y} \left(\frac{\partial u}{\partial x} + \frac{\partial v}{\partial y} \right) + \frac{F_y}{u} = 0 \tag{C.37}$$

where

$$\nabla^2 = \frac{\partial^2}{\partial x^2} + \frac{\partial^2}{\partial y^2}$$

with boundary conditions

$$T_x^* = \lambda \psi l + \mu \left[2 \frac{\partial u}{\partial x} l + \left(\frac{\partial u}{\partial y} + \frac{\partial v}{\partial x} \right) m \right]$$

$$T_y^* = \lambda \psi m + \mu \left[2 \frac{\partial v}{\partial y} m + \left(\frac{\partial v}{\partial x} + \frac{\partial u}{\partial y} \right) l \right] \tag{C.38}$$

Variational principles:
The matrix forms of these equations remain the same. Explicit forms may be recovered by substituting equations C.31–C.34.

Plane stress

Stress components:

$$\lfloor \sigma \rfloor = \lfloor \sigma_x \quad \sigma_y \quad \tau_{xy} \rfloor$$
$$\sigma_z = 0 \tag{C.39}$$

Strain components:

$$\lfloor \epsilon \rfloor = \lfloor \epsilon_x \quad \epsilon_y \quad \gamma_{xy} \rfloor$$

$$\epsilon_z = \frac{v}{1 - v}(\epsilon_x + \epsilon_y) \tag{C.40}$$

Hooke's law:

$$\{\sigma\} = [C]\{\epsilon\}$$

$$[C] = \frac{E}{1 - v^2}\begin{bmatrix} 1 & v & 0 \\ v & 1 & 0 \\ 0 & 0 & \dfrac{1 - v}{2} \end{bmatrix} \tag{C.41}$$

Static equilibrium:

$$\frac{\partial \sigma_x}{\partial x} + \frac{\partial \tau_{xy}}{\partial y} + F_x = 0$$

$$\frac{\partial \sigma_y}{\partial y} + \frac{\partial \tau_{xy}}{\partial x} + F_y = 0 \tag{C.42}$$

Compatibility:

$$\frac{\partial^2 \epsilon_y}{\partial z^2} + \frac{\partial^2 \epsilon_z}{\partial y^2} = 0$$

$$\frac{\partial^2 \epsilon_z}{\partial x^2} + \frac{\partial^2 \epsilon_x}{\partial z^2} = 0$$

$$\frac{\partial^2 \epsilon_x}{\partial y^2} + \frac{\partial^2 \epsilon_y}{\partial x^2} = 2\frac{\partial^2 \gamma_{xy}}{\partial x\, \partial y} \tag{C.43}$$

$$\frac{\partial^2 \epsilon_x}{\partial y\, \partial z} = \frac{\partial^2 \gamma_{xy}}{\partial x\, \partial z}$$

Differential equations for displacements:

$$\nabla^2 u + \frac{1 + v}{1 - v}\frac{\partial}{\partial x}\left(\frac{\partial u}{\partial x} + \frac{\partial v}{\partial y}\right) + \frac{F_x}{\mu} = 0$$

$$\nabla^2 v + \frac{1 + v}{1 - v}\frac{\partial}{\partial y}\left(\frac{\partial u}{\partial x} + \frac{\partial v}{\partial y}\right) + \frac{F_y}{\mu} = 0 \tag{C.44}$$

with the boundary conditions given by equations C.38.

Variational principles:

By substituting equations C.39–C.41 the explicit forms of these functionals may be obtained from the general forms given previously.

C.10   THE AIRY STRESS FUNCTION (TWO-DIMENSIONAL PROBLEMS)

The solution of plane strain and plane stress problems involves the determination of the three stress components, σ_x, σ_y, τ_{xy}. These components can be found by integrating the differential equations of equilibrium together with the compatibility and boundary conditions. One way to solve for the stress is to introduce a new function, the *Airy stress function*, and transform the governing equations into a single higher-order equation involving only one unknown. Then the higher-order equation is solved for the stress function only.

Under the assumption that the body forces in a given problem are derivable from a given potential function β^* as $\bar{F}^* = -\nabla\beta^*$, that is, $F_x^* = -(\partial\beta^*/\partial x)$, $F_y^* = -(\partial\beta^*/\partial y)$, the Airy stress function Φ is defined as

$$\sigma_x = \frac{\partial^2\Phi}{\partial y^2} + \beta$$

$$\sigma_y = \frac{\partial^2\Phi}{\partial x^2} + \beta \qquad\qquad (C.45)$$

$$\tau_{xy} = -\frac{\partial^2\Phi}{\partial x\,\partial y}$$

When the body force is simply the gravity force in say, the x direction, the potential is

$$\beta = -\rho g x$$

The definitions of equation C.45 identically satisfy the equations of static equilibrium, and when the compatibility conditions are invoked there result the following fourth-order equations for the Airy stress function.

Plain strain

$$\frac{\partial^4\Phi}{\partial x^4} + 2\frac{\partial^4\Phi}{\partial x^2\,\partial y^2} + \frac{\partial^4\Phi}{\partial y^4} + \frac{1-2v}{1-v}\left(\frac{\partial^2\beta^*}{\partial x^2} + \frac{\partial^2\beta^*}{2y^2}\right) \qquad (C.46)$$

or

$$\nabla^4\Phi + \frac{1-2v}{1-v}\nabla^2\beta^* = 0$$

Plain stress

$$\frac{\partial^4 \Phi}{\partial x^4} + 2 \frac{\partial^4 \Phi}{\partial x^2 \, \partial y^2} + \frac{\partial^4 \Phi}{\partial y^4} + (1 - v)\left(\frac{\partial^2 \beta^*}{\partial x^2} + \frac{\partial^2 \beta^*}{\partial y^2}\right) = 0 \qquad \text{(C.47)}$$

or

$$\nabla^4 \Phi + (1 - v)\nabla^2 \beta^* = 0$$

The boundary conditions for these equations are

$$X^* = l\sigma_x + m\tau_{xy}$$
$$Y^* = m\sigma_y + l\tau_{xy} \qquad \text{(C.48)}$$

The operators ∇^2 and ∇^4 are sometimes called the *harmonic* and *biharmonic* operators, respectively. Once the stress function has been found from either equation C.46 or C.47, the stresses may be obtained by differentiation according to equation C.45.

It is also possible to introduce stress functions for other types of elasticity problems, including three-dimensional ones, but discussion of these is beyond our present scope.

C.11 THERMAL STRESSES

The presence of thermal stresses due to nonuniform temperature distributions in a body does not alter the basic equations for stress and strain in Sections C.1 and C.2. However, the displacement field is modified. In general, the temperature distribution in the body is not known, but rather must be found from a solution of the thermal energy equation. Once the temperature distribution, T, is known relative to some datum, the coefficient of thermal expansion, α, is used to find an "equivalent" loading on the body. We define an equivalent loading as

$$\mathbf{F}_T = -\nabla(\alpha T) \qquad \text{and} \qquad P_T = \alpha T \qquad \text{(C.49)}$$

where $\mathbf{F}_T$ is a body force per unit volume and P_T is a local surface *pressure*. Then we can solve the thermal stress problem by noting that the displacement field in a nonuniformly heated body is the same as if the body were subjected to the body force $\mathbf{F}_T$ and the normal surface pressure P_T, together with the other forces and surface tractions already acting on it.

C.12 THIN-PLATE BENDING

We define a thin plate (see Figure 6.12) as a three-dimensional body of constant thickness h whose center plane is coincident with the x-y plane and whose deflections always remain small in relation to h. The classical theory of thin-plate bending incorporates three basic assumptions:

1. The center plane has no strain.
2. The stress component normal to the center plane, σ_z, is zero.
3. Normals to the center plane remain normal during bending.

According to these assumptions,

$$\epsilon_z = \gamma_{xz} = \gamma_{yz} = \sigma_z = 0 \qquad (C.50)$$

and the in-plane displacements are related to the deflection by

$$u = z \frac{\partial w}{\partial x} \qquad \text{and} \qquad v = z \frac{\partial w}{\partial y} \qquad (C.51)$$

These are the Kirchhoff constraints, which effectively reduce the plate-bending problem to that of finding only $w(x, y)$.

Instead of characterizing the plate-bending problem with the usual stress vector $\lfloor \sigma \rfloor = \lfloor \sigma_x \ \sigma_y \ \tau_{xy} \rfloor$ and the strain vector $\lfloor \epsilon \rfloor = \lfloor \epsilon_x \ \epsilon_y \ \gamma_{xy} \rfloor$, it is convenient to use other, analogous parameters to play roles similar to those of stress and strain. In place of the stress vector we introduce the *line moments* M_{xx}, M_{yy}, and M_{xy}, which are moments per unit length. The first subscript designates the axis normal to the plane on which the moment acts; the second subscript indicates the direction of the shear generating the moment. In place of the strain vector we shall use the *plate curvature* C_x, C_y, C_{xy}, defined as

$$C_x = -\frac{\partial^2 w}{\partial x^2}, \qquad C_y = -\frac{\partial^2 w}{\partial y^2}, \qquad C_{xy} = 2 \frac{\partial^2 w}{\partial x \, \partial y} \qquad (C.52)$$

Replacing stress components by line moments and strain resultants makes the problem formulation easier. In terms of line moments and curvatures, the constitutive equation for an orthotropic plate becomes

$$\begin{Bmatrix} M_{xx} \\ M_{yy} \\ M_{xy} \end{Bmatrix} = \begin{bmatrix} D_x & D_1 & 0 \\ D_1 & D_y & 0 \\ 0 & 0 & D_{xy} \end{bmatrix} \begin{Bmatrix} C_x \\ C_y \\ C_{xy} \end{Bmatrix} \qquad (C.53)$$

or

$$\{M\} = [D]\{C\}$$

where D_x, D_y, and D_1 are the flexural rigidities of the plate. If the plate is composed of an isotropic material, we have

$$[D] = \frac{Eh^3}{12(1-v^2)} \begin{bmatrix} 1 & v & 0 \\ v & 1 & 0 \\ 0 & 0 & \dfrac{1-v}{2} \end{bmatrix} \qquad (C.54)$$

When the transverse loading on the plate per unit area is $q(x, y)$, it can be shown that the differential equation governing the plate deflection is for an orthotropic solid

$$D_x \frac{\partial^4 w}{\partial x^2} + 2(D_1 + 2D_{xy}) \frac{\partial^4 w}{\partial x^2 \, \partial y^2} + D_y \frac{\partial^4 w}{\partial y^4} = q(x, y) \qquad \text{(C.55)}$$

and for an isotropic solid this reduces to

$$\nabla^4 w = \frac{\partial^4 w}{\partial x^2} + 2 \frac{\partial^4 w}{\partial x^2 \, dy^2} + \frac{\partial^4 w}{\partial y^4} = \frac{12q(x, y)(1 - v^2)}{Eh^3} \qquad \text{(C.56)}$$

The boundary conditions for equations C.55 and C.56 involve w and its derivatives up to and inclusive of the third order, and they depend on the type of edge conditions, for example, simply supported, built-in, free, and specified deflection.

Variational principles

Minimum potential energy:

$$\delta \Pi_p = 0 \qquad \text{(C.57)}$$

$$\Pi_p = \frac{1}{2} \iint_A \lfloor c \rfloor [D] \{c\} h \, dA - \int_{S_q} q^* w \, dS - \Sigma F_i^* w_i - \Sigma M_i^* \left\{ \begin{array}{c} -\dfrac{\partial w}{\partial x} \\[2mm] \text{or} \\[2mm] \dfrac{\partial w}{\partial y} \end{array} \right\}_i \qquad \text{(C.58)}$$

where Π_p = potential energy functional,
q^* = specified surface loading on surface portion S_q
F_i^* = specified concentrated forces in the z direction,
w_i = displacements corresponding to forces F_i,
M_i^* = specified concentrated moments,

$$\left\{ \begin{array}{c} -\dfrac{\partial w}{\partial x} \\[2mm] \text{or} \\[2mm] \dfrac{\partial w}{\partial y} \end{array} \right\}_i = \text{angular displacements corresponding to moments } M_i^*.$$

Note that for an isotropic solid equation C.56 is the Euler-Lagrange equation for the functional of equation C.58.

Minimum complementary energy (isotropic solid):

$$\Pi_c = \frac{1}{2} \iint_A \frac{12}{Eh^3} \left[\left(\frac{\partial U}{\partial x} \right)^2 + \left(\frac{\partial V}{\partial y} \right)^2 - 2v \frac{\partial U}{\partial x} \frac{\partial V}{\partial y} + \frac{1+v}{2} \left(\frac{\partial U}{\partial y} + \frac{\partial V}{\partial x} \right)^2 \right] dA$$

$$= + \iint_A \frac{12(1-v)}{Eh^3} \left[\Omega^2 - \Omega \left(\frac{\partial U}{\partial x} + \frac{\partial V}{\partial y} \right) \right] dA$$

$$+ \int_S \left(\frac{\partial w}{\partial n} M_n^* - w^* V_n \right) dS - \Sigma w_i^* F_i$$

where $U, V \equiv$ Southwell stress functions [8]

Ω = quantity related to the distributed load by $q = \nabla^2 \Omega$,

M_n^* = normal bending moment at the boundary,

V_n = normal shearing force at the boundary,

w^* = prescribed boundary displacement,

F_i = concentrated nodal forces.

Reissner's principle

$$\delta \Pi_R = 0$$

$$\Pi_R = \iint_A \left\{ \lfloor M \rfloor \{c\} - \frac{1}{2EI} \right.$$

$$\times [M_{xx}^2 + M_{yy}^2 - 2vM_{xx} + 2(1+v)M_{xy}^2] - qw \right\} h \, dA$$

$$- \int_{S_1} \left[\lfloor Q_n^*, M_n^*, M_{ns}^* \rfloor \left[w, \frac{\partial w}{\partial n}, \frac{\partial w}{\partial s} \right]^T \right] dS_1$$

$$+ \int_{S_2} \left[\lfloor Q_n, M_n, M_{ns} \rfloor \left[w - w^*, \frac{\partial w}{\partial n} - \frac{\partial w^*}{\partial n}, \frac{\partial w}{\partial s} - \frac{\partial w^*}{\partial s} \right]^T \right] dS_2$$

where I = moment of inertia about the centroidal axis,

n, s = normal and tangential directions on the boundary,

S_1 = portion of boundary where stress is specified,

Q = shear load per unit length,

S_2 = portion of boundary where displacement is specified.

REFERENCES

1. S. Timoshenko and J. N. Goodier, *Theory of Elasticity*, McGraw-Hill Book Company, New York, 1951.

2. A. E. H. Love, *A Treatise on the Mathematical Theory of Elasticity*, Dover Publications, New York, 1944.

3. Y. C. Fung, *Foundations of Solid Mechanics*, Prentice-Hall, Englewood Cliffs, N.J., 1965.
4. I. S. Sokolnikoff, *Mathematical Theory of Elasticity*, 2nd ed., McGraw-Hill Book Company, New York, 1956.
5. E. E. Sechler, *Elasticity in Engineering*, John Wiley and Sons, New York, 1952.
6. R. V. Southwell, *An Introduction to the Theory of Elasticity*, Oxford University Press, 1936.
7. E. Reissner, "On a Variational Theorem in Elasticity," *J. Math. Phys.*, Vol. 29, 1950, pp. 90–95.
8. R. V. Southwell, "On the Analogues Relating Flexure and Extension of Flat Plates," *Quart. J. Mech. Appl. Math.*, Vol. 3, 1950, pp. 257–270.
9. H. Kraus, *Thin Elastic Shells*, John Wiley and Sons, New York, 1967.
10. W. Flügge, *Stresses in Shells*, Springer-Verlag, Berlin, 1960.

APPENDIX D

BASIC EQUATIONS FROM
FLUID MECHANICS

D.1 INTRODUCTION

Here we review the introductory concepts and governing equations for the general flow of an isotropic single-species fluid. Again we assume that the continuum assumption holds. This means that the dimensions for a given problem are everywhere large compared to the mean-free-molecular path of the fluid substance.† Hence a fluid particle is taken as the smallest lump of fluid having enough molecules to allow the continuum assumption.

Since most engineers and scientists use the Eulerian description to formulate their problems, this description is used here exclusively. The various equations for different kinds of flow are given in Cartesian coordinates without derivation or proof. Readers desiring more details or statements of the equations in different coordinate systems should consult the selected references given at the end of this appendix.

D.2 DEFINITIONS AND CONCEPTS [1–3]

A *fluid* is a substance (either a liquid or a gas) that continuously deforms under the action of applied surface stresses. This implies that shear stresses do not exist in a fluid unless it is deforming. As a continuum, a fluid possesses the following physical properties: density, ρ; viscosity, μ; thermal conductivity, k; thermal diffusivity, a; specific heats, c_v and c_p; electrical conductivity, σ; dielectric constant, ϵ_R, magnetic permeability, ϵ_m; and expansivity factor, β.

† For example, the mean-free-molecular path for air at standard temperature and pressure is about 2×10^{-6} inch.

471

The *state of stress* is characterized in the same way in a fluid as it is in a solid (see Section C.2). For the special case of a fluid at rest, the off-diagonal terms of the stress tensor are zero (no shear stresses), and the diagonal terms are all the same and are equal to the negative of the hydrostatic pressure. For a fluid in motion, all terms of the stress tensor may be nonzero, though the stress tensor remains symmetric.

A flow field is characterized by a velocity vector that is a continuous function of space and time. To represent a fluid motion graphically, it is convenient to introduce the concepts of streamlines, pathlines, and streaklines. An imaginary line in a flow field along which flow velocity vectors are everywhere tangent at any instant is called a *streamline*. The trajectory or locus of points through which a fluid particle of fixed identity passes as it moves in a flow field is known as a *pathline*. A *streakline* is defined as a line that connects all fluid particles which at some time have passed through a fixed point in space. For an unsteady flow, streamlines, pathlines, and streaklines are, in general, different; but, for a steady flow, they are identical.

In this book we classify a flow field as one-, two-, or three-dimensional, depending on the number of independent space coordinates needed to specify the velocity field. In addition to classifying flow fields, we may also classify the various types of flow. Admittedly, such a classification is somewhat arbitrary because there can be much overlap between categories, but a given classification scheme can serve to highlight major divisions of the subject in spite of the hazards of oversimplification.

As Figure D.1 indicates, a flow may first be classified as either inviscid or viscous. Inviscid flows are frictionless flows characterized by zero viscosity. No real flows are inviscid, but there are numerous flow situations in which viscous effects can be neglected. The viscosity of the fluid and the magnitude

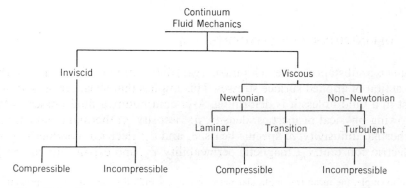

Figure D.1.   Some major categories or types of flow in continuum fluid mechanics.

of the gradient of the velocity field throughout the flow are the two primary factors that determine when viscous effects are insignificant. Often viscous effects are confined to a thin region or boundary layer near flow boundaries, and the rest of the flow can be considered frictionless.

Inviscid flows may further be classified as either compressible or incompressible, depending on whether density variations are large or relatively unimportant. High-speed gas flows are an example of compressible flows, whereas liquid flows are *usually* assumed to be incompressible.

Since all real fluids have a finite nonzero viscosity, the subject of viscous flow is of paramount importance in fluid mechanics. A viscous flow is said to be Newtonian if Stokes's† law of friction prevails for the fluid in question, and non-Newtonian if some other law governs the relation between stress and rate of strain. Most fluids are assumed to be Newtonian or purely viscous, but some real fluids exhibit more complex behavior. We shall limit our review here to Newtonian fluids only.

The dynamic macroscopic behavior of a flow determines whether it is classified as laminar or turbulent, or whether it is in transition between the the two. Fluid motion is said to be *laminar* if the fluid flows in imaginary laminas or layers and there is no macroscopic mixing of adjacent fluid layers. A flow is said to be *turbulent*, however, if such mixing occurs. The mixing of fluid particles from adjacent layers is caused by small velocity fluctuations superimposed on the mean flow velocity field. These velocity fluctuations have a random or irregular variation with the space and time coordinates, but they can be characterized by statistically distinct averages. Clearly, laminar flows are more amenable to detailed analysis than turbulent flows. *Transition* flow occurs when a laminar flow becomes unstable and approaches a turbulent flow. In other words, the type of flow that exists during the time span in which a flow changes from fully developed laminar to fully developed turbulent is known as transition flow. In such a flow, there can be patches of turbulent flow where irregularity and disorderliness prevail, and regions of laminar flow characterized by orderliness. Essentially, the flow changes in time by alternating between being laminar and being turbulent.

Any of these types of viscous flows can be compressible or incompressible. Furthermore, inviscid or viscous flows can be named according to the type of flow boundaries. We can have *internal flows* (i.e., pipe flows), *external flows* (i.e., flows over airfoils), or *internal-external flows* (i.e., flows around a center body in pipe). Flows may also be given a temporal description by labeling them as steady, periodic, quasi-steady, or transient.

Here we have indicated only the major types of flow and the adjectives used to describe them. Obviously, there are many other special types of flow

---

† Stokes's law is presented in detail in the next section.

which we have not considered. For example, under the heading of compressible flow, we can have subsonic, transonic, supersonic, or hypersonic flow; and viscous flows may be cavitated, separated, or stratified.

D.3 LAWS OF MOTION [4,5]

Solving fluid mechanics problems ultimately involves solving in some form the equations expressing the three laws of motion of the fluid:

1. Conservation of mass.
2. Conservation of momentum.
3. Conservation of energy.

In addition to these laws, it is sometimes necessary to include an equation of state and a viscosity equation for the fluid. These provide the pointwise relations among pressure, temperature, density, and viscosity regardless of whether the fluid is flowing or static. The conservation laws can be expressed in either integral or differential form. The integral form is most useful for determining the gross or global character of a flow as it interacts with a structure or other device. The differential form provides the relations that hold at any point in the flow; hence they are most useful for studying the detailed character. In the following, we shall summarize the differential form of the governing equations since finite element analyses of fluid flow problems rely on this form.

Differential continuity equation

By writing a mass balance for a differential control volume, we can show that the conservation-of-mass principle is expressed by

$$\frac{\partial \rho}{\partial t} + \frac{\partial(\rho u)}{\partial x} + \frac{\partial(\rho v)}{\partial y} + \frac{\partial(\rho w)}{\partial z} = 0 \qquad (D.1a)$$

or, in vector form,

$$\frac{\partial \rho}{\partial t} + \nabla \cdot (\rho \mathbf{V}) = 0 \qquad (D.1b)$$

where ρ = mass density,
 u, v, w = velocity components in the x, y, and z directions, respectively,
 t = time,
 $\mathbf{V} = u\hat{i} + v\hat{j} + w\hat{k}$,
 $\nabla = \hat{i}(\partial/\partial x) + \hat{j}(\partial/\partial y) + \hat{k}(\partial/\partial z)$.

Equations D.1 hold for the motion of any fluid satisfying the continuum assumption.

Differential momentum equation (Navier-Stokes equations)

Considering a differential volume (a fluid particle) and summing the forces acting on the particle results in the following differential equations of motion:

$$\rho\left(\frac{\partial u}{\partial t} + u\frac{\partial u}{\partial x} + v\frac{\partial u}{\partial y} + w\frac{\partial u}{\partial z}\right) = \rho B_x + \frac{\partial \sigma_{xx}}{\partial x} + \frac{\partial \tau_{yx}}{\partial y} + \frac{\partial \tau_{zx}}{\partial z} \quad \text{(D.2a)}$$

$$\rho\left(\frac{\partial v}{\partial t} + u\frac{\partial v}{\partial x} + v\frac{\partial v}{\partial y} + w\frac{\partial v}{\partial z}\right) = \rho B_y + \frac{\partial \tau_{xy}}{\partial x} + \frac{\partial \sigma_{yy}}{\partial y} + \frac{\partial \tau_{zy}}{\partial z} \quad \text{(D.2b)}$$

$$\rho\left(\frac{\partial w}{\partial t} + u\frac{\partial w}{\partial x} + v\frac{\partial w}{\partial y} + w\frac{\partial w}{\partial z}\right) = \rho B_z + \frac{\partial \tau_{xz}}{\partial x} + \frac{\partial \tau_{yz}}{\partial y} + \frac{\partial \sigma_{zz}}{\partial z} \quad \text{(D.2c)}$$

Here B_x, B_y, and B_z are the x, y, z components of the body force vector per unit volume:

$$\mathbf{B} = B_x \hat{i} + B_y \hat{j} + B_z \hat{k}$$

The left-hand side of equations D.2 is the substantial or material derivative of the velocity components; for example, the substantial derivative of u is:

$$\frac{Du}{Dt} = \frac{\partial u}{\partial t} + u\frac{\partial u}{\partial x} + v\frac{\partial u}{\partial y} + w\frac{\partial u}{\partial z}$$

The normal stresses σ_{xx}, σ_{yy}, and σ_{zz} include the fluid pressure P. Hence we can define new normal stresses of the form

$$\sigma_{xx} = -P + \sigma_x, \qquad \sigma_{yy} = -P + \sigma_y, \qquad \sigma_{zz} = -P + \sigma_z$$

Then, for instance, equation D.2a becomes

$$\rho\frac{Du}{Dt} = \rho B_x - \frac{\partial P}{\partial x} + \frac{\partial \sigma_x}{\partial x} + \frac{\partial \tau_{yx}}{\partial y} + \frac{\partial \tau_{zx}}{\partial z}$$

Equations D.2 are valid for any type of fluid. To use these equations to find velocity and pressure distributions for a particular flow and type of fluid, we must express the various stress components in terms of derivatives of velocity and the properties of the fluid. For Newtonian fluids, we have *Stokes's law of friction*, which is

$$
\left.\begin{array}{l}
\sigma_x = 2\mu \dfrac{\partial u}{\partial x} - \tfrac{2}{3}\mu \nabla \cdot \mathbf{V} \\[2mm]
\sigma_y = 2\mu \dfrac{\partial v}{\partial y} - \tfrac{2}{3}\mu \nabla \cdot \mathbf{V} \\[2mm]
\sigma_z = 2\mu \dfrac{\partial w}{\partial z} - \tfrac{2}{3}\mu \nabla \cdot \mathbf{V} \\[2mm]
\tau_{xy} = \tau_{yx} = \mu\left(\dfrac{\partial u}{\partial y} + \dfrac{\partial v}{\partial x}\right) \\[2mm]
\tau_{yz} = \tau_{zy} = \mu\left(\dfrac{\partial v}{\partial z} + \dfrac{\partial w}{\partial y}\right) \\[2mm]
\tau_{zx} = \tau_{xz} = \mu\left(\dfrac{\partial w}{\partial x} + \dfrac{\partial u}{\partial z}\right)
\end{array}\right\}
\tag{D.3}
$$

Substitution of equations D.3 into equations D.2 gives the general form of the Navier-Stokes equations:

$$
\rho \frac{Du}{Dt} = B_x - \frac{\partial P}{\partial x}
$$

$$
+ \frac{\partial}{\partial x}\left[2\mu\left(\frac{\partial u}{\partial x} - \tfrac{1}{3}\nabla \cdot \mathbf{V}\right)\right] + \frac{\partial}{\partial y}\left[\mu\left(\frac{\partial u}{\partial y} + \frac{\partial v}{\partial x}\right)\right] + \frac{\partial}{\partial z}\left[\mu\left(\frac{\partial w}{\partial x} + \frac{\partial u}{\partial z}\right)\right]
\tag{D.4a}
$$

$$
\rho \frac{Dv}{Dt} = B_y - \frac{\partial P}{\partial y}
$$

$$
+ \frac{\partial}{\partial y}\left[2\mu\left(\frac{\partial v}{\partial y} - \tfrac{1}{3}\nabla \cdot \mathbf{V}\right)\right] + \frac{\partial}{\partial z}\left[\mu\left(\frac{\partial v}{\partial z} + \frac{\partial w}{\partial y}\right)\right] + \frac{\partial}{\partial x}\left[\mu\left(\frac{\partial u}{\partial y} + \frac{\partial v}{\partial x}\right)\right]
\tag{D.4b}
$$

$$
\underbrace{\rho \frac{Dw}{Dt}}_{\substack{\text{inertia}\\\text{force}\\\text{terms}}} = \underbrace{B_z}_{\substack{\text{body}\\\text{force}\\\text{terms}}} - \underbrace{\frac{\partial P}{\partial z}}_{\substack{\text{pressure}\\\text{force}\\\text{terms}}}
$$

$$
\underbrace{+ \frac{\partial}{\partial z}\left[2\mu\left(\frac{\partial w}{\partial z} - \tfrac{1}{3}\nabla \cdot \mathbf{V}\right)\right] + \frac{\partial}{\partial x}\left[\mu\left(\frac{\partial w}{\partial x} + \frac{\partial u}{\partial z}\right)\right] + \frac{\partial}{\partial y}\left[\mu\left(\frac{\partial v}{\partial z} + \frac{\partial w}{\partial y}\right)\right]}_{\text{viscous force terms}}
\tag{D.4c}
$$

Thermal energy equation

For nonisothermal flows with temperature-dependent viscosity, the equations of continuity and momentum are coupled to the thermal energy equation, and all four equations must be solved simultaneously. In its most general form, the thermal energy equation is

$$\rho C_v \frac{DT}{Dt} + P\nabla \cdot \mathbf{V} + \nabla \cdot \mathbf{q}_r - \frac{\partial Q}{\partial t}$$

$$= \frac{\partial}{\partial x}\left(k \frac{\partial T}{\partial x}\right) + \frac{\partial}{\partial y}\left(k \frac{\partial T}{\partial y}\right) + \frac{\partial}{\partial z}\left(k \frac{\partial T}{\partial z}\right) + \mu\tilde{\Phi}(x, y, z, t) \quad (D.5)$$

where C_v = specific heat at constant volume,
k = thermal conductivity,
T = temperature,
$\mathbf{q}_r$ = radiation heat flux vector,
Q = internal heat generation,

and

$\tilde{\Phi}$ = viscous dissipation function

$$= \left(\frac{\partial u}{\partial z}\right)^2 + \left(\frac{\partial u}{\partial y}\right)^2 + \left(\frac{\partial v}{\partial x}\right)^2 + \left(\frac{\partial v}{\partial z}\right)^2 + \left(\frac{\partial w}{\partial x}\right)^2 + \left(\frac{\partial w}{\partial y}\right)^2$$

$$+ 2\left(\frac{\partial u}{\partial y}\frac{\partial v}{\partial x} + \frac{\partial v}{\partial z}\frac{\partial w}{\partial y} + \frac{\partial w}{\partial x}\frac{\partial u}{\partial z}\right)$$

$$+ \frac{4}{3}\left[\frac{\partial u}{\partial x}\left(\frac{\partial u}{\partial x} - \frac{\partial v}{\partial y}\right) + \frac{\partial v}{\partial y}\left(\frac{\partial v}{\partial y} - \frac{\partial w}{\partial z}\right) + \frac{\partial w}{\partial z}\left(\frac{\partial w}{\partial z} - \frac{\partial u}{\partial x}\right)\right]$$

Supplementary equations

The complete specification of a fluid flow problem requires two additional equations accompanying the continuity, momentum, and energy equations. These are the equation of state and the viscosity equation. Since the form of these equations is nonunique, we represent them generally as

Equation of state: $\rho = \rho(P, T)$ \hfill (D.6)

Viscosity equation: $\mu = \mu(P, T)$ \hfill (D.7)

Problem statement

The solution of general flow problems may be summarized as follows:

Given: (1) The geometry of the flow boundaries, (2) the physical properties of the fluid, (3) the seven governing equations: D.1, D.4a, D.4b, and D.4c, and D.5–D.7, and (4) a complete set of boundary and initial conditions for the governing equations.

Find: The distributions of velocity, u, v, and w; pressure, P; temperature, T; density, ρ; viscosity, μ.

The problem reduces to solving seven equations with seven unknowns. In addition to the obvious complexity of the problem as we have stated it thus far, other factors may cause difficulty. If the fluid forces acting on the bounding solids cause the shape of the boundary to deform, we must also simultaneously solve the elasticity equations for the solids. Thermal distortion of the boundaries may also be important. In this case, a complete solution would involve the solution of the heat conduction equation in the solids (equation D.5 with $u = v = w = 0$).

The governing equations hold at any instant of time and apply to either laminar or turbulent flows. However, in a turbulent flow, the field variables are fluctuating randomly about their mean values, and the mathematical nature of the problem becomes hopelessly complex. Because the instantaneous equations cannot be solved for a turbulent flow, the standard approach is to time-average the equations and obtain new ones which describe the temporally averaged distributions of the field variables. Discussion of the theoretical methods for predicting the behavior of turbulent flows is far beyond our purpose here. The interested reader can find treatments of this topic in specialized textbooks such as refs. 6 and 7.

Even if attention is restricted to laminar flows, a solution of the full set of equations has never been obtained. But, for many problems of practical interest, the governing equations simplify considerably and the mathematical difficulties become more tractable. In subsequent sections we shall present the governing equations for a number of special classes of problems. Often the governing equations and the form of a particular flow pattern can be described concisely in terms of a stream function. We consider this concept in the next section.

D.4 STREAM FUNCTIONS AND VORTICITY

Two-dimensional flows (both compressible and incompressible) can be conveniently characterized by introducing a mathematical artifice known as a stream function—$\psi(x, y, t)$. The stream function relates the concept of streamlines to the principle of mass conservation.

For a two-dimensional incompressible flow, the continuity equation reduces to

$$\frac{\partial u}{\partial x} + \frac{\partial v}{\partial y} = 0$$

This equation is identically satisfied if we define

$$u = \frac{\partial \psi}{\partial y} \quad \text{and} \quad v = -\frac{\partial \psi}{\partial x} \tag{D.8}$$

since

$$\frac{\partial u}{\partial x} + \frac{\partial v}{\partial y} = \frac{\partial^2 \psi}{\partial x\, \partial y} - \frac{\partial^2 \psi}{\partial y\, \partial x} = 0$$

In the flow field, lines of constant ψ and streamlines are identical and the flow rate between two streamlines is proportional to the numerical difference between the two stream functions corresponding to the two streamlines.

For a two-dimensional compressible flow, we have the analogous definitions

$$u = \frac{\rho_0}{\rho}\frac{\partial \psi}{\partial y}, \quad v = -\frac{\rho_0}{\rho}\frac{\partial \psi}{\partial x} \tag{D.9}$$

where ρ_0 is some reference density, and ρ is the local fluid density. Clearly, if we can find $\psi(x, y, t)$ for a given problem, we have essentially solved the problem. Equations governing the behavior of ψ for different types of flow problems will be given in the following sections as these problems are discussed.

Fluid rotation or *vorticity* is defined as the average angular velocity of any two mutually perpendicular line elements of a fluid particle. For a three-dimensional flow, vorticity is a three-component vector given by

$$\boldsymbol{\omega} = \omega_x \hat{i} + \omega_y \hat{j} + \omega_z \hat{k} \tag{D.10a}$$

where ω_x, ω_y, ω_z are the rotations† about the x, y, and z axes, respectively, and are defined by

$$\omega_x = \frac{1}{2}\left(\frac{\partial w}{\partial y} - \frac{\partial v}{\partial z}\right)$$

$$\omega_y = \frac{1}{2}\left(\frac{\partial u}{\partial z} - \frac{\partial w}{\partial x}\right) \tag{D.10b}$$

$$\omega_z = \frac{1}{2}\left(\frac{\partial v}{\partial x} - \frac{\partial u}{\partial y}\right)$$

† By convention, positive rotation follows the right-hand screw rule.

In vector notation, we have

$$2\boldsymbol{\omega} = \nabla \times \mathbf{V} \tag{D.10c}$$

The action of viscous forces in a flow field develops rotation. Hence any viscous flow is a rotational flow with $\boldsymbol{\omega} \neq 0$ [8]. In an initially irrotational flow field, rotation cannot be developed by the action of body forces or pressure forces. Only the action of shearing stress can produce rotationality. For this reason, inviscid ($\mu = 0$) flows initially irrotational can be assumed to remain irrotational, that is, $\boldsymbol{\omega} = 0$.

When $\mu = $ constant for rotational flows, the momentum equations may be written in terms of the vorticity as follows. The x component becomes

$$\frac{\partial \omega_x}{\partial t} + u \frac{\partial \omega_x}{\partial x} + v \frac{\partial \omega_x}{\partial y} + w \frac{\partial \omega_x}{\partial z}$$

$$= \omega_x \frac{\partial u}{\partial x} + \omega_y \frac{\partial u}{\partial y} + \omega_z \frac{\partial u}{\partial z} + \frac{u}{\rho} \left(\frac{\partial^2 \omega_x}{\partial x^2} + \frac{\partial^2 \omega_x}{\partial y^2} + \frac{\partial^2 \omega_x}{\partial z^2} \right)$$

Equations for the y and z components have similar forms. Using matrix notation, we can express all three equations in the concise form

$$\frac{D\boldsymbol{\omega}}{Dt} = (\boldsymbol{\omega} \cdot \nabla)\mathbf{V} + \frac{u}{\rho} \nabla^2 \boldsymbol{\omega} \tag{D.11}$$

For a two-dimensional flow with $\mu = $ constant, the stream function formulation can be used to express the combined continuity and momentum equations in the form

$$\frac{\partial}{\partial t} (\nabla^2 \psi) + \frac{\partial \psi}{\partial y} \frac{\partial}{\partial x} (\nabla^2 \psi) - \frac{\partial \psi}{\partial x} \frac{\partial}{\partial y} (\nabla^2 \psi) = \frac{\mu}{\rho} \nabla^4 \psi \tag{D.12}$$

where

$$\nabla^4 = \frac{\partial^4}{\partial x^4} + \frac{\partial^4}{\partial y^4} + 2 \frac{\partial^4}{\partial x^2 \, \partial y^2}$$

D.5 POTENTIAL FLOW [8,9]

Inviscid irrotational flow is called potential flow because the velocity field in the flow can be derived from a *potential function*. A well-known vector identity tells us that, for any function $\Phi(x, y, z, t)$ having continuous first and second derivatives,

$$\nabla \times (\nabla \Phi) = 0 \tag{D.13}$$

But the condition of irrotationality from equation D.12 states that $\nabla \times \mathbf{V} = 0$. Hence, for an irrotational flow, we must have $\mathbf{V} = \pm\nabla\Phi$. We choose the minus sign so that the positive direction of flow is in the direction of decreasing Φ. The potential function is Φ, and potential flow is characterized by the relation

$$\mathbf{V} = -\nabla\Phi \tag{D.14}$$

or

$$u = -\frac{\partial\Phi}{\partial x}, \qquad v = -\frac{\partial\Phi}{\partial y}, \qquad w = -\frac{\partial\Phi}{\partial z}$$

For a two-dimensional, incompressible, irrotational flow, we can show (using the continuity equation and equations D.12) that

$$\nabla^2\psi = \nabla^2\Phi = 0 \tag{D.15}$$

Also, the equations of momentum conservation become, from equations D.4,

$$\rho\frac{Du}{Dt} - B_x - \frac{\partial P}{\partial x} \tag{D.16a}$$

$$\rho\frac{Dv}{Dt} = B_y - \frac{\partial P}{\partial y} \tag{D.16b}$$

Note that, because of the convective inertia terms on the left-hand side, these momentum equations are nonlinear in the velocity components u and v. But, if the stream function or potential function formulation is used, the problem simply involves the solution of the linear Laplace equation—equation D.15. Since $\mu = 0$ for these flows, the question of whether the flow is isothermal is immaterial because the thermal energy equation is no longer coupled to the other equations. Our definition of the stream function applies only to one- and two-dimensional flows (viscid or inviscid). But the potential function can be used for any *inviscid* flow. From the continuity equation, we find

For an incompressible flow:

$$\nabla^2\Phi = 0 \tag{D.17a}$$

For a compressible flow;

$$\frac{\partial^2\Phi}{\partial x^2}\left[c^2 - \left(\frac{\partial\Phi}{\partial x}\right)^2\right] + \frac{\partial^2\Phi}{\partial y^2}\left[c^2 - \left(\frac{\partial\Phi}{\partial y}\right)^2\right] + \frac{\partial^2\Phi}{\partial z^2}\left[c^2 - \left(\frac{\partial\Phi}{\partial z}\right)^2\right]$$

$$-2\left(\frac{\partial^2\Phi}{\partial x\,\partial y}\frac{\partial\Phi}{\partial x}\frac{\partial\Phi}{\partial y} - \frac{\partial^2\Phi}{\partial y\,\partial z}\frac{\partial\Phi}{\partial y}\frac{\partial\Phi}{\partial z} - \frac{\partial^2\Phi}{\partial z\,\partial x}\frac{\partial\Phi}{\partial z}\frac{\partial\Phi}{\partial x}\right) = 0 \quad \text{(D.17b)}$$

where c is the local velocity of sound.

Though we have discussed only irrotational potential flows thus far, it is also possible to have *rotational* potential flows. These are often called *vortex flows*. Associated with vortex flows is the concept of *circulation*,[†] which is defined as the line integral about a closed path at some instant of time of the component of velocity tangent to the path. For any closed curve C, the circulation is

$$\Gamma = \oint_C \mathbf{V} \cdot d\mathbf{S} \tag{D.18a}$$

When the body forces in the flow are either zero or derivable from a potential function, and when the fluid density depends only on the pressure, we have an important relation known as *Kelvin's theorem*:

$$\frac{D\Gamma}{Dt} = \frac{D}{Dt} \oint \mathbf{V} \cdot d\mathbf{S} = 0 \tag{D.18b}$$

For the more general case of a viscous fluid, we have

$$\frac{D\Gamma}{Dt} = \frac{\mu}{\rho} \oint \nabla^2 \mathbf{V} \cdot d\mathbf{S} + \frac{1}{\rho} \oint \mathbf{B} \cdot d\mathbf{S} \tag{D.18c}$$

The magnitude of the lift force acting on a body immersed in a flow is directly related to the circulation by the equation

$$L = -\rho U \Gamma \tag{D.18d}$$

where ρ is the fluid density and U is the free stream velocity, whose direction is perpendicular to the direction of the lift force.

D.6 SLOW VISCOUS FLOW [10]

In contrast to potential flow, where the convective inertia terms are predominant, we have slow viscous flow or creeping motion, where only the viscous terms are important. For this type of flow, the momentum equations become

$$\rho \frac{\partial u}{\partial t} = \rho B_x - \frac{\partial P}{\partial x} + \frac{\partial}{\partial x}\left[2\mu\left(\frac{\partial u}{\partial x} - \tfrac{1}{3}\nabla \cdot \mathbf{V}\right)\right]$$

$$+ \frac{\partial}{\partial y}\left[\mu\left(\frac{\partial u}{\partial y} + \frac{\partial v}{\partial x}\right)\right] + \frac{\partial}{\partial z}\left[\mu\left(\frac{\partial w}{\partial x} + \frac{\partial u}{\partial z}\right)\right] \tag{D.19a}$$

† Circulation in the counterclockwise direction is taken as positive.

$$\rho \frac{\partial v}{\partial t} = \rho B_y - \frac{\partial P}{\partial y} + \frac{\partial}{\partial y}\left[2\mu\left(\frac{\partial v}{\partial y} - \tfrac{1}{3}\nabla \cdot \mathbf{V}\right)\right]$$

$$+ \frac{\partial}{\partial z}\left[\mu\left(\frac{\partial v}{\partial z} + \frac{\partial w}{\partial y}\right)\right] + \frac{\partial}{\partial x}\left[\mu\left(\frac{\partial u}{\partial y} + \frac{\partial v}{\partial x}\right)\right] \qquad \text{(D.19b)}$$

$$\rho \frac{\partial w}{\partial t} = \rho B_z - \frac{\partial P}{\partial z} + \frac{\partial}{\partial z}\left[2\mu\left(\frac{\partial w}{\partial z} - \tfrac{1}{3}\nabla \cdot \mathbf{V}\right)\right]$$

$$+ \frac{\partial}{\partial x}\left[\mu\left(\frac{\partial w}{\partial x} + \frac{\partial u}{\partial z}\right)\right] + \frac{\partial}{\partial y}\left[\mu\left(\frac{\partial v}{\partial z} + \frac{\partial w}{\partial y}\right)\right] \qquad \text{(D.19c)}$$

The continuity and energy equations are the same as those given in equations D.1 and D.5. If the viscosity, μ, is constant, equations D.19 may be written succinctly as

$$\rho \frac{\partial}{\partial t}\mathbf{V} = \rho\mathbf{B} - \nabla P + \mu\nabla^2\mathbf{V} \qquad \text{(D.20)}$$

For an incompressible steady flow, we have

$$\nabla P = \rho\mathbf{B} + \mu\nabla^2\mathbf{V} \qquad \text{(D.21)}$$

Taking the divergence of both sides of this equation gives

$$\nabla^2 P = \rho\nabla \cdot \mathbf{B} + \mu\nabla^2(\nabla \cdot \mathbf{V})$$

But, from the continuity equation, $\nabla \cdot \mathbf{V} = 0$; hence

$$\nabla^2 P = \rho\nabla \cdot \mathbf{B} \qquad \text{(D.22)}$$

And, in the absence of body forces,

$$\nabla^2 P = 0 \qquad \text{(D.23)}$$

The pressure $P(x, y, z)$ for an incompressible creeping flow without body forces is seen to be a potential function. If we further simplify the problem to two dimensions, the stream function previously defined applies, and the momentum equations, when cross differentiated to eliminate the pressure, may be written as

$$\nabla^4 \psi = 0 \qquad \text{(D.24)}$$

Note that the general problem of slow viscous flow is linear in the pressure P and the velocity components u, v, and w.

D.7 BOUNDARY LAYER FLOW [10]

Most fluid mechanics problems are neither potential flow nor creeping flow problems. Instead, they involve to some degree inertia effects as well as viscous effects. The flow of a viscous fluid over a flat plate provides a convenient example of flow in which both inertia terms and viscous terms appear.

In the thin-flow region immediately adjacent to the plate, viscous effects are most important. This region is called the boundary layer. Outside of the boundary layer, the fluid may be treated as inviscid. Determining the boundary layer thickness and the velocity distribution within the boundary layer becomes the central aspect of the problem. If the plate lies in the x-y plane and the free stream velocity is in the x direction, the governing equations for a general compressible fluid are as follows:

Continuity:

$$\frac{\partial \rho}{\partial t} + \frac{\partial(\rho u)}{\partial x} + \frac{\partial(\rho v)}{\partial z} = 0 \tag{D.25}$$

Momentum:

$$\frac{\partial u}{\partial t} + u\frac{\partial u}{\partial x} + v\frac{\partial u}{\partial z} = -\frac{1}{\rho}\frac{\partial P}{\partial x} + \frac{1}{\rho}\frac{\partial}{\partial z}\left(\mu\frac{\partial u}{\partial z}\right) + \rho B_x \beta(T - T_\infty) \tag{D.26a}$$

$$0 = \frac{\partial P}{\partial z} \tag{D.26b}$$

or, in terms of the stream function which satisfies continuity,

$$\frac{\partial^2 \psi}{\partial z\,\partial t} + \frac{\partial \psi}{\partial z}\frac{\partial^2 \psi}{\partial x\,\partial z} - \frac{\partial \psi}{\partial x}\frac{\partial^2 \psi}{\partial z^2} = -\frac{1}{\rho}\frac{\partial P}{\partial x} + \frac{\partial}{\partial z}\left(\mu\frac{\partial^2 \psi}{\partial z^2}\right) \tag{D.27}$$

Energy:

$$\rho c_p\left(u\frac{\partial T}{\partial x} + v\frac{\partial T}{\partial z}\right) = \frac{\partial}{\partial z}\left(k\frac{\partial T}{\partial z}\right) + \mu\left(\frac{\partial u}{\partial z}\right)^2 + u\frac{\partial P}{\partial x} \tag{D.28}$$

Supplementary equations:

$$\rho = \rho(P, T), \qquad \mu = \mu(P, T)$$

More complex equations apply for the boundary layers that form on curved walls. Note that even these simple boundary layer equations are nonlinear because of the presence of the convective inertia terms. Schlichting [10] provides a comprehensive discussion of numerous boundary layer problems and their solutions.

D.8   CLASSICAL VARIATIONAL PRINCIPLES

Since variational principles sometimes provide a very convenient basis for deriving finite element equations, we discuss here some of the useful variational principles of fluid mechanics. There are numerous variational principles governing various kinds of particular fluid mehcanics problems. These include principles for non-Newtonian fluids, water waves, rarefied gas flows, flows around suspended droplets, and so forth. Generally, though, all the principles fall into one of two categories. Either they apply to flows of an inviscid fluid ($\mu = 0$), or they apply to slow, viscous flow where inertia effects are negligible. There are no variational principles for the full Navier-Stokes equations containing all the inertia terms and all the viscous terms [11]. We will now summarize two variational principles: one for frictionless flows and the other for viscous flow without inertia.

Potential flow

For a steady, incompressible, irrotational flow the boundary value problem is described by the following equations:

$$\left.\begin{array}{l} \nabla^2\Phi = 0 \\ \mathbf{V} = \nabla\Phi \end{array}\right\} \quad \text{in domain } \Omega \tag{D.29}$$

$$\rho\hat{n}\cdot\mathbf{V} = \rho\hat{n}\cdot\nabla\Phi = f(x, y, z) \quad \text{on surface } S \tag{D.30}$$

The potential function $\Phi(x, y, z)$ must satisfy Laplace's equation in domain Ω, while the mass flux across the bounding surface S is specified according to equation D.30.

The function Φ satisfying equations D.29 and D.30 also minimizes the functional

$$I(\Phi) = \rho \int_\Omega \nabla\Phi\cdot\nabla\Phi \, d\Omega - 2\int_S \Phi f \, dS \tag{D.31}$$

Bateman [12] has given a variational principle for steady, compressible flows, and Herivel [13] derived a variational principle for unsteady, compressible flows.

Slow, viscous flow

A variational principle for steady, incompressible, isoviscous flow in which inertia forces are negligible was first stated by Helmholtz [14,15]. The principle states that the excess of the dissipation in the flow over twice the rate at which work is done by the specified surface tractions is a minimum. The functional to be minimized is

$$I(u, v, w, P) = \frac{1}{2} \int_{\Omega} \sigma_{ij} e_{ij} \, d\Omega - 2 \int_{S_2} (\sigma^x u + \sigma^y v + \sigma^z w) \, dS_2 \quad \text{(D.32a)}$$

where e_{ij} is the strain rate tensor, $i = 1, 2, 3, j = 1, 2, 3,$

$$e_{11} = 2\frac{\partial u}{\partial x}, \qquad e_{22} = 2\frac{\partial v}{\partial y}, \qquad e_{33} = 2\frac{\partial w}{\partial z}$$

$$e_{12} = \frac{\partial u}{\partial y} + \frac{\partial v}{\partial x} = e_{21}$$

$$e_{13} = \frac{\partial u}{\partial z} + \frac{\partial w}{\partial x} = e_{31}$$

$$e_{23} = \frac{\partial v}{\partial z} + \frac{\partial w}{\partial y} = e_{32}$$

$$\sigma_{ij} = -P\delta_{ij} + \mu e_{ij}, \qquad j = 1, 2, 3, \quad j = 1, 2, 3$$

$$\delta_{ij} = \begin{cases} 1, & i = j \\ 0, & i \neq j \end{cases}$$

and σ^x, σ^y, and σ^z are the specified surface tractions in the x, y, and z directions. Expanding the tensor notation in equation D.32 gives

$$I(u, v, w, p) = \frac{1}{2} \int_{\Omega} (\sigma_{11}e_{11} + 2e_{12}e_{12} + 2e_{13}e_{13}$$

$$+ 2e_{23}e_{23} + \sigma_{22}e_{22} + \sigma_{33}e_{33}) \, d\Omega$$

$$- 2 \int_{S_2} (\sigma^x u + \sigma^y v + \sigma^z w)_{\text{on boundary}} \, dS_2 \quad \text{(D.32b)}$$

The bounding surface of the solution domain is composed of two parts: S_1 and S_2. On part S_1 of the boundary, velocity is specified; and on the remaining part, S_2, the surface stresses are specified. The velocity and pressure distributions which satisfy the boundary conditions and minimize the functional of equation D.32b also satisfy the continuity equation and the three equations of momentum conservation.

Other flows

For a comprehensive discussion of the kinds of variational principles that have proved useful in the solution of fluid mechanics problems, the reader is encouraged to see the work of Finlayson [16,17]. In addition to summarizing the various classical variational principles, Finlayson introduces the formalism of Fréchet differentials to show that no variational principle exists for continuity equations and the Navier-Stokes equations.

REFERENCES

1. I. Shames, *Mechanics of Fluids*, McGraw-Hill Book Company, New York, 1962.

2. V. L. Streeter, *Fluid Dynamics*, McGraw-Hill Book Company, New York, 1948.

3. A. H. Jameson, *An Introduction to Fluid Mechanics*, Longmans, Green, London, 1942.

4. R. C. Binder, *Advanced Fluid Mechanics*, Prentice-Hall, Englewood Cliffs, N.J., 1958.

5. H. Rouse (ed.), *Advanced Mechanics of Fluids*, John Wiley and Sons, New York, 1959.

6. J. O. Hinze, *Turbulence*, McGraw-Hill Book Company, New York, 1959.

7. A. A. Townsend, *The Structure of Turbulent Shear Flow*, Cambridge, 1956.

8. H. R. Vallentine, *Applied Hydrodynamics*, Butterworth and Company, 1959.

9. L. M. Milne-Thomson, *Theoretical Hydrodynamics*, 4th ed., Macmillan Company, London, 1960.

10. H. Schlichting, *Boundary Layer Theory*, McGraw-Hill Book Company, New York, 1960.

11. C. B. Millikan, "On the Steady Motion of Viscous Incompressible Fluids, with Particular Reference to a Variational Principle," *Phil. Mag.*, Vol. 7, No. 44, April 1929.

12. H. Bateman, "Notes on a Differential Equation Which Occurs in the Two-Dimensional Motion of a Compressible Fluid and the Associated Variational Problems," *Proc. Roy. Soc. (London)*, Vol. 125A, 1929.

13. J. W. Herivel, "On a General Variational Principle for Dissipative Systems," *Proc. Roy. Irish Acad.*, Vol. 56, Sect. A, No. 4, 1954.

14. H. Helmholtz, "Zur Theorie der Stationären Ströme in Reibenden Flüssigkeiten," *Verh. Naturhist.-Med. Ver. Heidelberg*, Vol. V, October 1868.

15. H. Lamb, *Hydrodynamics*, Dover Publications, New York, 1945, Art. 344.

16. B. A. Finlayson, *The Method of Weighted Residuals and Variational Principles*, Academic Press, New York, 1972, Chapter 8.

17. B. A. Finlayson, "Existence of Variational Principles for the Navier-Stokes Equation," *Phys. Fluids*, Vol. 15, No. 6, June 1972.

AUTHOR INDEX

Abel, J.F., 13, 157, 230, 376
Abramowitz, M., 418
Adini, A., 176
Ahlberg, J.H., 10
Ahmad, S., 182, 185, 189
Akin, J.E., 13
Allan, T., 281, 288
Anderson, J., 365
Anderton, G.L., 157, 158, 159
Aral, M.M., 118, 267, 364
Argyris, J.H., 11, 169, 174,
  176, 181, 236, 281, 288, 315,
  324, 364
Arlett, P.L., 254
Atkinson, B., 340, 341, 346
Aubin, J.P., 10
Aziz, A.K., 10

Babuska, I., 10
Bahrani, A.K., 254
Bai, K.J., 315, 324
Baker, A.J., 119, 352, 364
Balasubramanian, R., 364
Bamford, R.M., 423
Barfield, W.D., 415
Bateman, H., 326, 327, 485
Bazeley, G.P., 81, 176
Beckenbach, E.F., 66
Bell, K., 13, 174
Belytschko, T., 365
Berard, G.P., 328
Berke, L., 13
Besseling, J.J., 12
Biggs, J.M., 270
Bijlaar, P.P., 179

Binder, R.C., 474
Birkhoff, G., 10
Birkhoff, M.H., 174
Blair, P., 365
Bogner, F.K., 173, 226, 227, 229
Booker, J.F., 281, 288
Bowley, W.W., 365
Boyd, W.W., 366
Bramble, J.H., 10
Bramlette, T.T., 119, 365
Bratanow, T., 352
Brebbia, C., 13, 176
Brocklebank, M.P., 340, 346
Bronlund, O.E., 236
Brown, C.B., 366
Buck, K.E., 169, 181
Buell, W.R., 415
Bug, G., 365
Bush, B.A., 415
Butlin, G.A., 174, 176

Card, C.C.H., 340, 341, 346
Carey, G.F., 13
Carlson, R.E., 10
Carnahan, B., 125, 263, 273
Carson, W.W., 253
Cea, J., 10
Chakrabarti, S., 131
Chang, G.Z., 351
Cheng, R.T., 352, 357, 365
Cherry, T.M., 327
Cheung, Y.K., 13, 81, 171, 176,
  177, 238
Chu, T.Y., 281, 299, 308, 309
Chu, W., 361, 365

SUBJECT INDEX